George Salmon

Lessons Introductory to the Modern Higher Algebra

Second Edition

George Salmon

Lessons Introductory to the Modern Higher Algebra
Second Edition

ISBN/EAN: 9783744775762

Printed in Europe, USA, Canada, Australia, Japan

Cover: Foto ©berggeist007 / pixelio.de

More available books at **www.hansebooks.com**

104, GRAFTON-STREET, DUBLIN,
November, 1866.

MATHEMATICAL WORKS

PUBLISHED BY

HODGES, SMITH AND CO.

A Treatise on the Analytic Geometry of Three
Dimensions. By GEORGE SALMON, D.D., F.R.S., Fellow and Senior
Tutor of Trinity College, Dublin. Second Edition, revised. 8vo. cloth
boards, price 12s.

By the same Author.

A Treatise on Higher Plane Curves: intended as
a Sequel to a Treatise on Conic Sections. By GEORGE SALMON, D.D.
8vo. cloth boards, price 12s.

By the same Author.

Lessons Introductory to the Modern Higher Algebra.
By GEORGE SALMON, D.D. Second Edition, revised.

Chapters on the Modern Geometry of the Point,
Line, and Circle; being the Substance of Lectures delivered in the
University of Dublin to the Candidates for Honours in the first year
of Arts. By the Rev. RICHARD TOWNSEND, M.A., Fellow and Tutor
of Trinity College. 2 vols. 8vo. price 22s. 6d.

Observations made at the Magnetical and Meteoro-
logical Observatory at Trinity College, Dublin: under the Direction
of the Rev. HUMPHREY LLOYD, D.D., D.C.L., Fellow of Trinity College,
Dublin, and formerly Professor of Natural Philosophy in the University.
Vol. I. 1840—1843, 4to. price 21s.

Principles of Modern Geometry, with numerous
Applications to Plane and Spherical Figures; and an Appendix con-
taining questions for Exercise. Intended chiefly for the use of Junior
Students. By JOHN MULCAHY, LL.D., late Professor of Mathematics,
Queen's College, Galway. Second Edition, revised. 8vo. cloth boards,
price 9s.

Lectures on Quaternions: containing a Systematic
Statement of a New Mathematical Method. With numerous Illustrative Diagrams, and some Geometrical and Physical Applications. By SIR WILLIAM ROWAN HAMILTON, LL.D., M.R.I.A., Andrews' Professor of Astronomy in the University of Dublin; Royal Astronomer of Ireland, &c., &c. 8vo. cloth boards, price 21s.

A Treatise on Heat. The Thermometer, Dilation,
Change of State, and Laws of Vapours. By the Rev. R. V. DIXON, D.D., Rector of Clogherny, late Fellow of Trinity College, &c., &c. 8vo. cloth boards, price 12s. 6d.

The Elements of Plane and Spherical Trigonometry,
with the Nature and Properties of Logarithms, &c., for the use of Students in the University. By THOMAS LUBY, D.D., Senior Fellow of Trinity College, Dublin. Part I. Third Edition, corrected and enlarged, price 6s.

The Elements of Plane Astronomy. By JOHN
BRINKLEY, D.D., late Bishop of Cloyne; edited by THOMAS LUBY, D.D., M.R.I.A., Senior Fellow of Trinity College, Dublin. 8vo. cloth boards, price 12s.

Two Geometrical Memoirs of the General Properties
of Cones of the Second Degree and of Spherical Cones. By M. CHASLES. Translated from the French, with Notes and Additions, and an Appendix on the Application of Analysis to Spherical Geometry. By CHARLES GRAVES, D.D., Dean of the Chapel Royal and Fellow of Trinity College, Dublin. 8vo. price 6s.

Elements of Optics. By HUMPHREY LLOYD, D.D.,
D.C.L., Vice-Provost of Trinity College, Dublin. 8vo. cloth boards, price 6s. 6d.

Tables for Facilitating the Calculation of Earthwork,
in the cutting of Embankments of Railways, Canals, and other Public Works. By SIR JOHN MACNEILL, LL.D., F.R.S., M.R.I.A., Professor of Practical Engineering, Trinity College, Dublin. 8vo. cloth boards, price 31s. 6d.

DUBLIN: HODGES, SMITH, & CO., GRAFTON STREET,
PUBLISHERS TO THE UNIVERSITY.

LESSONS

INTRODUCTORY TO THE

MODERN HIGHER ALGEBRA.

BY

THE REV. GEORGE SALMON, D.D.,

FELLOW AND SENIOR TUTOR, TRINITY COLLEGE, DUBLIN.

SECOND EDITION.

Dublin:
HODGES, SMITH, AND CO., GRAFTON STREET,
BOOKSELLERS TO THE UNIVERSITY.

1866.

CAMBRIDGE:
Printed by W. Metcalfe, Trinity Street, corner of Green Street.

TO

A. CAYLEY, ESQ., AND J. J. SYLVESTER, ESQ.,

I BEG TO INSCRIBE

THIS ATTEMPT TO RENDER SOME OF THEIR DISCOVERIES BETTER KNOWN,

In Acknowledgment

OF

THE OBLIGATIONS I AM UNDER, NOT ONLY TO THEIR PUBLISHED WRITINGS,

BUT ALSO

TO THEIR INSTRUCTIVE CORRESPONDENCE.

PREFACE.

I FEAR I cannot defend the title of this volume as very accurately describing its contents. The title "Lessons Introductory" applied well enough to the former edition, which grew out of Lectures given some years ago to my Class, and which was not intended to do more than to supply students with the preliminary information necessary to enable them to read with advantage the original memoirs whence my materials were derived. That edition has, however, been for a long time out of print, and in reprinting it now I found so many additions necessary to bring it up to the present state of science, that the book has become doubled in size, and might fairly assume the less modest title of a "Treatise" on the subjects with which it deals. Neither does the name "Modern Higher Algebra" very precisely define the nature of these subjects. The Theory of Elimination, and that of Determinants cannot be said to be very modern; and I do not meddle with some parts of Higher Algebra, for which much has been done in modern times; as, for instance, the

Theory of Numbers, or the General Theory of the Resolution of Equations. But it is no great abuse of language to give, in a special sense, the name "Modern Higher Algebra" to that which forms the principal subject of this volume—the Theory of Linear Transformations. Since Mr. Cayley's discovery of Invariants, quite a new department of Algebra has been created; and there is no part of Mathematics in which an able mathematician, who had turned his attention to other subjects some twenty years ago, would find more difficulty in reading a memoir of the present day, and would more feel the want of an elementary guide to inform him of the meaning of the terms employed, and to establish the truth of the theorems assumed to be known.

With respect to the use of new words I have tried to steer a middle course. In this part of Algebra combinations of ideas require to be frequently spoken of which were not of important use in the older Algebra. This has made it necessary to employ some new words, in order to avoid an intolerable amount of circumlocution. But feeling that every strange term makes the science more repulsive to a beginner, I have generally preferred the use of a periphrasis to the introduction of a new word which I was not likely often to have occasion to employ. Students who may be disappointed by not finding in this volume the explanation of some words which occur in modern algebraical memoirs, will be likely to find the desired information in the Glossary added to Mr. Sylvester's paper (*Philosophical Transactions*, 1853, p. 543).

The first four or five Lessons in this volume were printed a year or two ago: it having been at that time my intention to publish separately the Lessons on Determinants as a manual for the use of Students. At the time when these Lessons were written I had not met Baltzer's *Treatise on Determinants*, a work remarkable for the rigorous and scientific manner in which its principles are evolved. But I most regretted not to have met with it earlier, on account of his careful indication of the original authorities for the several theorems. Although very sensible of the value of these historical notices, I have, in the text of these Lessons, too often omitted to assign the theorems to their original authors, because my knowledge not having been obtained by any recent course of study, I did not find it easy to name the sources whence I had derived it, nor had I mathematical learning enough to be able to tell whether these sources were the originals. I have now tried to supply the references omitted in the text by adding a few historical notes; following Baltzer's guidance as far as it would serve me. Where I have had only my own reading to trust to, it is only too likely that I have in several cases failed to trace theorems back to their first discoverers, and I must ask the indulgence of any living authors to whom I have in this way unwittingly done injustice.

I have to thank my friends Dr. Hart, Mr. Traill, and Mr. Burnside, for help given me at various times in the revision of the proofs of this work, though, in justice to these gentlemen, I must add that there is a considerable part of it which was printed under circumstances where I could not have the benefit of their assistance, and for the errors in which they are not responsible. I have already intimated that my obligations to Messrs. Cayley and Sylvester are not merely those which every one must owe who writes on a branch of Algebra which they have done so much to create. I was in constant correspondence with them at the time when some of their most important discoveries were made, and I owe my knowledge of these discoveries as much to their letters as to any printed papers. I must also express my thanks to M. Hermite for his obliging readiness to remove by letter difficulties which occurred to me in my study of his published memoirs.

TRINITY COLLEGE, DUBLIN,
October 16th, 1866.

CONTENTS.

LESSON I.
DETERMINANTS.—PRELIMINARY ILLUSTRATIONS AND DEFINITIONS.

	PAGE
Rule of signs	5
Sylvester's umbral notation	7

LESSON II.
REDUCTION AND CALCULATION OF DETERMINANTS.

Minors [called by Jacobi partial determinants]	9
Examples of reduction	12
Product of differences of n quantities expressed as a determinant	13
Reduction of bordered Hessians	15

LESSON III.
MULTIPLICATION OF DETERMINANTS.

The theorem stated as one of linear transformation	18
Extension of the theorem	18
Examples of multiplication of determinants	20
Product of squares of differences of n quantities	20
Radius of sphere circumscribing a tetrahedron	20
Relation connecting mutual distances of points on a circle or sphere	21
Of five points in space	22

LESSON IV.
MINOR AND RECIPROCAL DETERMINANTS.

Solution of a system of linear equations	23
Reciprocal systems [called by Cauchy *adjoint* systems]	24
Minors of reciprocal system expressed in terms of those of the original	25
Minors of a determinant which vanishes	26

b

LESSON V.

SYMMETRICAL AND SKEW SYMMETRICAL DETERMINANTS.

	PAGE
Differentials of a determinant with respect to its coefficients	27
If a determinant vanishes, the same bordered is a perfect square	29
Skew symmetric determinants of odd degree vanish	30
Of even degree are perfect squares [Cayley]	31
Nature of the square root [see Jacobi, *Crelle*, II. 354; XXIX. 236]	32
Orthogonal substitutions	33

LESSON VI.

SYMMETRICAL DETERMINANTS.

Equation of secular inequalities has always real roots	36
Sylvester's proof	36
Another new proof	37
Borchardt's proof	37
Sylvester's expressions for Sturm's functions in terms of the roots	38

LESSON VII.

SYMMETRIC FUNCTIONS.

Newton's formulæ for sums of powers of roots	45
Rules for weight and order of a symmetric function	46
Rule for sum of powers of differences of roots	46
Differential equation of function of differences	47
Symmetric functions of homogeneous equations	49
Differential equation where binomial coefficients are used	50
Serret's notation	51
Brioschi's expression for the operation in terms of the roots	52

LESSON VIII.

ORDER AND WEIGHT OF ELIMINANTS.

Elimination by symmetric functions	54
Order and weight of resultant of two equations	56
Symmetric functions of common values of system of two equations	59
Extension of principles to any number of equations	62

LESSON IX.

EXPRESSION OF ELIMINANTS AS DETERMINANTS.

Elimination by process for greatest common measure	63
Euler's method	64
Conditions that two equations should have two common factors	65
Sylvester's dialytic method	65
Bezout's method	66
Cayley's statement of it	68
Jacobians defined	69
Expression by determinants, in particular cases, of resultant of three equations	70
Cayley's method of expressing resultants as quotients of determinants	71

LESSON X.

DETERMINATION OF COMMON ROOTS.

	PAGE
Expression of roots common to a system of equations by the differentials of the resultant	75
Equations connecting these differentials when the resultant vanishes	76
Expressions by the minors of Bezout's matrix	77
General expression for differentials of resultants with respect to any quantities entering into the equations	81
General conditions that a system may have two common roots	82

LESSON XI.

DISCRIMINANTS.

Order and weight of discriminants	84
Discriminant expressed in terms of the roots	86
Discriminant of product of two or more functions	86
Discriminant is of form $a_0\phi + a_1^2\psi$	86
Formation of discriminants by the differential equation	87
Method of finding the equal roots when the discriminant vanishes	88
Extension to any number of variables	90
Discriminant of a quadratic function	92

LESSON XII.

LINEAR TRANSFORMATIONS.

Invariance of discriminants	93
Number of independent invariants	95
Invariants of systems of quantics	96
Covariants	98
Every invariant of a covariant is an invariant of the original	98
Invariants of emanants are covariants	99
Contravariants	101
$x\xi + y\eta + $ &c. absolutely unaltered by transformation	103
Mixed concomitants	104
Evectants	105
Evectant of discriminant of a quantic whose discriminant vanishes	106

LESSON XIII.

FORMATION OF INVARIANTS AND COVARIANTS.

Method by symmetric functions	107
Invariants, &c. which vanish when two or more roots are equal	109
Mutual differentiation of covariants and contravariants	109
Differential coefficients substituted for the variables in a contravariant give covariants	111
Every binary quantic has an invariant of even degree	112
Cubinvariant of a quartic	112
Every quantic of odd order has invariant of the 4th order	113
Weight of an invariant of given order	113
The differential equation	113
Binary quantics of even degree cannot have invariants of odd order	114

CONTENTS.

	PAGE
Skew invariants	115
Determination of numbers of independent invariants by the differential equation	116
Coefficients of covariants determined by the differential equation	116
Source of product of two covariants is product of their sources	118
Extension to any number of variables	119

LESSON XIV.
SYMBOLICAL REPRESENTATION OF INVARIANTS AND COVARIANTS.

Formation of derivative symbols	121
Order of derivative in coefficients and in the variables	123
Table of invariants of the third order	124
Hermite's law of reciprocity	125
Derivative symbols for ternary quantics	128
Symbols for evectants	129
Method of Aronhold and Clebsch	130
Clebsch's proof that every invariant can be thus expressed symbolically	132

LESSON XV.
CANONICAL FORMS.

Generality of a form determined by its number of constants	134
Reduction of a quadratic function to a sum of squares	135
Principle that the number of negative squares is unaffected by real substitution	135
Reduction of cubic to its canonical form	136
Discriminant of a cubic and its Hessian differ only in sign	136
General reduction of quantic of odd degree	137
Methods of forming canonizant	138
Condition that quantic of order $2n$ should be reducible to a sum of n, $2n^{th}$ powers [called by Sylvester *catalecticant*]	140
Canonical forms for quantics of even order	141
Canonical forms for sextic	142
For ternary and quaternary cubic	144

LESSON XVI.
SYSTEMS OF QUANTICS.

Combinants defined, differential equation satisfied by them	144
Number of double points in an involution	145
Factor common to two quantics is also factor in Jacobian	146
Order of condition that $u + kv$ may have cubic factor	147
Nature of discriminant of Jacobian	148
Discriminant of discriminant of $u + kv$	148
Proof that resultant is a combinant	150
Discriminant with respect to x, y, of a function of u, v	151
Discriminant of discriminant of $u + kv$ for ternary quantics	152
Tact-invariant of two curves	153
Tact-invariant of complex curves	155
Of $k-1$, k-ary quantics	155
Of two quaternary quantics	156
Osculants	156

CONTENTS.

LESSON XVII.
APPLICATIONS TO BINARY QUANTICS.

	PAGE
Invariants when said to be distinct	158
Number of independent covariants	159
THE QUADRATIC	159
Resultant of two quadratics	160
Condition that three should form a system in involution	160
THE CUBIC	160
Geometric meaning of covariant cubic	161
Square of this cubic expressed in terms of the other covariants	162
Solution of cubic	162
SYSTEM OF CUBIC AND QUADRATIC	162
SYSTEM OF TWO CUBICS	164
Resultant of the system	165
Condition that $u + \lambda v$ may have a cubic factor	165
Mode of dealing with equations which contain a superfluous variable	166
Invariants of invariants of $u + \lambda v$ are combinants	167
THE QUARTIC	169
Catalecticants	169
Every invariant of a quartic is a rational function of S and T	170
Discriminant of a quartic	170
The same derived from theory of two cubics	171
Relation between covariants of a cubic derived from that of invariants of a quartic	172
Conditions that a quartic should have two square factors	173
Reduction of quartic to its canonical form	174
General solution of quartic	175
Criteria for real and imaginary roots	175
The quartic can be brought to its canonical form by real substitutions	176
Sextic covariant of quartic	177
Invariants of system of quartic and its Hessian	177
SYSTEM OF TWO QUARTICS	179
Their resultant	180
Condition that $u + \lambda v$ should be perfect square	180
Condition that $u + \lambda v$ should have cubic factor	181
Special form when both quartics are sum of two fourth powers	182
THE QUINTIC	183
Condition that two quartics should be capable of being got by differentiation from the same quintic	184
Discriminant of quintic	184
Its fundamental invariants	188
Conditions for two pairs of equal roots	189
All invariants of a quantic vanish if more than half its roots be all equal	189
Hermite's canonical form	189
Hermite's skew invariant	189
Covariants of quintic	190
Sign of discriminant of any quantic determines whether it has an odd or even number of pairs of imaginary roots	192
Criteria furnished by Sturm's theorem for a quintic	192
If roots all real, canonizant has imaginary factors	193
Invariantive expression of criteria for real roots	193
Sylvester's criteria	197

	PAGE
Conditions involving a constant varying within certain limits	198
Sextic resolvent of a quintic	200
Harley and Cockle's resolvent	201
Expression of invariants in terms of roots	202
THE SEXTIC—its invariants	202
Conditions for cubic factor or for two square factors	204
The discriminant	205
Simpler invariant of tenth order	207
The skew invariant expressed in terms of other invariants	210
Functions likely to afford criteria for real roots	211

LESSON XVIII.

ON THE ORDER OF RESTRICTED SYSTEMS OF EQUATIONS.

Order and weight of systems defined	214
Restricted systems	215
Determinant system, k rows, $k+1$ columns	217
Order and weight of conditions that two equations should have two common roots	219
System of conditions that three equations should have a common root	221
Intersection of quantics having common curves	222
Case of distinct curves	225
Problem to find the number of quadrics which pass through five points and touch four planes	226
Rank of curve represented by a system of k rows, $k+1$ columns	227
System of k rows, $k+2$ columns	228
System of conditions that two equations should have three common roots	229
System of quantics having a surface common	230
Having common two surfaces	233
System of conditions that three ternary quantics should have two common points	235, 238
Rule when the constants in systems of equations are connected by relations	237

LESSON XIX.

APPLICATIONS OF SYMBOLICAL METHODS.

When a figure disappears from a derivative symbol, U is a factor	240
Result of substituting in reciprocal, differentials for the variables	241
Symbols for derivatives of derivatives	243
Symbolical expression for resultant of quadratic and equation of nth degree	245
Formula for discriminant of quartic	247
Clebsch's investigation of resultant of quadratic and n-tic	249

APPENDIX.

Value of skew invariant of a sextic	253

NOTES.

	PAGE
History of determinants	266
Commutants	267
Hessians	268
Symmetric functions	268
Elimination	268
Discriminants	269
Linear transformations	269
On the number of invariants or covariants of a binary quantic	271
Canonical forms	273
Combinants	273
Applications to binary quantics	274
The quintic	274
Every skew invariant vanishes when alternate terms in a quantic are wanting	274
Hermite's forme-types	274
The Tschirnhausen transformation	275
On the order of systems of equations	280
Mr. S. Roberts' methods of investigation	280
Bezoutiants	282
Tables of resultants	283
Hirsch and Cayley's tables of symmetric functions	285
Mr. M. Roberts' tables of sums of powers of differences	293
Index	294

ERRATA.

p. 27, line 11, *for* $\Sigma_s a_{rs} a'_{rs} = \Delta$, *read* $\Sigma_s a_{rs} a_{rs} = \Delta$.

p. 140, lines 2 and 3 from bottom, *for* $\mathfrak{X}$, *read* $\mathfrak{X}$.

p. 168, line 16, *for* "invariants," *read* "covariants."

p. 194, note, line 3 from bottom, *for* "is of the recurring form," *read* "is capable of being linearly transformed to the recurring form."

p. 197, line 17, the word "not" has been omitted.

p. 200, note, line 11 from bottom, *for* D, *read* JD.

p. 211, lines 4 and 5, interchange the words positive and negative.

p. 211, line 3 from bottom, *for* $A^3 - 100B^2$, *read* $A^2 - 100B$.

LESSONS ON HIGHER ALGEBRA.

LESSON I.

DETERMINANTS.—PRELIMINARY ILLUSTRATIONS.

1. If we are given n homogeneous equations of the first degree between n variables, we can eliminate the variables, and obtain a result involving the coefficients only, which is called the *determinant* of those equations. We shall, in what follows, give rules for the formation of these determinants, and shall state some of their principal properties; but we think that the general theory will be better understood if we first give illustrations of its application to the simplest examples.

Let us commence, then, with two equations between two variables
$$a_1 x + b_1 y = 0, \quad a_2 x + b_2 y = 0.$$
The variables are eliminated by adding the first equation multiplied by b_2 to the second multiplied by $-b_1$, when we get $a_1 b_2 - a_2 b_1 = 0$, the left-hand member of which is the determinant required. The ordinary notation for this determinant is
$$\begin{vmatrix} a_1, & b_1 \\ a_2, & b_2 \end{vmatrix}.$$
We shall, however, often, for brevity, write $(a_1 b_2)$ to express this determinant, leaving the reader to supply the term with the negative sign; and in this notation it is obvious that $(a_1 b_2) = -(a_2 b_1)$. The coefficients a_1, b_1, &c., which enter into the expression of a determinant, are called the *constituents* of that determinant, and the products $a_1 b_2$, &c., are called the *elements* of the determinant.

B

2. It can be verified at once that we should have obtained the same result if we had eliminated the variables between the equations
$$a_1x + a_2y = 0, \quad b_1x + b_2y = 0.$$
In other words
$$\begin{vmatrix} a_1, & b_1 \\ a_2, & b_2 \end{vmatrix} = \begin{vmatrix} a_1, & a_2 \\ b_1, & b_2 \end{vmatrix}$$
or the value of the determinant is not altered if we write the horizontal rows vertically, and *vice versâ*.

3. If we are given two homogeneous equations between three variables,
$$a_1x + a_2y + a_3z = 0, \quad b_1x + b_2y + b_3z = 0;$$
these equations are sufficient to determine the mutual ratios of x, y, z. Thus, by eliminating y and x alternately, we can express x and y in terms of z; when we find
$$(a_1b_2)\,x = (a_2b_3)\,z\,; \quad (a_1b_2)\,y = (a_3b_1)\,z.$$
In other words x, y, z, are proportional respectively to (a_2b_3), (a_3b_1), (a_1b_2). Substituting these values in the original equations, we obtain the idéntical relations
$$a_1(a_2b_3) + a_2(a_3b_1) + a_3(a_1b_2) = 0, \quad b_1(a_2b_3) + b_2(a_3b_1) + b_3(a_1b_2) = 0;$$
relations which are verified at once by writing them at full length, as for instance
$$a_1(a_2b_3 - a_3b_2) + a_2(a_3b_1 - a_1b_3) + a_3(a_1b_2 - a_2b_1) = 0.$$
The notation
$$\begin{Vmatrix} a_1, & a_2, & a_3 \\ b_1, & b_2, & b_3 \end{Vmatrix}$$
(where the number of columns is greater than the number of rows) is used to express the three determinants which can be obtained by suppressing in turn each one of the columns; viz. the three determinants of which we have been speaking, (a_2b_3), (a_3b_1), (a_1b_2).

4. Let us now proceed to a system of three equations
$$a_1x + b_1y + c_1z = 0, \quad a_2x + b_2y + c_2z = 0, \quad a_3x + b_3y + c_3z = 0.$$
Then, if we multiply the first by (a_2b_3), the second by (a_3b_1), the third by (a_1b_2), and add; the coefficients of x and y will vanish

in virtue of the identical relations of Art. 3, and the determinant required is
$$c_1(a_2b_3) + c_2(a_3b_1) + c_3(a_1b_2);$$
or, writing at full length,
$$c_1a_2b_3 - c_1a_3b_2 + c_2a_3b_1 - c_2a_1b_3 + c_3a_1b_2 - c_3a_2b_1.$$
It may also be written in either of the forms,
$$a_1(b_2c_3) + a_2(b_3c_1) + a_3(b_1c_2) = 0, \quad b_1(c_2a_3) + b_2(c_3a_1) + b_3(c_1a_2).$$
This determinant is expressed by the notation
$$\begin{vmatrix} a_1, & b_1, & c_1 \\ a_2, & b_2, & c_2 \\ a_3, & b_3, & c_3 \end{vmatrix}$$
though we shall often use for it the abbreviation $(a_1b_2c_3)$.

It is useful to observe that
$$(a_2b_3c_1) = (a_1b_2c_3), \text{ but } (a_1b_3c_2) = -(a_1b_2c_3).$$
For by analogy of notation,
$(a_2b_3c_1) = a_2(b_3c_1) + a_3(b_1c_2) + a_1(b_2c_3)$, which is the same as $(a_1b_2c_3)$, while
$(a_1b_3c_2) = a_1(b_3c_2) + a_3(b_2c_1) + a_2(b_1c_3)$, which is the same as $-(a_1b_2c_3)$.

5. We should have obtained the same result of elimination if we had eliminated between the three equations
$$a_1x + a_2y + a_3z = 0, \quad b_1x + b_2y + b_3z = 0, \quad c_1x + c_2y + c_3z = 0.$$
For if we proceed on the same system as before, multiplying the first equation by (b_2c_3), the second by (c_2a_3), and the third by (a_2b_3), and add, then the coefficients of y and z vanish, and the determinant is obtained in the form
$$a_1(b_2c_3) + b_1(c_2a_3) + c_1(a_2b_3),$$
which, expanded, is found to be identical with $(a_1b_2c_3)$. Hence
$$\begin{vmatrix} a_1, & b_1, & c_1 \\ a_2, & b_2, & c_2 \\ a_3, & b_3, & c_3 \end{vmatrix} = \begin{vmatrix} a_1, & a_2, & a_3 \\ b_1, & b_2, & b_3 \\ c_1, & c_2, & c_3 \end{vmatrix}$$
or the determinant is not altered by writing the horizontal rows vertically, and *vice versâ;* a property which will be proved to be true of every determinant.

6. Using the notation
$$\begin{Vmatrix} a_1, & a_2, & a_3, & a_4 \\ b_1, & b_2, & b_3, & b_4 \\ c_1, & c_2, & c_3, & c_4 \end{Vmatrix}$$
to denote the system of determinants obtained by omitting in turn each one of the columns; these four determinants are connected by the relations

$$a_1(a_2b_3c_4) - a_2(a_3b_4c_1) + a_3(a_4b_1c_2) - a_4(a_1b_2c_3) = 0,$$
$$b_1(a_2b_3c_4) - b_2(a_3b_4c_1) + b_3(a_4b_1c_2) - b_4(a_1b_2c_3) = 0,$$
$$c_1(a_2b_3c_4) - c_2(a_3b_4c_1) + c_3(a_4b_1c_2) - c_4(a_1b_2c_3) = 0.$$

These relations may be either verified by actual expansion of the determinants, or else may be proved by a method analogous to that used in Art. 3. Take the three equations

$$a_1x + a_2y + a_3z + a_4w = 0,$$
$$b_1x + b_2y + b_3z + b_4w = 0,$$
$$c_1x + c_2y + c_3z + c_4w = 0.$$

Then (as in Art. 5) we can eliminate y and z by multiplying the equations by (b_2c_3), (c_2a_3), (a_2b_3), respectively, and adding, when we get
$$(a_1b_2c_3)x + (a_4b_2c_3)w = 0.$$
In like manner, multiplying by (b_3c_1), (c_3a_1), (a_3b_1) respectively, we get
$$(a_2b_3c_1)y + (a_4b_3c_1)w = 0.$$
And in like manner,
$$(a_3b_1c_2)z + (a_4b_1c_2)w = 0.$$

Now, attending to the remarks about signs (Art. 4), these equations are equivalent to
$$(a_1b_2c_3)x = -(a_2b_3c_4)w, \quad (a_1b_2c_3)y = (a_3b_4c_1)w, \quad (a_1b_2c_3)z = -(a_4b_1c_2)w.$$
or x, y, z, w are respectively proportional to $(a_2b_3c_4)$, $-(a_3b_4c_1)$, $(a_4b_1c_2)$, $-(a_1b_2c_3)$; substituting which values in the original equations, we obtain the identities already written.

7. If now we have to eliminate between the four equations
$$a_1x + b_1y + c_1z + d_1w = 0,$$
$$a_2x + b_2y + c_2z + d_2w = 0,$$
$$a_3x + b_3y + c_3z + d_3w = 0,$$
$$a_4x + b_4y + c_4z + d_4w = 0,$$
we have only to multiply the first by $(a_2b_3c_4)$, the second by $-(a_3b_4c_1)$, the third by $(a_4b_1c_2)$, the fourth by $-(a_1b_2c_3)$, and add,

when the coefficients of x, y, z vanish identically, and the determinant is found to be

$$d_1(a_2b_3c_4) - d_2(a_3b_4c_1) + d_3(a_4b_1c_2) - d_4(a_1b_2c_3);$$

or, writing it full length,

$$a_1b_2c_3d_4 - a_1b_3c_2d_4 + a_2b_3c_1d_4 - a_2b_1c_3d_4 + a_3b_1c_2d_4 - a_3b_2c_1d_4 + a_1b_4c_2d_3$$
$$- a_1b_2c_4d_3 + a_4b_2c_1d_3 - a_4b_1c_2d_3 + a_2b_1c_4d_3 - a_2b_4c_1d_3 + a_3b_4c_1d_2$$
$$- a_3b_1c_4d_2 + a_4b_1c_3d_2 - a_4b_3c_1d_2 + a_1b_3c_4d_2 - a_1b_4c_3d_2 + a_3b_4c_2d_1$$
$$- a_2b_3c_4d_1 + a_4b_3c_2d_1 - a_4b_2c_3d_1 + a_2b_4c_3d_1 - a_3b_4c_2d_1.$$

8. There is no difficulty in extending to any number of equations the process here employed; and the reader will observe that the general expression for a determinant is $\Sigma \pm a_1 b_2 c_3 d_4$, &c., where each product must include all the varieties of the n letters and of the n suffixes, without repetition or omission, and the determinant contains all possible such products which can be formed. With regard to the sign to be affixed to each element of the determinant, the following is the rule: We give the sign + to the term $a_1b_2c_3d_4$, &c. obtained by reading the determinant from the left-hand top to the right-hand bottom corner; and then "*the sign + or − is affixed to each other product according as it is derived from this leading term by an even or odd number of permutations of suffixes.*" Thus, in the last example, the second term $a_1b_3c_2d_4$ differs from the first only by a permutation of the suffixes of b and c; it therefore has an opposite sign. The third term, $a_2b_3c_1d_4$, differs from the second by a permutation of the suffixes of a and c; it therefore has an opposite sign: but it has the same sign with the first term, since it can only be derived from it by *twice* permuting suffixes.

Ex. In the determinant $(a_1b_2c_3d_4e_5)$, what sign is to be affixed to the element $a_3b_5c_2d_1e_4$?

From the first term, permuting the suffixes of a and c, we get $a_3b_2c_1d_4e_5$, the first constituent of which is the same as that in the given term: next permuting the suffixes of b and e, we get $a_3b_5c_1d_4e_2$, which has two constituents the same as the given term: next, permuting c and e, we get $a_3b_5c_2d_4e_1$: lastly, permuting d and e, we get the given term $a_3b_5c_2d_1e_4$. Since, then, there has been an even number (four) of permutations, the sign of the term is +. In fact, the signs of the series of terms are

$$a_1b_2c_3d_4e_5 - a_3b_2c_1d_4e_5 + a_3b_5c_1d_4e_2 - a_3b_5c_2d_4e_1 + a_3b_5c_2d_1e_4.*$$

---

* Comparing the elements $a_1b_2c_3d_4e_5$, $a_3b_5c_2d_1e_4$, it will be seen that the suffix 1 which came first in the former element, is in the latter preceded by three constituents; that the suffix 2 is preceded by two which came after it before, and the suffix 4 by one. The total number of displacements is therefore six. The rule of signs is

6   DETERMINANTS.

9. *A cyclic interchange of suffixes alters the sign when the number of terms in the product is even; but not so when the number of terms is odd.* Thus $a_2 b_1$, being got from $a_1 b_2$ by one interchange of suffixes, has a different sign; but $a_2 b_3 c_1$ has the same sign with $a_1 b_2 c_3$ from which it is derived by a double permutation. For, changing the suffixes of a and b, $a_1 b_2 c_3$ becomes $a_2 b_1 c_3$, and changing the suffixes of b and c, this again becomes $a_2 b_3 c_1$. In like manner $a_2 b_3 c_4 d_1$ has an opposite sign to $a_1 b_2 c_3 d_4$, being derived from it by a triple permutation, viz. through the steps $a_2 b_1 c_3 d_4$, $a_2 b_3 c_1 d_4$, $a_2 b_3 c_4 d_1$.

10. We are now in a position to replace our former definition of a determinant by another, which we make the foundation of the subsequent theory. In fact, since a determinant is only a function of its constituents a_1, b_1, c_1, &c., and does not contain the variables x, y, z, &c., it is obviously preferable to give a definition which does not introduce any mention of equations between these quantities x, y, z.

*Let there be n^2 quantities arrayed in a square of n columns and n rows, then the sum with proper signs (as explained, Art. 9) of all possible products of n constituents, one constituent being taken from each horizontal and each vertical row is called the determinant of these quantities and is said to be of the nth order. Constituents are said to be *conjugate* to each other, when the place which each occupies in the horizontal rows is the same as that which the other occupies in the vertical rows. A determinant is said to be *symmetrical* when the conjugate constituents are equal to each other; for example,

$$\begin{vmatrix} a, & h, & g \\ h, & b, & f \\ g, & f, & c \end{vmatrix}.$$

---

sometimes given in the form that the sign of an element is +, when the total number of displacements as compared with the order in the leading term is even, and *vice versâ*.

* We might have commenced with this definition of a determinant, the preceding articles being unnecessary to the scientific development of the theory. We have thought, however, that the illustrations there given would make the general theory more intelligible; and also that the importance of the study of determinants would more clearly appear, when it had been shown that every elimination of the variables from a system of equations of the first degree, and every solution of such a system, gives rise to determinants, such systems of equations being of constant occurrence in every department of pure and applied mathematics.

11. In these first lessons, as in the previous examples, we write all the constituents in the same row with the same letter, and those in the same column with the same suffix. The most common notation however is to write the constituents of a determinant with a double suffix, one suffix denoting the row, and the other the column, to which the constituent belongs. Thus the determinant of the third order would be written

$$\begin{vmatrix} a_{1,1} & a_{1,2} & a_{1,3} \\ a_{2,1} & a_{2,2} & a_{2,3} \\ a_{3,1} & a_{3,2} & a_{3,3} \end{vmatrix},$$

or else $\quad \Sigma \pm a_{1,1} a_{2,2} a_{3,3}$,

where in the sum the suffixes are interchanged in all possible ways. The preceding notation is occasionally modified by the omission of the letter a, and the determinant is written

$$\begin{vmatrix} (1, 1), & (1, 2), & (1, 3) \\ (2, 1), & (2, 2), & (2, 3) \\ (3, 1), & (3, 2), & (3, 3) \end{vmatrix}.$$

Lastly, Mr. Sylvester has suggested what he calls an *umbral* notation. Consider, for example, the determinant

$$\begin{vmatrix} a\alpha, & b\alpha, & c\alpha, & d\alpha \\ a\beta, & b\beta, & c\beta, & d\beta \\ a\gamma, & b\gamma, & c\gamma, & d\gamma \\ a\delta, & b\delta, & c\delta, & d\delta \end{vmatrix}$$

the constituents of which are $a\alpha$, $b\alpha$, &c., and where a, b, c, &c. are not quantities, but as it were shadows of quantities; that is to say, they have no meaning separately, except in combination with one of the other class of umbræ α, β, γ, &c. Thus, for example, if α, β, γ, δ represent the suffixes 1, 2, 3, 4, the constituents in the notation we have ourselves employed, are all formed by combining one of the letters a, b, c, d with one of the figures 1, 2, 3, 4. Now the determinant above written is written by Mr. Sylvester more compactly

$$\begin{matrix} a, & b, & c, & d, \\ \alpha, & \beta, & \gamma, & \delta, \end{matrix}$$

which denotes the sum of all possible products of the form $a\alpha . b\beta . c\gamma . d\delta$, obtained by giving the terms in the second line every possible permutation, and changing sign according to the ordinary rule with every permutation.

LESSON II.

REDUCTION AND CALCULATION OF DETERMINANTS.

12. WE have in the last Lesson given the rule for the formation of determinants, and exemplified some of their properties in particular cases. We shall in this Lesson prove these properties in general, together with some others, which are most frequently used in the reduction and calculation of determinants.

The value of a determinant is not altered if the vertical rows be written horizontally, and vice versâ (see Arts. 2, 5).

This follows immediately from the law of formation (Art. 10), which is perfectly symmetrical with respect to the columns and rows. One of the principal advantages of the notation with double suffixes is that it exhibits most distinctly the symmetry which exists between the horizontal and vertical lines.

13. *If any two rows (or two columns) be interchanged, the sign of the determinant is altered.*

For the effect of the change is evidently a single permutation of two of the letters (or of two of the suffixes), which by the law of formation causes a change of sign.

14. *If two rows (or if two columns) be identical, the determinant vanishes.*

For these two rows being interchanged, we ought (Art. 13) to have a change of sign: but the interchange of two identical lines can produce no change in the value of the determinant. Its value, then, does not alter when its sign is changed; that is to say, it is $= 0$.

This theorem also follows immediately from the definition of a determinant, as the result of elimination between n linear equations. For that elimination is performed by solving for the variables from $n-1$ of the equations, and substituting the values so found in the nth. But if this n^{th} equation be the same as one of the others, it must vanish identically when these values are substituted in it.

15. *If every constituent in any row (or in any column) be multiplied by the same factor, then the determinant is multiplied by that factor.*

This follows at once from the fact that every term in the expansion of the determinant contains as a factor, one, and but one, constituent belonging to the same row or to the same column.

Thus, for example, since every element of the determinant

$$\begin{vmatrix} a_1, & b_1, & c_1 \\ a_2, & b_2, & c_2 \\ a_3, & b_3, & c_3 \end{vmatrix}$$

contains either $a_1, a_2,$ or a_3, the determinant can be written in the form $a_1A_1 + a_2A_2 + a_3A_3$ (where neither $A_1, A_2,$ nor A_3 contains any constituent from the a column); and if a_1, a_2, a_3 be each multiplied by the same factor k, the determinant will be multiplied by that factor.

COR. If the constituents in one row or column differ only by a constant multiplier from those in another row or column, the determinant vanishes. Thus

$$\begin{vmatrix} ka_2, & a_2, & a_3 \\ kb_2, & b_2, & b_3 \\ kc_2, & c_2, & c_3 \end{vmatrix} = k \begin{vmatrix} a_2, & a_2, & a_3 \\ b_2, & b_2, & b_3 \\ c_2, & c_2, & c_3 \end{vmatrix} = 0 \text{ (Art. 14)}.$$

16. If in any determinant we erase any number of rows and the same number of columns, the determinant formed with the remaining rows and columns is called a *minor* of the given determinant. The minors formed by erasing one row and one column may be called first minors; those formed by erasing two rows and two columns, second minors, and so on.

We have, in the last article, observed that if the constituents of one column of a determinant be $a_1, a_2, a_3,$ &c., the determinant may be written in the form $a_1A_1 + a_2A_2 + a_3A_3 + $ &c. And it is evident that A_1 is the minor obtained by erasing the line and column which contain a_1, &c. For every element of the determinant which contains a_1 can contain no other constituent from the column a or the line (1); and a_1 must be multiplied by all possible combinations of products of $n-1$ constituents, taken one from each of the other rows and columns. But the aggregate of these form the minor A_1.

10 REDUCTION AND CALCULATION OF DETERMINANTS.

Compare Art. 7. In like manner the determinant may be written $a_1 A_1 + b_1 B_1 + c_1 C_1 +$ &c., where B_1 is the minor formed by erasing the row and column which contain b_1.

17. *If all the constituents but one vanish in any row or column of a determinant of the nth order, its calculation is reduced to the calculation of a determinant of the $n-1^{th}$ order.* For, evidently, if a_2, a_3, &c. all vanish, the determinant $a_1 A_1 + a_2 A_2 +$ &c. reduces to the single term $a_1 A_1$; and A_1 is a determinant having one row and one column less than the given determinant.

18. *If every constituent in any row (or in any column) be resolvable into the sum of two others, the determinant is resolvable into the sum of two others.*

This follows from the principle used in Art. 16. Thus, if in the Example there given, we write $a_1 + a_1$ for a_1; $b_1 + \beta_1$ for b_1; $c_1 + \gamma_1$ for c_1; then the determinant becomes

$(a_1 + a_1) A_1 + (b_1 + \beta_1) B_1 + (c_1 + \gamma_1) C_1$

$= \{a_1 A_1 + b_1 B_1 + c_1 C_1\} + \{a_1 A_1 + \beta_1 B_1 + \gamma_1 C_1\}.$

Thus we have

$$\begin{vmatrix} a_1 + a_1, & a_2, & a_3 \\ b_1 + \beta_1, & b_2, & b_3 \\ c_1 + \gamma_1, & c_2, & c_3 \end{vmatrix} = \begin{vmatrix} a_1, & a_2, & a_3 \\ b_1, & b_2, & b_3 \\ c_1, & c_2, & c_3 \end{vmatrix} + \begin{vmatrix} a_1, & a_2, & a_3 \\ \beta_1, & b_2, & b_3 \\ \gamma_1, & c_2, & c_3 \end{vmatrix}.$$

In like manner, if the terms in any one column were each the sum of any number of others, the determinant could be resolved into the same number of others.

19. If again, in the preceding, the terms in the second column were also each the sum of others (if, for instance, we were to write for a_2, $a_2 + a_2$; for b_2, $b_2 + \beta_2$; for c_2, $c_2 + \gamma_2$), then each of the determinants on the right-hand side of the last equation could be resolved into the sum of others; and we see, without difficulty, that

$(a_1 + a_1, b_2 + \beta_2, c_3) = (a_1 b_2 c_3) + (a_1 \beta_2 c_3) + (a_1 b_2 c_3) + (a_1 \beta_2 c_3).$

And if each of the constituents in the first column could be resolved into the sum of m others, and each of those of the second into the sum of n others, then the determinant could be resolved into the sum of mn others. For we should first, as in the last Article, resolve the determinant into the sum of m others, by taking, instead of the first column, each one of the m partial columns; and then, in like manner, resolve each of these into n others, by dealing similarly with the second column. And so, in general, if each of the constituents of a determinant consist of the sum of a number of terms, so that each of the columns can be resolved into the sum of a number of partial columns (the first into m partial columns, the second into n, the third into p, &c.), then the determinant is equal to the sum of all the determinants which can be formed by taking, instead of each column, one of its partial columns; and the number of such determinants will be the product of the numbers m, n, p, &c.

20. *If the constituents of one row or column are respectively equal to the sum of the corresponding constituents of other rows or columns, multiplied respectively by constant factors, the determinant vanishes.* For in this case the determinant can be resolved into the sum of others which separately vanish. Thus

$$\begin{vmatrix} ka_2 + la_3, & a_2, & a_3 \\ kb_2 + lb_3, & b_2, & b_3 \\ kc_2 + lc_3, & c_2, & c_3 \end{vmatrix} = \begin{vmatrix} ka_2, & a_2, & a_3 \\ kb_2, & b_2, & b_3 \\ kc_2, & c_2, & c_3 \end{vmatrix} + \begin{vmatrix} la_3, & a_2, & a_3 \\ lb_3, & b_2, & b_3 \\ lc_3, & c_2, & c_3 \end{vmatrix}.$$

But the last two determinants vanish (Cor., Art. 15).

21. *A determinant is not altered if we add to each constituent of any row or column the corresponding constituents of any of the other rows or columns multiplied respectively by constant factors.* Thus

$$\begin{vmatrix} a_1 + ka_2 + la_3, & a_2, & a_3 \\ b_1 + kb_2 + lb_3, & b_2, & b_3 \\ c_1 + kc_2 + lc_3, & c_2, & c_3 \end{vmatrix} = \begin{vmatrix} a_1, & a_2, & a_3 \\ b_1, & b_2, & b_3 \\ c_1, & c_2, & c_3 \end{vmatrix} + \begin{vmatrix} ka_2 + la_3, & a_2, & a_3 \\ kb_2 + lb_3, & b_2, & b_3 \\ kc_2 + lc_3, & c_2, & c_3 \end{vmatrix}.$$

But the last determinant vanishes (Art. 20).* The following examples will show how the principles just explained are applied to simplify the calculation of determinants.

---

* The beginner will be careful to observe that though the determinant is not altered if we substitute in the first row $a_1 + ka_2 + la_3$ for a_1, &c.; yet if we make the same substitution in the second row for a_2, &c. we multiply the determinant by k; and if for a_3, &c. we multiply it by l.

12 REDUCTION AND CALCULATION OF DETERMINANTS.

Ex. 1. Let it be required to calculate the following determinant:

$$\begin{vmatrix} 9, & 13, & 17, & 4 \\ 18, & 28, & 33, & 8 \\ 30, & 40, & 54, & 13 \\ 24, & 37, & 46, & 11 \end{vmatrix} = \begin{vmatrix} 1, & 1, & 1, & 4 \\ 2, & 4, & 1, & 8 \\ 4, & 1, & 2, & 13 \\ 2, & 4, & 2, & 11 \end{vmatrix} = \begin{vmatrix} 1, & 1, & 1, & 1 \\ 2, & 4, & 1, & 1 \\ 4, & 1, & 2, & 6 \\ 2, & 4, & 2, & 3 \end{vmatrix}.$$

The second determinant is derived from the first by subtracting from the constituents of the first, second, and third columns, twice, three times, and four times, the corresponding constituents of the last column. The third determinant is derived from the second by subtracting the sum of the first three columns from the last. Whenever we have, as now, a determinant for which all the constituents of one row are equal, we can get by subtraction one for which all the constituents but one of one row vanish, and so reduce the calculation to that of a determinant of lower order (Art. 17). Thus subtracting the first column from each of those following, the determinant last written becomes

$$\begin{vmatrix} 1, & 0, & 0, & 0 \\ 2, & 2, & -1, & -1 \\ 4, & -3, & -2, & 2 \\ 2, & 2, & 0, & 1 \end{vmatrix} = \begin{vmatrix} 2, & -1, & -1 \\ -3, & -2, & 2 \\ 2, & 0, & 1 \end{vmatrix} = \begin{vmatrix} 4, & -1, & -1 \\ -7, & -2, & 2 \\ 0, & 0, & 1 \end{vmatrix} = \begin{vmatrix} 4, & -1 \\ -7, & -2 \end{vmatrix}.$$

The third of these follows from the one preceding by subtracting double the last column from the first, when we have a determinant of only the second order, whose value is $-8-7 = -15$.

Ex. 2. The calculation of the following determinant is necessary (*Solid Geometry*, p. 175):

$$\begin{vmatrix} 5, & -10, & 11, & 0 \\ -10, & -11, & 12, & 4 \\ 11, & 12, & -11, & 2 \\ 0, & 4, & 2, & -6 \end{vmatrix} = \begin{vmatrix} 5, & -10, & 11, & 0 \\ -32, & -35, & 34, & 0 \\ 11, & 12, & -11, & 2 \\ 1, & 5, & 3, & 0 \end{vmatrix} = -2 \begin{vmatrix} 5, & -10, & 11 \\ -32, & -35, & 34 \\ 1, & 5, & 3 \end{vmatrix}$$

$$= 10 \begin{vmatrix} 5, & -2, & 1 \\ 32, & 7, & 1 \\ 1, & 1, & 8 \end{vmatrix} = 10 \begin{vmatrix} 5, & -2, & 1 \\ 27, & 9, & 0 \\ -39, & 17, & 0 \end{vmatrix} = 90 \begin{vmatrix} 3, & 1 \\ -39, & 17 \end{vmatrix} = 90 \, (51 + 39) = 8100.$$

The first transformation is made by subtracting double the third row from the second, and adding the sum of the second and third to the fourth. In the next step it will be observed that since the sign of the term $a_1 b_2 c_4 d_3$ is opposite to that of $a_1 b_2 c_3 d_4$, when c_4 is the only constituent of the last column which does not vanish, the determinant becomes $-c_4 \, (a_1 b_2 d_3)$. In the next step, we add the second and third columns, we take out the factor 5 common to the second column, and the sign – common to the second row. We then subtract the first row from the second, and eight times the first row from the last, and the remainder is obvious.

Ex. 3.
$$\begin{vmatrix} 7, & -2, & 0, & 5 \\ -2, & 6, & -2, & 2 \\ 0, & -2, & 5, & 3 \\ 5, & 2, & 3, & 4 \end{vmatrix} = -972 \; (Solid \; Geometry, \text{ p. 168}).$$

Ex. 4.
$$\begin{vmatrix} 25, & -15, & 23, & -5 \\ -15, & -10, & 19, & 5 \\ 23, & 19, & -15, & 9 \\ -5, & 5, & 9, & -5 \end{vmatrix} = 194400 \; (Solid \; Geometry, \text{ p. 175}).$$

REDUCTION AND CALCULATION OF DETERMINANTS. 13

Ex. 5. Given n quantities a, β, γ, &c., to find the value of

$$\begin{vmatrix} 1, & 1, & 1, & 1, & \&c. \\ a, & \beta, & \gamma, & \delta, & \&c. \\ a^2, & \beta^2, & \gamma^2, & \delta^2, & \&c. \\ \cdots & \cdots & \cdots & \cdots & \cdots \\ a^{n-1}, & \beta^{n-1}, & \gamma^{n-1}, & \delta^{n-1}, & \&c. \end{vmatrix}.$$

It is evident (Art. 14) that this determinant would vanish if $a = \beta$, therefore $a - \beta$ is a factor in it. In like manner so is every other difference between any two of the quantities a, β, &c. The determinant is therefore

$$= \pm (a - \beta)(a - \gamma)(a - \delta)(\beta - \gamma)(\beta - \delta)(\gamma - \delta) \&c.$$

For the determinant is either equal to this product, or to the product multiplied by some factor. But there can be no factor containing a, β, &c., since the product contains a^{n-1}, β^{n-1}, &c.; and the determinant can contain no higher power of a, β, &c.; and by comparing the coefficients of a^n [1] it will be seen that the determinant contains no numerical factor. This example may also be treated in the same way as the next example.

Ex. 6. To calculate
$$\begin{vmatrix} 1, & 1, & 1, & 1 \\ a, & \beta, & \gamma, & \delta \\ a^2, & \beta^2, & \gamma^2, & \delta^2 \\ a^4, & \beta^4, & \gamma^4, & \delta^4 \end{vmatrix}.$$

Subtract the last column from each of the first three, and the determinant becomes divisible by $(a - \delta)(\beta - \delta)(\gamma - \delta)$ the quotient being

$$\begin{vmatrix} 1, & 1, & 1 \\ a + \delta, & \beta + \delta, & \gamma + \delta \\ a^3 + a^2\delta + a\delta^2 + \delta^3, & \beta^3 + \beta^2\delta + \beta\delta^2 + \delta^3, & \gamma^3 + \gamma^2\delta + \gamma\delta^2 + \delta^3 \end{vmatrix}.$$

Subtract again the last column from the two preceding and the determinant is seen to be divisible by $(a - \gamma)(\beta - \gamma)$, and its value is thus at once found to be

$$(a - \delta)(\beta - \delta)(\gamma - \delta)(a - \gamma)(\beta - \gamma)(a - \beta)(a + \beta + \gamma + \delta).$$

Ex. 7. In the solution of a geometrical problem it became necessary to determine λ from the equation

$$\begin{vmatrix} a^3, & b^3, & c^3 \\ (a + \lambda)^3, & (b + \lambda)^3, & (c + \lambda)^3 \\ (2a + \lambda)^3, & (2b + \lambda)^3, & (2c + \lambda)^3 \end{vmatrix} = 0.$$

Subtract the first row from the second, and divide by λ; subtract 8 times the first row from the last and divide by λ; then subtract the second row from the third and divide by 3; and lastly, subtract this last row from the second and divide by λ, when the determinant becomes

$$\begin{vmatrix} a^3, & b^3, & c^3 \\ 2a + \lambda, & 2b + \lambda, & 2c + \lambda \\ 3a^2 + a\lambda, & 3b^2 + b\lambda, & 3c^2 + c\lambda \end{vmatrix} = 0.$$

Again, subtract the first column from the second and third, and then the second from the third, divide by $b - a$, $c - a$, $c - b$; and then from the first column subtract a times the second and add ab times the last; and from the second column take $(a + b)$ times the last, and we have finally

$$\begin{vmatrix} abc, & -(ab + bc + ca), & a + b + c \\ \lambda, & 2, & 0 \\ 0, & \lambda, & 3 \end{vmatrix} = 0,$$

which reduced is

$$(a + b + c)\lambda^2 + 3(ab + bc + ca)\lambda + 6abc = 0.$$

Ex. 8.
$$\begin{vmatrix} (b+c)^2, & a^2, & a^2 \\ b^2, & (c+a)^2, & b^2 \\ c^2, & c^2, & (a+b)^2 \end{vmatrix} = 2abc\,(a+b+c)^3.$$

Ex. 9.
$$\begin{vmatrix} 1, & 1, & 1 \\ \sin\alpha, & \sin\beta, & \sin\gamma \\ \cos\alpha, & \cos\beta, & \cos\gamma \end{vmatrix} = 4\sin\tfrac{1}{2}(\alpha-\beta)\sin\tfrac{1}{2}(\beta-\gamma)\sin\tfrac{1}{2}(\alpha-\gamma).$$

Ex. 10.
$$\begin{vmatrix} \cos\tfrac{1}{2}(\alpha-\beta), & \cos\tfrac{1}{2}(\beta-\gamma), & \cos\tfrac{1}{2}(\gamma-\alpha) \\ \cos\tfrac{1}{2}(\alpha+\beta), & \cos\tfrac{1}{2}(\beta+\gamma), & \cos\tfrac{1}{2}(\gamma+\alpha) \\ \sin\tfrac{1}{2}(\alpha+\beta), & \sin\tfrac{1}{2}(\beta+\gamma), & \sin\tfrac{1}{2}(\gamma+\alpha) \end{vmatrix} = 2\sin\tfrac{1}{2}(\alpha-\beta)\sin\tfrac{1}{2}(\beta-\gamma)\sin\tfrac{1}{2}(\alpha-\gamma).$$

Ex. 11.
$$\begin{vmatrix} \sin\alpha, & \sin\beta, & \sin\gamma \\ \cos\alpha, & \cos\beta, & \cos\gamma \\ \sin\alpha\cos\alpha, & \sin\beta\cos\beta, & \sin\gamma\cos\gamma \end{vmatrix}$$
$$= 2\sin\tfrac{1}{2}(\alpha-\beta)\sin\tfrac{1}{2}(\beta-\gamma)\sin\tfrac{1}{2}(\alpha-\gamma)\{\sin(\alpha+\beta)+\sin(\beta+\gamma)+\sin(\gamma+\alpha)\}.$$

Ex. 12. Many of these examples may be applied to the calculation of areas of triangles, it being remembered that the double area of the triangle formed by three points is
$$\begin{vmatrix} 1, & 1, & 1 \\ x', & x'', & x''' \\ y', & y'', & y''' \end{vmatrix},$$
and by three lines $ax+by+c$, &c. is $\begin{vmatrix} a, & b, & c \\ a', & b', & c' \\ a'', & b'', & c'' \end{vmatrix}^2$ divided by $(ab'-a'b)(ac'-ca')(bc'-cb')$
(see *Conic Sections*, p. 33). For example, the area of the triangle formed by the centres of curvature of three points on a parabola is (the co-ordinates of a centre of curvature being $\tfrac{1}{2}p + 3x'$, $-\dfrac{4y'^3}{p^2}$)

$$\dfrac{6}{p^3}\begin{vmatrix} 1, & 1, & 1 \\ y'^2, & y''^2, & y'''^2 \\ y'^3, & y''^3, & y'''^3 \end{vmatrix} = \dfrac{6}{p^3}(y'-y'')(y''-y''')(y'''-y')(y'y''+y''y'''+y'''y').$$

In like manner may be investigated the area of the triangle formed by three normals, or any other three lines connected with the curve.

Ex. 13. $\begin{vmatrix} 0, & c, & b \\ c, & 0, & a \\ b, & a, & 0 \end{vmatrix} = 2abc;$ $\begin{vmatrix} 0, & c, & b, & d \\ c, & 0, & a, & e \\ b, & a, & 0, & f \\ d, & e, & f, & 0 \end{vmatrix} = a^2d^2+b^2e^2+c^2f^2-2abde-2bcef-2adcf.$

Ex. 14. Prove $\begin{vmatrix} 0, & 1, & 1, & 1 \\ 1, & 0, & z^2, & y^2 \\ 1, & z^2, & 0, & x^2 \\ 1, & y^2, & x^2, & 0 \end{vmatrix} = \begin{vmatrix} 0, & x, & y, & z \\ x, & 0, & z, & y \\ y, & z, & 0, & x \\ z, & y, & x, & 0 \end{vmatrix}$
$$= -(x+y+z)(y+z-x)(z+x-y)(x+y-z).$$

Ex. 15.
$$\begin{vmatrix} a, & \lambda, & \lambda, & \lambda, & \&c. \\ \lambda, & b, & \lambda, & \lambda, & \&c. \\ \lambda, & \lambda, & c, & \lambda, & \&c. \\ \lambda, & \lambda, & \lambda, & d, & \&c. \\ & & \&c. & & \end{vmatrix},$$

REDUCTION AND CALCULATION OF DETERMINANTS. 15

where all the constituents are equal except those in the principal diagonal, is $\phi(\lambda) - \lambda \frac{d\phi}{d\lambda}$; where $\phi(\lambda)$ is the continued product $(a - \lambda)(b - \lambda)$, &c.

Ex. 16. Let u be a homogeneous function of the nth order in any number of variables; and let u_1, u_2, u_3, &c. denote its differential coefficients with regard to the variables x_1, x_2, x_3, &c.; and, in like manner, let u_{11}, u_{12}, u_{13} denote the differential coefficients of u_1, &c. Then, by Euler's theorem of homogeneous functions, we have

$$nu = u_1 x_1 + u_2 x_2 + u_3 x_3 + \&c., \quad (n-1) u_1 = x_1 u_{11} + x_2 u_{12} + x_3 u_{13} + \&c., \quad \&c.$$

We shall hereafter speak at length of the determinant (called the Hessian) formed with the second differential coefficients, whose rows are u_{11}, u_{12}, u_{13}, &c.; u_{21}, u_{22}, u_{23}, &c.; &c. At present our object is to show how to reduce a class of determinants of frequent occurrence, viz. those which are formed by bordering the matrix of the Hessian, either with the first differential coefficients, or with other quantities, as for example

$$\begin{vmatrix} u_{11}, & u_{12}, & u_{13}, & u_1, & a_1 \\ u_{21}, & u_{22}, & u_{23}, & u_2, & a_2 \\ u_{31}, & u_{32}, & u_{33}, & u_3, & a_3 \\ u_1, & u_2, & u_3, & 0, & 0 \\ a_1, & a_2, & a_3, & 0, & 0 \end{vmatrix}.$$

In this example we only take three variables, but the processes which we shall employ are applicable to the case of any number of variables. In this example the determinant formed by the first three rows and columns is the Hessian which we shall call H. We shall denote the determinant written above by the abbreviation $\begin{pmatrix} u & a \\ u & a \end{pmatrix}$, while we use $\begin{pmatrix} u \\ u \end{pmatrix}$, $\begin{pmatrix} a \\ a \end{pmatrix}$, $\begin{pmatrix} u \\ a \end{pmatrix}$ to denote the determinants of four rows, formed by bordering the matrix of the Hessian with a single row and column, either both u's or both a's, or one u and the other a. We also write $a_1 x_1 + a_2 x_2 + a_3 x_3 = a$. If now we multiply the first column of the above written determinant by x_1, the second by x_2, the third by x_3, and subtract from $n-1$ times the fourth column, the first three terms vanish, the fourth becomes $-nu$, and the fifth $-a$. Again, multiply the fourth row by $n-1$, and subtract in like manner the first, second, and third rows multiplied by x_1, x_2, x_3 respectively, and the first three terms vanish, the fourth remains unchanged, and the last becomes $-a$. Thus then $(n-1)^2$ times the determinant originally written is proved to be equal to

$$\begin{vmatrix} u_{11}, & u_{12}, & u_{13}, & 0, & a_1 \\ u_{21}, & u_{22}, & u_{23}, & 0, & a_2 \\ u_{31}, & u_{32}, & u_{33}, & 0, & a_3 \\ 0, & 0, & 0, & -n(n-1)u, & -a \\ a_1, & a_2, & a_3, & -a, & 0 \end{vmatrix}.$$

But now since (Art. 15) a determinant which has only two terms d_4, d_5 of the fourth row, which do not vanish, is expressible in the form $d_4 D_4 + d_5 D_5$; the above determinant may be resolved into the sum of two others, and we find that the originally given determinant

$$\begin{pmatrix} u\ a \\ u\ a \end{pmatrix} = -\frac{n}{n-1} u \begin{pmatrix} a \\ a \end{pmatrix} - \frac{1}{(n-1)^2} a^2 H.$$

In like manner it is proved that $\begin{pmatrix} u \\ u \end{pmatrix} = -\frac{n}{n-1} Hu$. Or, again, if there be four variables, and if the matrix of the Hessian be triply bordered, we prove in the same way that

$$\begin{pmatrix} u & \alpha & \beta \\ u & \alpha & \beta \end{pmatrix} = -\frac{n}{n-1} u \begin{pmatrix} \alpha & \beta \\ \alpha & \beta \end{pmatrix} - \frac{1}{(n-1)^2} \alpha^2 \begin{pmatrix} \beta \\ \beta \end{pmatrix} - \frac{1}{(n-1)^2} \beta^2 \begin{pmatrix} \alpha \\ \alpha \end{pmatrix} + \frac{2}{(n-1)^2} \alpha\beta \begin{pmatrix} \alpha \\ \beta \end{pmatrix}.$$

When u is of the second degree, it is to be noted that, in the case of three variables, $\begin{pmatrix} \alpha \\ \alpha \end{pmatrix}$ expresses the condition that the line α should touch the conic u; and $\begin{pmatrix} \alpha & \beta \\ \alpha & \beta \end{pmatrix}$ expresses the condition that the intersection of the lines α, β should be on the conic. In like manner for four variables $\begin{pmatrix} \alpha \\ \alpha \end{pmatrix}$, $\begin{pmatrix} \alpha & \beta \\ \alpha & \beta \end{pmatrix}$, $\begin{pmatrix} \alpha & \beta & \gamma \\ \alpha & \beta & \gamma \end{pmatrix}$ respectively are the conditions that a plane should touch, that the intersection of two planes should touch, and that the intersection of three planes should be on the quadric u. The equation then $\begin{pmatrix} u & \alpha & \beta \\ u & \alpha & \beta \end{pmatrix} = 0$ expresses that the polar plane of a point passes through one or other of the two points where the line $\alpha\beta$ meets the quadric. But points having this property lie only on the tangent planes at these two points. The transformation therefore that we have given for $\begin{pmatrix} u & \alpha & \beta \\ u & \alpha & \beta \end{pmatrix}$ expresses the equation of the tangent planes at the points where $\alpha\beta$ meets the quadric, and the transformation for $\begin{pmatrix} u & \alpha \\ u & \alpha \end{pmatrix}$ gives the equation of the tangent cone where α meets the quadric.

LESSON III.

MULTIPLICATION OF DETERMINANTS.

22. WE shall in this lesson show that the product of two determinants may be expressed as a determinant.

The product of two determinants is the determinant whose constituents are the sums of the products of the constituents in any row of one by the corresponding constituents in any row of the other.

For example, the product of the determinants $(a_1 b_2 c_3)$ and $(\alpha_1 \beta_2 \gamma_3)$ is

$$\begin{vmatrix} a_1\alpha_1 + b_1\beta_1 + c_1\gamma_1, & a_1\alpha_2 + b_1\beta_2 + c_1\gamma_2, & a_1\alpha_3 + b_1\beta_3 + c_1\gamma_3 \\ a_2\alpha_1 + b_2\beta_1 + c_2\gamma_1, & a_2\alpha_2 + b_2\beta_2 + c_2\gamma_2, & a_2\alpha_3 + b_2\beta_3 + c_2\gamma_3 \\ a_3\alpha_1 + b_3\beta_1 + c_3\gamma_1, & a_3\alpha_2 + b_3\beta_2 + c_3\gamma_2, & a_3\alpha_3 + b_3\beta_3 + c_3\gamma_3 \end{vmatrix}.$$

The proofs which we shall give for this particular case will apply equally in general. Since the constituents of the determinant just written are each the sum of three terms, the determinant can (by Art. 19) be resolved into the sum of the 27 determinants, obtained by taking any one partial column of the

MULTIPLICATION OF DETERMINANTS. 17

first, second, and third column. We need not write down the whole 27, but give two or three specimen terms:

$$\begin{vmatrix} a_1\alpha_1, & a_1\alpha_2, & a_1\alpha_3 \\ a_2\alpha_1, & a_2\alpha_2, & a_2\alpha_3 \\ a_3\alpha_1, & a_3\alpha_2, & a_3\alpha_3 \end{vmatrix} + \begin{vmatrix} a_1\alpha_1, & b_1\beta_2, & c_1\gamma_3 \\ a_2\alpha_1, & b_2\beta_2, & c_2\gamma_3 \\ a_3\alpha_1, & b_3\beta_2, & c_3\gamma_3 \end{vmatrix} + \begin{vmatrix} a_1\alpha_1, & c_1\gamma_2, & b_1\beta_3 \\ a_2\alpha_1, & c_2\gamma_2, & b_2\beta_3 \\ a_3\alpha_1, & c_3\gamma_2, & b_3\beta_3 \end{vmatrix} + \&c.$$

Now it will be observed that in all these determinants every column has a common factor, which (Art. 15) may be taken out as a multiplier of the entire determinant. The specimen terms already given may therefore be written in the form

$$\alpha_1\alpha_2\alpha_3 \begin{vmatrix} a_1, & a_1, & a_1 \\ a_2, & a_2, & a_2 \\ a_3, & a_3, & a_3 \end{vmatrix} + \alpha_1\beta_2\gamma_3 \begin{vmatrix} a_1, & b_1, & c_1 \\ a_2, & b_2, & c_2 \\ a_3, & b_3, & c_3 \end{vmatrix} + \alpha_1\gamma_2\beta_3 \begin{vmatrix} a_1, & c_1, & b_1 \\ a_2, & c_2, & b_2 \\ a_3, & c_3, & b_3 \end{vmatrix}.$$

But the first of these determinants vanishes, since two columns are the same; the second is the determinant $(a_1b_2c_3)$; and the third (Art. 13) is $= -(a_1b_2c_3)$. In like manner, every other partial determinant will vanish which has two columns the same; and it will be found that every determinant which does not vanish will be $(a_1b_2c_3)$, while the factors which multiply it will be the elements of the determinant $(\alpha_1\beta_2\gamma_3)$.

It would have been equally possible to break up the determinant into a series of terms, every one of which would have been the determinant $(\alpha_1\beta_2\gamma_3)$ multiplied by one of the elements of $(a_1b_2c_3)$.

23. On account of the importance of this theorem, we give another proof, founded on our first definition of a determinant.

The determinant which we examined in the last Article is the result of elimination between the equations

$$(a_1\alpha_1 + b_1\beta_1 + c_1\gamma_1)x + (a_1\alpha_2 + b_1\beta_2 + c_1\gamma_2)y + (a_1\alpha_3 + b_1\beta_3 + c_1\gamma_3)z = 0,$$
$$(a_2\alpha_1 + b_2\beta_1 + c_2\gamma_1)x + (a_2\alpha_2 + b_2\beta_2 + c_2\gamma_2)y + (a_2\alpha_3 + b_2\beta_3 + c_2\gamma_3)z = 0,$$
$$(a_3\alpha_1 + b_3\beta_1 + c_3\gamma_1)x + (a_3\alpha_2 + b_3\beta_2 + c_3\gamma_2)y + (a_3\alpha_3 + b_3\beta_3 + c_3\gamma_3)z = 0.$$

Now if we write

$$\alpha_1 x + \alpha_2 y + \alpha_3 z = X,$$
$$\beta_1 x + \beta_2 y + \beta_3 z = Y,$$
$$\gamma_1 x + \gamma_2 y + \gamma_3 z = Z,$$

the three preceding equations may be written

$$a_1 X + b_1 Y + c_1 Z = 0,$$
$$a_2 X + b_2 Y + c_2 Z = 0,$$
$$a_3 X + b_3 Y + c_3 Z = 0,$$

from which eliminating X, Y, Z, we see at once that $(a_1 b_2 c_3)$ must be a factor in the result. But also a system of values of x, y, z can be found to satisfy the three given equations, provided that a system can be found to satisfy simultaneously the equations $X = 0$, $Y = 0$, $Z = 0$. Hence $(\alpha_1 \beta_2 \gamma_3) = 0$, which is the condition that the latter should be possible, is also a factor in the result. And since we can see without difficulty that the degree of the result in the coefficients is exactly the same as that of the product of these quantities, the result is $(a_1 b_2 c_3)(\alpha_1 \beta_2 \gamma_3)$.

It appears from the present Article that the theorem concerning the multiplication of determinants can be expressed in the following form, in which we shall frequently employ it: *If a system of equations*

$$a_1 X + b_1 Y + c_1 Z = 0, \quad a_2 X + b_2 Y + c_2 Z = 0, \quad a_3 X + b_3 Y + c_3 Z = 0$$

be transformed by the substitutions

$$X = \alpha_1 x + \alpha_2 y + \alpha_3 z, \quad Y = \beta_1 x + \beta_2 y + \beta_3 z, \quad Z = \gamma_1 x + \gamma_2 y + \gamma_3 z,$$

then the determinant of the transformed system will be equal to $(a_1 b_2 c_3)$ the determinant of the original system, multiplied by $(\alpha_1 \beta_2 \gamma_3)$, which we shall call the modulus of transformation.

24. The theorems of the last Articles may be extended as follows: We might have two sets of constituents, the number of rows being different from the number of columns; for example

$$\left\| \begin{array}{ccc} a_1, & b_1, & c_1 \\ a_2, & b_2, & c_2 \end{array} \right\| \quad \left\| \begin{array}{ccc} \alpha_1, & \beta_1, & \gamma_1 \\ \alpha_2, & \beta_2, & \gamma_2 \end{array} \right\|,$$

and from these we could form, in the same manner as in the last Articles, the determinant

$$\left| \begin{array}{cc} a_1 \alpha_1 + b_1 \beta_1 + c_1 \gamma_1, & a_2 \alpha_1 + b_2 \beta_1 + c_2 \gamma_1 \\ a_1 \alpha_2 + b_1 \beta_2 + c_1 \gamma_2, & a_2 \alpha_2 + b_2 \beta_2 + c_2 \gamma_2 \end{array} \right|,$$

whose value we purpose to investigate.

MULTIPLICATION OF DETERMINANTS.

Now, first, let the number of columns be greater than the number of rows, as in the example just written; so that each constituent of the new determinant is the sum of a number of terms greater than the number of rows: then proceeding as in Art. 22, the value of the determinant is

$$\begin{vmatrix} a_1\alpha_1, & a_2\alpha_1 \\ a_1\alpha_2, & a_2\alpha_2 \end{vmatrix} + \begin{vmatrix} a_1\alpha_1, & b_2\beta_1 \\ a_1\alpha_2, & b_2\beta_2 \end{vmatrix} + \&c.$$

$$= (a_1 b_2)(\alpha_1 \beta_2) + (a_1 c_2)(\alpha_1 \gamma_2) + (b_1 c_2)(\beta_1 \gamma_2).$$

That is to say, *the new determinant is the sum of the products of every possible determinant which can be formed out of the one set of constituents by the corresponding determinant formed out of the other set of constituents.*

25. But in the second place, let the number of rows exceed the number of columns. Thus, from the two sets of constituents,

$$\begin{Vmatrix} a_1, & b_1 \\ a_2, & b_2 \\ a_3, & b_3 \end{Vmatrix} \quad \begin{Vmatrix} \alpha_1, & \beta_1 \\ \alpha_2, & \beta_2 \\ \alpha_3, & \beta_3 \end{Vmatrix},$$

let us form the determinant

$$\begin{vmatrix} a_1\alpha_1 + b_1\beta_1, & a_2\alpha_1 + b_2\beta_1, & a_3\alpha_1 + b_3\beta_1 \\ a_1\alpha_2 + b_1\beta_2, & a_2\alpha_2 + b_2\beta_2, & a_3\alpha_2 + b_3\beta_2 \\ a_1\alpha_3 + b_1\beta_3, & a_2\alpha_3 + b_2\beta_3, & a_3\alpha_3 + b_3\beta_3 \end{vmatrix}.$$

Then when we proceed to break this up into partial determinants in the manner already explained, it will be found impossible to form any partial determinant which shall not have two columns the same. *The determinant therefore will vanish identically.* Or this may be seen immediately by adding a column of cyphers to each matrix and then multiplying; when we get the determinant last written as the product of two factors each equal cypher.

26. A useful particular case of Art. 22 is, that *the square of a determinant is a symmetrical determinant* (see Art. 10). Thus the square of $(a_1 b_2 c_3)$ is

$$\begin{vmatrix} a_1^2 + b_1^2 + c_1^2, & a_1 a_2 + b_1 b_2 + c_1 c_2, & a_1 a_3 + b_1 b_3 + c_1 c_3 \\ a_1 a_2 + b_1 b_2 + c_1 c_2, & a_2^2 + b_2^2 + c_2^2, & a_2 a_3 + b_2 b_3 + c_2 c_3 \\ a_1 a_3 + b_1 b_3 + c_1 c_3, & a_2 a_3 + b_2 b_3 + c_2 c_3, & a_3^2 + b_3^2 + c_3^2 \end{vmatrix}.$$

Again, it appears by Art. 24 that the sum of the squares of the determinants $(a_1 b_2)^2 + (b_1 c_2)^2 + (c_1 a_2)^2$ is the determinant

$$\begin{vmatrix} a_1^2 + b_1^2 + c_1^2, & a_1 a_2 + b_1 b_2 + c_1 c_2 \\ a_1 a_2 + b_1 b_2 + c_1 c_2, & a_2^2 + b_2^2 + c_2^2 \end{vmatrix}.$$

Ex. 1. If a_1, b_1, c_1; a_2, b_2, c_2 be the direction-cosines of two lines in space, and θ their inclination to each other, $\cos \theta = a_1 a_2 + b_1 b_2 + c_1 c_2$; and the identity last proved gives $\sin^2 \theta = (a_1 b_2)^2 + (b_1 c_2)^2 + (c_1 a_2)^2$.

Ex. 2. In the theory of equations it is important to express the product of the squares of the differences of the roots; now the product of the differences of n quantities has been expressed as a determinant (Ex. 5, p. 13), and if we form the square of this determinant we obtain

$$\begin{vmatrix} s_0, & s_1, & s_2 & \ldots & s_{n-1} \\ s_1, & s_2, & s_3 & \ldots & s_n \\ s_2, & s_3, & s_4 & \ldots & s_{n+1} \\ \ldots\ldots\ldots\ldots\ldots\ldots\ldots \\ s_{n-1}, & s_n, & s_{n+1} \ldots s_{2n-2} \end{vmatrix},$$

where s_p denotes the sum of the pth powers of the quantities a, β, &c.

Ex. 3. In like manner it is proved by Art. 24 that the determinant

$$\begin{vmatrix} s_0, & s_1 \\ s_1, & s_2 \end{vmatrix} = \Sigma (a - \beta)^2,$$

$$\begin{vmatrix} s_0, & s_1, & s_2 \\ s_1, & s_2, & s_3 \\ s_2, & s_3, & s_4 \end{vmatrix} = \Sigma (a - \beta)^2 (\beta - \gamma)^2 (\gamma - a)^2.$$

We thus form a series of determinants, the last of which is the product of the squares of the differences of a, β, &c.; all similar determinants beyond this vanishing identically by Art. 25. This series of determinants is of great importance in the theory of algebraic equations.

Ex. 4. Let the origin be taken at the centre of the circle circumscribing a triangle, whose radius is R; and let M be the area of the triangle, then

$$2MR = \begin{vmatrix} x', & y', & R \\ x'', & y'', & R \\ x''', & y''', & R \end{vmatrix} \text{ and } -2MR = \begin{vmatrix} x', & y', & -R \\ x'', & y'', & -R \\ x''', & y''', & -R \end{vmatrix}.$$

Multiply these determinants according to the rule, and the first term $x'^2 + y'^2 - R^2$ vanishes; the second $x'x'' + y'y'' - R^2 = -\frac{1}{2}\{(x' - x'')^2 + (y' - y'')^2\} = -\frac{1}{2}c^2$, where c is a side of the triangle. Hence then

$$-4M^2 R^2 = -\frac{1}{8} \begin{vmatrix} 0, & c^2, & b^2 \\ c^2, & 0, & a^2 \\ b^2, & a^2, & 0 \end{vmatrix} = -\frac{1}{4} a^2 b^2 c^2,$$

whence $R = \dfrac{abc}{4M}$ as is well known.

Ex. 5. The same process may be used to find an expression for the radius of the sphere circumscribing a tetrahedron. Starting with the expression for the volume of the tetrahedron

$$6V = \begin{vmatrix} x', & y', & z', & 1 \\ x'', & y'', & z'', & 1 \\ x''', & y''', & z''', & 1 \\ x'''', & y'''', & z'''', & 1 \end{vmatrix}.$$

We find, as before, if a, d; b, e; c, f are pairs of opposite edges of the tetrahedron

$$-36R^2V^2 = \tfrac{1}{16}\begin{vmatrix} 0, & c^2, & b^2, & d^2 \\ c^2, & 0, & a^2, & e^2 \\ b^2, & a^2, & 0, & f^2 \\ d^2, & e^2, & f^2, & 0 \end{vmatrix},$$

whence if $ad + be + cf = 2S$; by Ex. 13, p. 14, we find

$$36R^2V^2 = S(S - ad)(S - be)(S - cf).$$

Ex. 6. The above proofs by Mr. Burnside were suggested by the following proof by Joachimsthal of an expression for the area of a triangle inscribed in an ellipse. Multiply the equations

$$\frac{2M}{ab} = \begin{vmatrix} \frac{x'}{a}, & \frac{y'}{b}, & 1 \\ \frac{x''}{a}, & \frac{y''}{b}, & 1 \\ \frac{x'''}{a}, & \frac{y'''}{b}, & 1 \end{vmatrix}; \quad -\frac{2M}{ab} = \begin{vmatrix} \frac{x'}{a}, & \frac{y'}{b}, & -1 \\ \frac{x''}{a}, & \frac{y''}{b}, & -1 \\ \frac{x'''}{a}, & \frac{y'''}{b}, & -1 \end{vmatrix}.$$

And the product is a symmetrical determinant, of which the leading terms, such as $\frac{x'^2}{a^2} + \frac{y'^2}{b^2} - 1$, vanish when the points $x'y'$, $x''y''$, $x'''y'''$ are on the curve, while the other terms are $\frac{x'x''}{a^2} + \frac{y'y''}{b^2} - 1$, &c. Now it can easily be proved that if γ be a side of the triangle and b''' the parallel semi-diameter

$$\frac{\gamma^2}{b'''^2} = \frac{(x' - x'')^2}{a^2} + \frac{(y' - y'')^2}{b^2} = 2\left(1 - \frac{x'x''}{a^2} - \frac{y'y''}{b^2}\right).$$

Thus we have

$$-4\frac{M^2}{a^2b^2} = -\tfrac{1}{8}\begin{vmatrix} 0, & \frac{\gamma^2}{b'''^2}, & \frac{\beta^2}{b''^2} \\ \frac{\gamma^2}{b'''^2}, & 0, & \frac{a^2}{b'^2} \\ \frac{\beta^2}{b''^2}, & \frac{a^2}{b'^2}, & 0 \end{vmatrix} = -\frac{a^2\beta^2\gamma^2}{4b'^2b''^2b'''^2}.$$

Ex. 7. The following investigation of the relation connecting the mutual distances of four points on a circle (or five points on a sphere) is given by Mr. Cayley (*Cambridge and Dublin Mathematical Journal*, Vol. II., p. 270).

Substitute the co-ordinates of each point in the general equation of a circle

$$x^2 + y^2 - 2Ax - 2By + C = 0,$$

and eliminate A, B, C; when we get a determinant with four rows such as

$$x'^2 + y'^2, \quad -2x', \quad -2y', \quad 1.$$

Multiply this by the determinant (which only differs by a numerical factor from the preceding) whose four rows are such as $1, x', y', x'^2 + y'^2$; and the first term of the product determinant vanishes, the second being $(x' - x'')^2 + (y' - y'')^2$. If then the square of the distance between two of the points be $(12)^2$, the product determinant is

$$\begin{vmatrix} 0, & (12)^2, & (13)^2, & (14)^2 \\ (21)^2, & 0, & (23)^2, & (24)^2 \\ (31)^2, & (32)^2, & 0, & (34)^2 \\ (41)^2, & (42)^2, & (43)^2, & 0 \end{vmatrix} = 0,$$

which is the relation required. As has been already seen, this determinant expanded gives the well-known relation $(12)(34) \pm (13)(24) \pm (14)(23) = 0$. The relation connecting five points on a sphere is the corresponding determinant with five rows.

Ex. 8. To find a relation connecting the mutual distances of three points on a line, four points on a plane, or five points in space. We prefix a unit and cyphers to the two determinants which we multiplied in the last example, thus

$$\begin{vmatrix} 1, & 0, & 0, & 0 \\ x'^2+y'^2, & -2x', & -2y', & 1 \\ & \&c. & & \end{vmatrix} \times \begin{vmatrix} 0, & 0, & 0, & 1 \\ 1, & x', & y', & x'^2+y'^2 \\ & \&c. & & \end{vmatrix}.$$

We have now got five rows and only four columns, therefore the product formed as in Art. 25, will vanish identically. But this is the determinant

$$\begin{vmatrix} 0, & 1, & 1, & 1, & 1 \\ 1, & 0, & (12)^2, & (13)^2, & (14)^2 \\ 1, & (21)^2, & 0, & (23)^2, & (24)^2 \\ 1, & (31)^2, & (32)^2, & 0, & (34)^2 \\ 1, & (41)^2, & (42)^2, & (43)^2, & 0 \end{vmatrix} = 0,$$

which is the relation required. If we erase the outside row and column we have the relation connecting three points on a line; and if we add another row $1, (51)^2, (52)^2$, &c., we get the relation connecting the mutual distances of five points in space. We might proceed to calculate these determinants by subtracting the second column from each of these succeeding, and then the first row from those succeeding, when we get

$$\begin{vmatrix} 2(12)^2, & (12)^2+(13)^2-(23)^2, & (12)^2+(14)^2-(24)^2 \\ (12)^2+(13)^2-(23)^2, & 2(13)^2, & (13)^2+(14)^2-(34)^2 \\ (12)^2+(14)^2-(24)^2, & (13)^2+(14)^2-(34)^2, & 2(14)^2 \end{vmatrix} = 0.$$

Now the determinants might have been obtained directly in this reduced but unsymmetrical form by taking the origin at the point (1), and forming, as in Art. 25, with the constituents $x'y', x''y''$, &c., the determinant which vanishes identically,

$$\begin{vmatrix} x'^2+y'^2, & x'x''+y'y'', & x'x'''+y'y''' \\ x'x''+y'y'', & x''^2+y''^2, & x''x'''+y''y''' \\ x'x'''+y'y''', & x''x'''+y''y''', & x'''^2+y'''^2 \end{vmatrix},$$

which it will readily be seen is equivalent to that last written.

Ex. 9. To find the relation connecting the arcs which join four points on a sphere. Take the origin at the centre of the sphere, and form with the direction-cosines of the radii vectores to each point, $\cos a', \cos \beta', \cos \gamma'$; $\cos a''$, &c. a determinant which vanishes identically, and it will be

$$\begin{vmatrix} 1, & \cos ab, & \cos ac, & \cos ad \\ \cos ba, & 1, & \cos bc, & \cos bd \\ \cos ca, & \cos cb, & 1, & \cos cd \\ \cos da, & \cos db, & \cos dc, & 1 \end{vmatrix} = 0.$$

If we substitute for each cosine, $\cos ab, 1-\dfrac{(ab)^2}{2r^2}+$ &c., and then suppose r the radius of the sphere to be infinite, we derive from the determinant of this article, that of the last article connecting four points on a plane.

Ex. 10. If $\phi(\lambda) = \begin{vmatrix} a-\lambda, & h, & g, \\ h, & b-\lambda, & f, \\ g, & f, & c-\lambda \end{vmatrix}$, calculate $\phi(\lambda).\phi(-\lambda)$. The determinant is one of like form with λ^2 instead of λ, the first line being $A-\lambda^2$, H, G, &c., where

$$A = a^2+h^2+g^2, \quad B = b^2+f^2+h^2, \quad C = c^2+g^2+f^2,$$
$$F = gh+f(b+c), \quad G = hf+g(c+a), \quad H = fg+h(a+b),$$

and the expanded determinant equated to cypher gives $\lambda^6 - L\lambda^4 + M\lambda^2 - N = 0$, where
$L = a^2 + b^2 + c^2 + 2(f^2 + g^2 + h^2)$,
$M = (bc - f^2)^2 + (ca - g^2)^2 + (ab - h^2)^2 + 2(af - gh)^2 + 2(bg - hf)^2 + 2(ch - fg)^2$,
and N is the square of the original determinant with λ in it $= 0$; L, M, N are then all essentially positive quantities. In like manner if $\phi(\lambda)$ be formed similarly from any symmetrical determinant, $\phi(\lambda) \phi(-\lambda)$ equated to nothing, gives an equation for λ^2, whose signs are alternately positive and negative, which therefore by Des Cartes's rule cannot have a negative root. The above constitutes Mr. Sylvester's proof that the roots of the equation $\phi(\lambda) = 0$ are all real. It is evident, from what has been just said, that no root can be of the form $\beta \sqrt{(-1)}$, and in order to see that no root can be of the form $a + \beta \sqrt{(-1)}$, it is only necessary to write $a - a = a'$, $b - a = b'$, $c - a = c'$ when the case is reduced to the preceding.

LESSON IV.

MINOR AND RECIPROCAL DETERMINANTS.

27. We have seen (Art. 16) that the minors of any determinant are connected with the corresponding constituents by the relation

$$a_1 A_1 + a_2 A_2 + a_3 A_3 + \&c. = \Delta,$$

and these minors are connected with the other constituents by the identical relations

$$b_1 A_1 + b_2 A_2 + b_3 A_3 + \&c. = 0,$$
$$c_1 A_1 + c_2 A_2 + c_3 A_3 + \&c. = 0, \&c.$$

For since the determinant is equal to $a_1 A_1 + a_2 A_2 + \&c.$, and since $A_1, A_2, \&c.$, do not contain $a_1, a_2, \&c.$, therefore $b_1 A_1 + b_2 A_2 + \&c.$, is what the determinant would become if we were to make in it $a_1 = b_1$, $a_2 = b_2$, &c.; but the determinant would then have two columns identical, and would therefore vanish (Art. 14).

28. We can now briefly write the solution of a system of equations

$$a_1 x + b_1 y + c_1 z + \&c. = \xi,$$
$$a_2 x + b_2 y + c_2 z + \&c. = \eta,$$
$$a_3 x + b_3 y + c_3 z + \&c. = \zeta, \&c.,$$

for, multiply the first by A_1, the second by A_2, &c., and add, and the coefficients of y, z, &c., will vanish identically, while the coefficient of x will be $a_1A_1 + a_2A_2 + $ &c., which is the determinant formed out of the coefficients on the left-hand side of the equation, which we shall call Δ. Thus we get

$$\Delta x = A_1\xi + A_2\eta + A_3\zeta + \text{&c.},$$
$$\Delta y = B_1\xi + B_2\eta + B_3\zeta + \text{&c.},$$
$$\Delta z = C_1\xi + C_2\eta + C_3\zeta + \text{&c.}, \text{&c.}$$

29. The *reciprocal* of a given determinant is the determinant whose constituents are the minors corresponding to each constituent of the given one. Thus the reciprocal of $(a_1 b_2 c_3)$ is

$$\begin{vmatrix} A_1, & B_1, & C_1 \\ A_2, & B_2, & C_2 \\ A_3, & B_3, & C_3 \end{vmatrix},$$

where A_1, B_1, &c., have the meaning already explained. If we call this reciprocal Δ', and multiply it by the original determinant Δ, by the rule of Art. 22 we get

$$\begin{vmatrix} a_1A_1+b_1B_1+c_1C_1, & a_2A_1+b_2B_1+c_2C_1, & a_3A_1+b_3B_1+c_3C_1 \\ a_1A_2+b_1B_2+c_1C_2, & a_2A_2+b_2B_2+c_2C_2, & a_3A_2+b_3B_2+c_3C_2 \\ a_1A_3+b_1B_3+c_1C_3, & a_2A_3+b_2B_3+c_2C_3, & a_3A_3+b_3B_3+c_3C_3 \end{vmatrix}.$$

But (Art. 27) $a_1A_1 + b_1B_1 + c_1C_1 = \Delta$, $a_1A_2 + b_1B_2 + c_1C_2 = 0$, &c. This determinant, therefore, reduces to

$$\begin{vmatrix} \Delta, & 0, & 0 \\ 0, & \Delta, & 0 \\ 0, & 0, & \Delta \end{vmatrix} = \Delta^3.$$

Hence $(a_1 b_2 c_3)(A_1 B_2 C_3) = (a_1 b_2 c_3)^3$; therefore $(A_1 B_2 C_3) = (a_1 b_2 c_3)^2$. And in general, $\Delta'\Delta = \Delta^n$; therefore $\Delta' = \Delta^{n-1}$.

30. If we take the second system of equations in Art. 28, and solve these back again for ξ, η, &c., in terms of Δx, Δy, &c., we get

$$\Delta'\xi = \mathfrak{a}_1\Delta x + \mathfrak{b}_1\Delta y + \mathfrak{c}_1\Delta z + \text{&c.},$$

where $\mathfrak{a}_1$, $\mathfrak{b}_1$, $\mathfrak{c}_1$ are the minors of the reciprocal determinant. But these values for ξ, η, ζ, &c. must be identical with the expressions originally given; hence remembering that $\Delta' = \Delta^{n-1}$, we get, by comparison of coefficients,

$$\mathfrak{a}_1 = \Delta^{n-2}a_1, \quad \mathfrak{b}_1 = \Delta^{n-2}b_1, \quad \mathfrak{c}_1 = \Delta^{n-2}c_1, \text{&c.},$$

which express, in terms of the original coefficients, the first minors of the reciprocal determinant.

31. We have seen that, considering any one column a of a determinant, every element contains as a factor a constituent from that column, and therefore the determinant can be written in the form $\Sigma a_p A_p$. In like manner, considering any two columns a, b of the determinant, it can be written in the form $\Sigma (a_p b_q) A_{p, q}$, where the sum $\Sigma (a_p b_q)$ is intended to express all possible determinants which can be formed by taking two rows of the given two columns.

For every element of the determinant contains as factors, a constituent from the column a, and another from the column b; and any term $a_p b_q c_r d_s$, &c., must, by the rule of signs, be accompanied by another, $-a_q b_p c_r d_s$, &c. Hence we see that the form of the determinant is $\Sigma (a_p b_q) A_{p, q}$; and, by the same reasoning as in Art. 16, we see that the multiplier $A_{p, q}$ is the minor formed by omitting the two rows and columns in which a_p, b_q occur.

In like manner, considering any p columns of the determinant, it can be expressed as the sum of all possible determinants that can be formed by taking any p rows of the selected columns, and multiplying the minor formed with them, by the *complemental* minor; that is to say, the minor formed by erasing these rows and columns. For example,

$(a_1 b_2 c_3 d_4 e_5)$
$= (a_1 b_2)(c_3 d_4 e_5) - (a_1 b_3)(c_2 d_4 e_5) + (a_1 b_4)(c_2 d_3 e_5) - (a_1 b_5)(c_2 d_3 e_4)$
$+ (a_2 b_3)(c_1 d_4 e_5) - (a_2 b_4)(c_1 d_3 e_5) + (a_2 b_5)(c_1 d_3 e_4) + (a_3 b_4)(c_1 d_2 e_5)$
$- (a_3 b_5)(c_1 d_2 e_4) + (a_4 b_5)(c_1 d_2 e_3).$

The sign of each term in the above is determined without difficulty by the rule of signs (Art. 8).

It is evident, as in Art. 27, that if we write in the above a c for every b, the sum $\Sigma (a_1 c_2)(c_3 d_4 e_5)$ must vanish identically, since it is what the determinant would become if the c column were equal to the b column.

32. The theorem of Art. 30 may be extended as follows: *Any minor of the order p which can be formed out of the inverse constituents A_1, B_1, &c., is equal to the complementary of the cor-*

responding minor of the original determinant, multiplied by the $(p-1)^{\text{th}}$ power of that determinant.

For example, in the case where the original determinant is of the fifth order,

$$(A_1B_2) = \Delta\, (c_3d_4e_5), \quad (A_1B_2C_3) = \Delta^2\, (d_4e_5), \&c.$$

The method in which this is proved in general will be sufficiently understood from the proof of the first Example. We have

$$\Delta x = A_1\xi + A_2\eta + A_3\zeta + A_4\omega + A_5v,$$

$$\Delta y = B_1\xi + B_2\eta + B_3\zeta + B_4\omega + B_5v.$$

Therefore

$$\Delta B_2 x - \Delta A_2 y = (A_1B_2)\,\xi + (A_3B_2)\,\zeta + (A_4B_2)\,\omega + (A_5B_2)\,v.$$

But we can get another expression for x in terms of the same five quantities, y, ξ, ζ, ω, v. For, consider the original equations,

$$\xi = a_1x + b_1y + c_1z + d_1w + e_1u,$$
$$\zeta = a_3x + b_3y + c_3z + d_3w + e_3u,$$
$$\omega = a_4x + b_4y + c_4z + d_4w + e_4u,$$
$$v = a_5x + b_5y + c_5z + d_5w + e_5u,$$

and eliminate z, w, u, when we get

$$(a_1c_3d_4e_5)x + (b_1c_3d_4e_5)y = (c_3d_4e_5)\,\xi - (c_4d_5e_1)\,\zeta + (c_5d_1e_3)\,\omega - (c_1d_3e_4)\,v\,;$$

and since $(a_1c_3d_4e_5)$ is by definition $= B_2$, comparing these equations with those got already, we find $(A_1B_2) = \Delta\,(c_3d_4e_5)$, &c. Q.E.D.

Ex. 1. If a determinant vanish, its minors A_1, A_2, &c. are respectively proportional to B_1, B_2, &c. For we have just proved that $A_1B_2 - A_2B_1 = \Delta C$, where C is the second minor obtained by suppressing the first two rows and columns. If then $\Delta = 0$, we have $A_1 : A_2 :: B_1 : B_2$, &c.

Ex. 2. A particular example of the above, which is of frequent occurrence, is obtained by applying these principles to the determinant considered, Ex. 16, p. 15. We thus find, using the notation of that example $\begin{pmatrix}\alpha\\\alpha\end{pmatrix}\begin{pmatrix}\beta\\\beta\end{pmatrix} - \begin{pmatrix}\alpha\\\beta\end{pmatrix}^2 = \Delta \begin{pmatrix}\alpha\,\beta\\\alpha\,\beta\end{pmatrix}$; (see *Solid Geometry*, p. 48).

*LESSON V.

SYMMETRICAL AND SKEW SYMMETRICAL DETERMINANTS.

33. In this lesson it is convenient to employ the double suffix notation, and to write the constituents a_{11}, a_{12}, &c.; and we therefore begin by expressing in this notation some of the results already obtained. We denote the constituents of the reciprocal determinant by α_{11}, α_{12}, &c., where, if a_{rs} be any constituent of the original, α_{rs} is the minor obtained by erasing the row and column which contain that constituent. The equations of Art. 27 may then be written

$$a_{r1}\alpha_{r1} + a_{r2}\alpha_{r2} + a_{r3}\alpha_{r3} + \&c. = \Delta,$$
$$a_{r1}\alpha_{r'1} + a_{r2}\alpha_{r'2} + a_{r3}\alpha_{r'3} + \&c. = 0,$$

or more briefly $\Sigma_s a_{rs}\alpha_{rs} = \Delta$, $\Sigma_s a_{rs}\alpha_{r's} = 0$; that is to say, the sum of the products $a_{rs}\alpha_{r's}$ (where we give every value to s from 1 to n) is $= 0$, when r and r' are different, and $= \Delta$ when $r = r'$.

Since any constituent a_{rs} enters into the determinant only in the first degree, it is obvious that the factor α_{rs}, which multiplies it, is the differential coefficient of the determinant taken with respect to a_{rs}; similarly, that the second minor (Art. 31) which multiplies the product of two constituents a_{mn}, a_{rs} is the second differential coefficient of the determinant taken with respect to these two constituents, &c.

If any of the constituents be functions of any variable x, the entire differential of the determinant, with regard to that variable, is evidently $\alpha_{11} \dfrac{da_{11}}{dx} + \alpha_{12} \dfrac{da_{12}}{dx} + \&c.$

Ex. If u_1, v_1, &c. denote the first differentials of u, v, &c. with respect to x; u_2, v_2 the second differentials, &c., prove

$$\frac{d}{dx}\begin{vmatrix} u, & v, & w \\ u_1, & v_1, & w_1 \\ u_2, & v_2, & w_2 \end{vmatrix} = \begin{vmatrix} u, & v, & w \\ u_1, & v_1, & w_1 \\ u_3, & v_3, & w_3 \end{vmatrix}.$$

The differential is the sum of the nine products of the minor obtained by suppressing each term into the differential of that term, viz.

$$(a_{11}u_1 + a_{12}v_1 + a_{13}w_1) + (a_{21}u_2 + a_{22}v_2 + a_{23}w_2) + (a_{31}u_3 + a_{32}v_3 + a_{33}w_3).$$

But the first three terms denote the result of changing in the given determinant the

---

* Portions marked thus may be omitted on a first reading.

first row into u_1, v_1, w_1, and therefore vanish; the second three terms vanish as denoting the result of changing the second row into u_2, v_2, w_2; and there only remain the last three terms which denote the result of changing the last row into u_3, v_3, w_3. The same proof evidently applies to the similar determinant of the nth order formed with n functions.

34. The determinant is said to be symmetrical (Art. 10) when every two conjugate constituents are equal ($a_{rs} = a_{sr}$). In this case it is to be observed that the corresponding minors will also be equal ($\alpha_{rs} = \alpha_{sr}$); for it easily appears that the determinant got by suppressing the rth row and the sth column, differs only by an interchange of rows for columns, from that got by suppressing the sth row and the rth column. It appears from the last article that if any constituent a_{sr} were given as any function of its conjugate a_{rs}, the differential coefficient of the determinant, with regard to a_{rs}, would be $\alpha_{rs} + \alpha_{sr}\dfrac{da_{sr}}{da_{rs}}$. In the present case then where $a_{rs} = a_{sr}$, $\alpha_{rs} = \alpha_{sr}$, the differential coefficient of the determinant, with regard to a_{rs}, is $2\alpha_{rs}$. The differential coefficient, however, with respect to one of the terms in the leading diagonal a_{rr}, remains as before α_{rr}, since such a term has no conjugate distinct from itself.

35. If, as before, α_{rs} denote the first minor of any determinant answering to any constituent a_{rs}, and if β_{ik} denote the first minor of the determinant α_{rs} answering to any constituent a_{ik}, which will, of course, be a second minor of the original determinant, then this last may be written

$$\Delta = a_{rs}\alpha_{rs} - \Sigma_{ik} a_{rk} a_{is} \beta_{ik};$$

where we are to give i every value except r, and k every value except s. For any element of the determinant which does not contain the constituent a_{rs} must contain some other constituent from the rth row and some other from the sth column; that is to say, must contain a product such as $a_{rk}a_{is}$ where i and k are two numbers different from r and s respectively. But as we have already seen the aggregate of all the terms which multiply a_{rs} is α_{rs}; and the coefficient of $a_{rk}a_{is}$ (by Art. 31) differs only in sign from that of $a_{rs}a_{ik}$; that is to say, differs only in sign from the coefficient of a_{ik} in α_{rs}. Therefore $-\beta_{ik}$ is the value of the coefficient in question.

Thus then if we have calculated a symmetric determinant of the $n-1^{\text{th}}$ order, we can see what additional terms occur in the determinant of the nth order. Let Δ be the determinant, D that obtained by suppressing the outside row and column, β_{rs} any minor of the latter, and we have

$$\Delta = D a_{nn} - \Sigma_r a_{nr}^2 \beta_{rr} - 2\Sigma_{rs} a_{nr} a_{ns} \beta_{rs},$$

where r is supposed to be different from s, and every value is to be given to r and s from 1 to $n-1$.

Again, we have occasion often, as at p. 15, to deal with determinants such as

$$\begin{vmatrix} a_{11}, & a_{12}, & a_{13}, & \lambda_1 \\ a_{12}, & a_{22}, & a_{23}, & \lambda_2 \\ a_{13}, & a_{23}, & a_{33}, & \lambda_3 \\ \lambda_1, & \lambda_2, & \lambda_3, & \end{vmatrix},$$

obtained by bordering a symmetric determinant horizontally and vertically with the same constituents. This is in fact a symmetric determinant of the order one higher, the last term vanishing, and is

$$-\{a_{11}\lambda_1^2 + a_{22}\lambda_2^2 + a_{33}\lambda_3^2 + 2a_{23}\lambda_2\lambda_3 + 2a_{13}\lambda_1\lambda_3 + 2a_{12}\lambda_1\lambda_2\},$$

or generally $\quad -\Sigma_r a_{rr} \lambda_r^2 - 2\Sigma_{rs} a_{rs} \lambda_r \lambda_s.$

36. *If any symmetric determinant vanishes, the same determinant bordered as in the last article, is, with sign changed if need be, a perfect square, when considered as a function of* λ_1, λ_2, λ_3, *&c.* We saw (Art. 32, Ex. 1) that when the determinant vanishes $a_{11} a_{22} = a_{12}^2$, &c., whence it is evident that a_{11}, a_{22}, &c. must have all the same sign, and where we have generally $a_{rs} = \pm \sqrt{(a_{rr} a_{ss})}$. Further, since it was shewn in the same example that when a determinant vanishes, the constituents in the second row are proportional to those in the first, it follows that the signs to be given to the radicals are not all arbitrary. If, for instance, in the above we write $a_{12} = +\sqrt{(a_{11}a_{22})}$, $a_{13} = +\sqrt{(a_{11}a_{33})}$, then we are forced to give the positive sign also to the square root in $a_{23} = \sqrt{(a_{22}a_{33})}$. Substituting then these values in the result of the last article, it becomes, if a_{11}, &c. be positive, the negative square,

$$-\{\lambda_1 \sqrt{(a_{11})} + \lambda_2 \sqrt{(a_{22})} + \lambda_3 \sqrt{(a_{33})} + \&c.\}^2,$$

30 SYMMETRICAL AND SKEW SYMMETRICAL DETERMINANTS.

and if a_{11}, &c. be negative the determinant is a positive square. What has been just proved may be stated a little differently. We may consider the bordered determinant as the original determinant; of which that obtained by suppressing the row and column containing λ is a first minor, and a_{33} obtained by suppressing the next outside row and column is a second minor. And what we have proved with respect to any symmetrical determinant wanting the last term a_{nn}, is that if the first minor obtained by erasing the outside row and column vanish, then the determinant itself and the second minor, similarly obtained, must have opposite signs. And this will be equally true if a_{nn} does not vanish. For in the expansion of the determinant, a_{nn} is multiplied by the first minor, which vanishes by hypothesis, and therefore the presence or absence of a_{nn} does not affect the truth of the result.

37. A *skew symmetric* determinant is one in which every term is equal to its conjugate with its sign changed. The terms a_{rr} in the leading diagonal, being each its own conjugate, must in this case vanish; otherwise each could not be equal to itself with sign changed.

A skew symmetrical determinant of odd degree vanishes. For if we multiply each row by -1; in other words, if we change the sign of every term, it is easy to see that we get the same result as if we were to read the columns of the original determinant as rows and *vice versâ*. Thus then, a skew symmetrical determinant is not altered when multiplied by $(-1)^n$; and therefore when n is odd, such a determinant must vanish.

It is easy to see that the minor a_{rr} differs by the sign of every term from the minor a_{rs}, and therefore $a_{rs} = (-1)^{n-1} a_{rs}$. Hence $a_{sr} = a_{rs}$ when n is odd, and is equal with contrary sign when n is even. a_{rr} is itself a skew symmetric determinant and therefore vanishes when the original determinant is of even degree.

The differential coefficient of the determinant, with regard to any constituent a_{rs}, being $a_{rs} + a_{sr} \dfrac{da_{sr}}{da_{rs}}$ is $a_{rs} - a_{sr}$. When therefore n is even it is $= 2a_{rs}$ and when n is odd it vanishes.

SYMMETRICAL AND SKEW SYMMETRICAL DETERMINANTS. 31

38. *Every skew symmetrical determinant of even degree is a perfect square.*

We have seen (Art. 35) that any determinant is

$$= a_{nn}\alpha_{nn} - \Sigma_{rs} a_{ns} a_{rn} \beta_{rs},$$

and in the present case a_{nn} vanishes, as does also α_{nn}, which is a skew symmetric of odd degree. On this account therefore we have, as in Art. 36, $\beta_{rs}^2 = \beta_{rr}\beta_{ss}$; and therefore exactly as in that article, the determinant is shown to be

$$\{a_{1n} \sqrt{(\beta_{11})} + a_{2n} \sqrt{(\beta_{22})} + a_{3n} \sqrt{(\beta_{33})} + \&c.\}^2.$$

The determinant is therefore a perfect square if β_{11}, β_{22} are perfect squares, but these are skew symmetrics of the order $n-2$. Hence the theorem of this article is true for determinants of order n if true for those of order $n-2$, and so on. But it is evidently true for a determinant of the second order

$$\begin{vmatrix} 0, & a_{12} \\ -a_{12}, & 0 \end{vmatrix},$$ which is $= a_{12}^2$. Hence it is generally true. *

39. We have seen that the square root of the determinant contains one term $a_{n-1,n} \sqrt{(\beta_{n-1,n-1})}$, where $\beta_{n-1,n-1}$ contains no terms with either of the suffixes $n-1$ or n. But taking any two of the remaining suffixes, such as $n-3$, $n-2$, we see that $\sqrt{(\beta_{n-1,n-1})}$ contains a term $a_{n-3,n-2} \sqrt{(\gamma_{n-3,n-3})}$, where $\gamma_{n-3,n-3}$ contains no term with any of the four suffixes of which account has already been taken. Proceeding in this manner we see that the square root will be the sum of a number of terms such as $a_{12}a_{34}a_{56}\ldots a_{n-1,n}$; each of which is the product of $\tfrac{1}{2}n$ constituents, and in which no suffix is repeated twice. The form however obtained in the last article $\alpha_{1n} \sqrt{(\beta_{11})} \pm \alpha_{2n}\sqrt{(\beta_{22})} + \&c.$ does not show what sign is to be affixed to each term. Thus if the method of the last article be applied to the skew symmetric of the fourth order, its square root appears to be $a_{12}a_{34} \pm a_{13}a_{24} \pm a_{14}a_{23}$; but it has not been shown which signs we are to choose. This however will appear from the following considerations: If in the given determinant we interchange any two suffixes 1, 2; since this amounts to a transposition of the first and second row, and also of the first and second column, the determinant is not altered. Its square root then must be

a function, such that if we interchange any two suffixes it will remain unaltered, or at most change sign. But that it *will* change sign is evident on considering any term $a_{12}a_{34}$, &c. which, if we interchange the suffixes 1 and 2, becomes $a_{21}a_{34}$, &c.; that is to say, changes sign, since $a_{21} = -a_{12}$. It follows then, in the particular example just considered, that the signs of the terms are $a_{12}a_{34} - a_{13}a_{24} + a_{14}a_{23}$; for if we give the second term a positive sign, the interchange of 2 and 3, which alters the sign of the last term would leave the first two unchanged. And generally the rule is, that the square root is the sum of all possible terms derived from $a_{12}a_{34}...a_{n-1,n}$ by interchange of the suffixes 2, 3...n, where, as in determinants, we change sign with every permutation.

40. We can reduce to the calculation of skew symmetric determinants, the calculation of what Mr. Cayley calls a skew determinant, viz. where, though the conjugate terms are equal with opposite signs, $a_{ik} = -a_{ki}$, yet the leading terms a_{ii}, a_{kk}, &c. do not vanish. We shall suppose, for simplicity, that these leading terms all have a common value, λ. We prefix the following lemma: If in any determinant we denote by D, the result of making all the leading terms $= 0$, by D_i what the minor corresponding to a_{ii} becomes when the leading terms are all made $= 0$, by D_{ik} what the second minor corresponding to $a_{ii}a_{kk}$ becomes when the leading terms vanish, &c.; then the given determinant, expanded as far as the leading terms are concerned, is
$$\Delta = D + \Sigma a_{ii}D_i + \Sigma a_{ii}a_{kk}D_{ik} + ... + a_{11}a_{22}...a_{nn},$$
where in the first sum i is given every value from 1 to n, where in the second sum i, k are any binary combinations of these numbers, &c.

For the part of the determinant which contains no leading term is evidently D. Since the terms which contain a_{ii} are $a_{ii}A_{ii}$, where A_{ii} is the corresponding minor, the terms which contain a_{ii} and no other leading term are got by making the leading terms $= 0$ in A_{ii}, &c.

41. If this lemma be applied to the case of the skew determinant defined in the last article, all the terms D_i, D_{ii}, &c.

are skew symmetric determinants; of which, those of odd order vanish, while those of even order are perfect squares. The term $a_{11}a_{22}...a_{nn}$ is λ^n, and the determinant is

$$\Delta = \lambda^n + \lambda^{n-2}\Sigma D_2 + \lambda^{n-4}\Sigma D_4 + \&c.,$$

where D_2, D_4, &c. denote skew symmetrical determinants of the second, fourth, &c. orders formed from the original in the manner explained in the last article.

Ex. 1. $\begin{vmatrix} \lambda, & a_{12}, & a_{13} \\ a_{21}, & \lambda, & a_{23} \\ a_{31}, & a_{32}, & \lambda \end{vmatrix}$, where $a_{21} = -a_{12}$, &c. $= \lambda^3 + \lambda(a_{12}^2 + a_{13}^2 + a_{23}^2)$.

Ex. 2. The similar skew determinant of the fourth order expanded is
$$\lambda^4 + \lambda^2(a_{12}^2 + a_{13}^2 + a_{14}^2 + a_{23}^2 + a_{24}^2 + a_{34}^2) + (a_{12}a_{34} + a_{13}a_{42} + a_{14}a_{23})^2.$$

42. Mr. Cayley (see *Crelle*, Vol. XXXII., p. 119) has applied the theory of skew determinants to that of orthogonal substitutions, of which we shall here give some account. It is known (see *Solid Geometry*, p. 10) that when we transform from one set of three rectangular axes to another, if a, b, c, &c. be the direction-cosines of the new axes, we have

$$X = ax + by + cz, \quad Y = a'x + b'y + c'z, \quad Z = a''x + b''y + c''z;$$

that we have $\quad X^2 + Y^2 + Z^2 = x^2 + y^2 + z^2,$

whence $a^2 + a'^2 + a''^2 = 1$, &c., $ab + a'b' + a''b'' = 0$, &c.;
that also we have

$$x = aX + a'Y + a''Z, \quad y = bX + b'Y + b''Z, \quad z = cX + c'Y + c''Z,$$

and that we have the determinant formed by

$$a, b, c; \; a', b', c'; \; \&c. = \pm 1.$$

It is also useful (in studying the theory of rotation for example) instead of using nine quantities a, b, c, &c. connected by six relations, to express all in terms of three independent variables. Now all this may be generalized as follows: If we have a function of any number of variables, it can be transformed by a linear substitution by writing

$$x = a_{11}X + a_{12}Y + a_{13}Z + \&c., \quad y = a_{21}X + a_{22}Y + a_{23}Z + \&c., \&c.,$$

and the substitution is called orthogonal if we have

$$x^2 + y^2 + z^2 + \&c. = X^2 + Y^2 + Z^2 + \&c.,$$

which implies the equations

$$a_{11}^2 + a_{21}^2 + \&c. = 1, \quad a_{11}a_{12} + a_{21}a_{22} + \&c. = 0, \; \&c.$$

34 SYMMETRICAL AND SKEW SYMMETRICAL DETERMINANTS.

Thus the n^2 quantities a_{11}, &c. are connected by $\tfrac{1}{2}n(n+1)$ relations and there are only $\tfrac{1}{2}n(n-1)$ of them independent.

We have then conversely

$$X = a_{11}x + a_{21}y + a_{31}z + \&c., \quad Y = a_{12}x + a_{22}y + a_{32}z + \&c., \&c.,$$

equations which are immediately verified by substituting on the right-hand side of the equations for x, y, z, &c. their values. And hence, the equation $X^2 + Y^2 + \&c. = x^2 + y^2 + \&c.$ gives us the new system of relations

$$a_{11}^2 + a_{12}^2 + \&c. = 1, \quad a_{11}a_{21} + a_{12}a_{22} + \&c. = 0.$$

Lastly, forming by the ordinary rule for multiplication of determinants, the square of the determinant formed with the n^2 quantities a_{11}, &c., every term of the square vanishes except the leading terms which are all $=1$. The value of the square is therefore $=1$. Thus the theorems which we know to be true in the case of determinants of the third order are generally true, and it only remains to show how to express the n^2 quantities in terms of $\tfrac{1}{2}n(n-1)$ independent quantities.

43. Let us suppose that we have a skew determinant of the $(n-1)^{\text{th}}$ order, b_{11}, b_{12}, &c. where $b_{ik} = -b_{ki}$, and $b_{11} = b_{22} = b_{ii} = 1$; and let us suppose that we form with these constituents two different sets of linear substitutions, viz.

$$x = b_{11}\xi + b_{12}\eta + b_{13}\zeta + \&c., \quad X = b_{11}\xi + b_{21}\eta + b_{31}\zeta + \&c.,$$
$$y = b_{21}\xi + b_{22}\eta + b_{23}\zeta + \&c., \quad Y = b_{12}\xi + b_{22}\eta + b_{32}\zeta + \&c.,$$
$$z = b_{31}\xi + b_{32}\eta + b_{33}\zeta + \&c., \quad Z = b_{13}\xi + b_{23}\eta + b_{33}\zeta + \&c.,$$

from adding which equations we have, in virtue of the given relations between b_{11}, b_{12}, &c.,

$$x + X = 2\xi, \quad y + Y = 2\eta, \&c.$$

If now the first set of equations be solved for ξ, η, &c. in terms of x, y, &c., we find, by Art. 28,

$$\Delta \xi = \beta_{11}x + \beta_{21}y + \beta_{31}z + \&c., \quad \Delta \eta = \beta_{12}x + \beta_{22}y + \&c.,$$

(where β_{11}, β_{12}, &c. are minors of the determinant in question;) and putting for 2ξ, $x + X$, &c., these equations give

$$\Delta X = (2\beta_{11} - \Delta)x + 2\beta_{21}y + 2\beta_{31}z + \&c.,$$
$$\Delta Y = 2\beta_{12}x + (2\beta_{22} - \Delta)y + 2\beta_{32}z + \&c., \&c.,$$

SYMMETRICAL AND SKEW SYMMETRICAL DETERMINANTS. 35

which express X, Y, &c. in terms of x, y, &c. But if we had solved from the second set of equations ξ, η, &c. in terms of X and Y, we should have got

$$\Delta\xi = \beta_{11}X + \beta_{12}Y + \beta_{13}Z + \&c., \quad \Delta\eta = \beta_{21}X + \beta_{22}Y + \beta_{23}Z + \&c.,$$

whence, as before,

$$\Delta x = (2\beta_{11} - \Delta)X + 2\beta_{12}Y + 2\beta_{13}Z + \&c.,$$
$$\Delta y = 2\beta_{21}X + (2\beta_{22} - \Delta)Y + 2\beta_{23}Z + \&c.$$

Thus, then, if we write

$$\frac{2\beta_{11} - \Delta}{\Delta} = a_{11}, \quad \frac{2\beta_{ii} - \Delta}{\Delta} = a_{ii}; \quad \frac{2\beta_{12}}{\Delta} = a_{12}, \quad \frac{2\beta_{ik}}{\Delta} = a_{ik},$$

we have x, y, &c. connected with X, Y, &c. by the relations

$$x = a_{11}X + a_{12}Y + \&c., \quad y = a_{21}X + a_{22}Y + \&c., \quad \&c.,$$
$$X = a_{11}x + a_{21}y + \&c., \quad Y = a_{12}x + a_{22}y + \&c., \quad \&c.$$

We have then x, y, &c., X, Y, &c. connected by an orthogonal substitution, for if we substitute in the value of x, the values of X, Y, &c. given by the second set of equations, in order that our results may be consistent, we must have

$$a_{11}^2 + a_{12}^2 + a_{13}^2 + \&c. = 1, \quad a_{11}a_{21} + a_{12}a_{22} + a_{13}a_{23} + \&c. = 0, \quad \&c.$$

Thus then we have seen that taking arbitrarily the $\frac{1}{2}n(n-1)$ quantities, b_{12}, b_{13}, &c., we are able to express in terms of these the coefficients of a general orthogonal transformation of the nth order.

Ex. 1. To form an orthogonal transformation of the second order. Write

$$\Delta = \begin{vmatrix} 1, & \lambda \\ -\lambda, & 1 \end{vmatrix} = 1 + \lambda^2,$$

then $\beta_{11} = \beta_{22} = 1$, $\beta_{12} = \lambda$, $\beta_{21} = -\lambda$, and our transformation is

$$(1 + \lambda^2)\,x = (1 - \lambda^2)X + 2\lambda Y, \quad (1 + \lambda^2)\,y = -2\lambda X + (1 - \lambda^2)Y,$$
$$(1 + \lambda^2)\,X = (1 - \lambda^2)x - 2\lambda y, \quad (1 + \lambda^2)\,Y = 2\lambda x + (1 - \lambda^2)y.$$

Ex. 2. To form an orthogonal transformation of the third order. Write

$$\Delta = \begin{vmatrix} 1, & \nu, & -\mu \\ -\nu, & 1, & \lambda \\ \mu, & -\lambda, & 1 \end{vmatrix} = 1 + \lambda^2 + \mu^2 + \nu^2.$$

Then the constituents of the reciprocal system are

$$\begin{vmatrix} 1 + \lambda^2, & \nu + \lambda\mu, & -\mu + \lambda\nu \\ -\nu + \lambda\mu, & 1 + \mu^2, & \lambda + \mu\nu \\ \mu + \lambda\nu, & -\lambda + \mu\nu, & 1 + \nu^2 \end{vmatrix},$$

consequently the coefficients of the orthogonal substitution hence derived are

$1 + \lambda^2 - \mu^2 - \nu^2$, $\quad 2(\nu + \lambda\mu)$, $\quad 2(\lambda\nu - \mu)$,

$2(\lambda\mu - \nu)$, $\quad 1 + \mu^2 - \lambda^2 - \nu^2$, $\quad 2(\mu\nu + \lambda)$,

$2(\lambda\nu + \mu)$, $\quad 2(\mu\nu - \lambda)$, $\quad 1 + \nu^2 - \lambda^2 - \mu^2$,

where each term is to be divided by $1 + \lambda^2 + \mu^2 + \nu^2$.*

*LESSON VI.

SYMMETRICAL DETERMINANTS.

44. IF we add the quantity λ to each of the leading terms of a symmetrical determinant, and equate the result to 0, we have an equation of considerable importance in analysis.† We have already given one proof (Mr. Sylvester's) that the roots of this equation are all real (Ex. 10, p. 22), and we purpose in this Lesson to give another proof by M. Borchardt (see *Liouville*, Vol. XII., p. 50) chiefly because the principles involved in this proof are worth knowing for their own sakes. First, however, we may remark that a simple proof may be obtained by the application of a principle proved in Art. 36. Take the determinant

$$\begin{vmatrix} a_{11} + \lambda, & a_{12}, & a_{13}, & \&c. \\ a_{21}, & a_{22} + \lambda, & a_{23}, & \&c. \\ a_{31}, & a_{32}, & a_{33} + \lambda, & \&c. \\ \&c. & & & \end{vmatrix},$$

and form from it a minor, as in Art. 36, by erasing the outside line and column: form from this again another minor by the

---

* The geometric meaning of these coefficients may be stated as follows: Write $\lambda = a \tan\tfrac{1}{2}\theta$, $\mu = b \tan\tfrac{1}{2}\theta$, $\nu = c \tan\tfrac{1}{2}\theta$, then the new axes may be derived from the old by rotating the system through an angle θ round an axis whose direction-cosines are a, b, c. The theory of orthogonal substitutions was first investigated by Euler, (*Nov. Comm. Petrop.*, Vol. XV., p. 75, and Vol. XX., p. 217) who gave formulæ for the transformation as far as the fourth order. The quantities λ, μ, ν, in the case of the third order, were introduced by Rodrigues, *Liouville*, Vol. V., p. 405. The general theory, explained above, connecting linear transformations with skew determinants was given by Cayley, *Crelle*, Vol. XXXII., p. 119.

† It occurs in the determination of the secular inequalities of the planets (see Laplace, *Mecanique Celeste*, Part I., Book II., Art. 56).

same rule, and so on. We shall thus have a series of functions of λ, whose degrees regularly diminish from the nth to the 1st; and we may take any positive constant to complete the series. Now, if we substitute successively in this series any two values of λ, and count in each case the variations of sign as in Sturm's theorem, it is easy to see that the difference in the number of variations cannot exceed the number of roots of the equation of the nth degree which lie between the two assumed values of λ. This appears at once from what was proved in Art. 36, that if λ be taken so as to make any of these minors vanish, the two adjacent functions in the series will have opposite signs. It follows, then, precisely as in the proof of Sturm's theorem, that if we diminish λ regularly from $+\infty$ to $-\infty$, then as λ passes through a root of any of these minors, the number of variations in the series will not be affected; and that a change in the number of variations can only take place when λ passes through a root of the first equation, namely, that in which λ enters in the nth degree. The total number of variations, therefore, cannot exceed the number of real roots of this equation.

But obviously, in all these functions the sign of the highest power of λ is positive; hence, when we substitute $+\infty$, we get no variation; when we substitute $-\infty$, the terms become alternately positive and negative, and we get n variations; the equation we are discussing must, therefore, have n real roots. It is easy to see, in like manner, that the roots of every one of the series of functions are all real, and that the roots of each are interposed as limits between the roots of the function next above it in the series.

45. It will be perceived that in the preceding Article we have substituted, for the functions of Sturm's theorem, another series of functions possessing the same fundamental property, viz. that when one vanishes, the two adjacent have opposite signs. M. Borchardt's proof, however, which we now proceed to give, depends on a direct application of Sturm's theorem.

The first principle which it will be necessary to use is a theorem given by Mr. Sylvester (*Philosophical Magazine*, December, 1839), that the several functions in Sturm's series, expressed in terms of the roots of the given equation, differ only by positive square multipliers from the following. The first two (namely, the function itself, and its derived) are of course, $(x-\alpha)(x-\beta)(x-\gamma)$, &c., $\Sigma(x-\beta)(x-\gamma)$, &c.; and the remaining ones are

$\Sigma(\alpha-\beta)^2(x-\gamma)(x-\delta)$, &c.; $\Sigma(\alpha-\beta)^2(\beta-\gamma)^2(\gamma-\alpha)^2(x-\delta)$, &c., &c., where we take the product of any k factors of the given equation, and multiplying by the product of the squares of the differences of all the roots not contained in these factors, form the corresponding symmetric function. We commence by proving this theorem.*

46. In the first place, let U be the function, V its first derived, R_2, R_3, &c. the series of Sturm's remainders: then it is easy to see that any one of them can be expressed in the form $AV - BU$. For, from the fundamental equations

$$U = Q_1 V - R_2, \quad V = Q_2 R_2 - R_3, \quad R_2 = Q_3 R_3 - R_4, \text{ &c.,}$$

we have

$$R_2 = Q_1 V - U,$$

$$R_3 = Q_2 R_2 - V = (Q_2 Q_1 - 1) V - Q_2 U,$$

$$R_4 = (Q_3 Q_2 - 1) R_2 - Q_3 V = (Q_1 Q_2 Q_3 - Q_1 - Q_3) V - (Q_2 Q_3 - 1) U,$$

and so on. We have then in general† $R_k = AV - BU$, where, since all the Q's are of the first degree in x, it is easy to see that A is of the degree $k-1$, and B of the degree $k-2$, while R_k is of the degree $n-k$.

---

* I suppose that Mr. Sylvester must have originally divined the form of these functions from the characteristic property of Sturm's functions, viz. that if the equation has two equal roots $\alpha = \beta$, every one of them must become divisible by $x - \alpha$. Consequently, if we express any one of these functions as the sum of a number of products $(x-\alpha)(x-\beta)$, &c., every product which does not include either $x - \alpha$ or $x - \beta$ must be divisible by $(\alpha - \beta)^2$; and it is evident in this way that the theorem *ought* to be true. The method of verification here employed does not differ essentially from Sturm's proof, *Liouville*, Vol. VII., p. 356.

† The theory of continued fractions which we are virtually applying here shows that if we have $R_k = A_k V - B_k U$, $R_{k+1} = A_{k+1} V - B_{k+1} U$, then $A_k B_{k+1} - A_{k+1} B_k$ is constant and $= 1$. In fact, since $R_{k+1} = Q_k R_k - R_{k-1}$, we have

$$A_{k+1} = Q_k A_k - A_{k-1}, \quad B_{k+1} = Q_k B_k - B_{k-1},$$

whence $\qquad A_k B_{k+1} - A_{k+1} B_k = A_{k-1} B_k - A_k B_{k-1},$

and by taking the values in the first two equations above, namely, where $k = 2$, and $k = 3$, we see that the constant value $= 1$.

But now this property would suffice to determine R_2, R_3, &c. directly. Thus, if in the equation $R_2 = Q_1 V - U$ we assume $Q_1 = ax + b$, where a and b are unknown constants, the condition that the coefficients of the two highest powers of x on the right-hand side of the equation must vanish (since R_2 is only of the degree $n-2$) is sufficient to determine a and b. And so in general, if in the function $AV - BU$ we write for A the most general function of the $k-1^{st}$ degree containing k constants, and for B the most general function of the $k-2^{nd}$ degree, containing $k-1$ constants, we appear to have in all $2k-1$ constants at our disposal, and have in reality one less, since one of the coefficients may by division be made $=1$.* We have then just constants enough to be able to make the first $2k-2$ terms of the equation vanish, or to reduce it from the degree $n+k-2$ to the degree $n-k$. The problem, then, to form a function of the degree $n-k$, and expressible in the form $AV - BU$, where A and B are of the degrees $k-1$, $k-2$, is perfectly definite, and admits but of one solution. If, then, we have ascertained that any function R_k is expressible in the form $AV - BU$, where A and B are of the right degree, we can infer that R_k must be identical with the corresponding Sturm's remainder, or at least only differ from it by a constant multiplier. It is in this way that we shall identify with Sturm's remainders the expressions in terms of the roots, Art. 45.

47. Let us now, to fix the ideas, take any one of these functions, suppose

$$\Sigma (\alpha - \beta)^2 (\beta - \gamma)^2 (\gamma - \alpha)^2 (x - \delta)(x - \varepsilon), \&c.,$$

and we shall prove that it is of the form $AV - BU$, where in the example chosen A is to be of the second degree, and B of the first in x. Now we can immediately see what we are to assume for the form of A, by making $x = \alpha$ on both sides of the equation. The right-hand side of the equation will then become $A(\alpha - \beta)(\alpha - \gamma)(\alpha - \delta)(\alpha - \varepsilon)$, &c. since U vanishes; and the left-hand side will become

$$\Sigma (\alpha - \beta)^2 (\beta - \gamma)^2 (\gamma - \alpha)^2 (\alpha - \delta)(\alpha - \varepsilon), \&c.$$

---

* Just as the six constants in the most general equation of a conic are only equivalent to five independent constants, and only enable us to make the curve satisfy five conditions.

It follows, then, that the supposition $x = \alpha$ must reduce A to the form $\Sigma (\beta - \gamma)^2 (\alpha - \beta)(\alpha - \gamma)$, and it is at once suggested that we ought to take for A the symmetric function

$$\Sigma (\beta - \gamma)^2 (x - \beta)(x - \gamma).$$

And in like manner, in the general case, we are to take for A the symmetric function of the product of $k-1$ factors of the original equation multiplied by the product of the squares of the differences of all the roots which enter into these factors. It will not be necessary to our purpose actually to determine the coefficients in B, which we shall therefore write down in its most general form. Let us then write down

$$\Sigma(\alpha - \beta)^2 (\beta - \gamma)^2 (\gamma - \alpha)^2 (x - \delta), \&c. = \Sigma(\alpha - \beta)^2 (x - \alpha)(x - \beta)$$
$$\times \Sigma (x - \beta)(x - \gamma), \&c. + (ax + b)(x - \alpha)(x - \beta), \&c.,$$

which we are to prove is an identical equation. Now, since an equation of the pth degree can only have p roots, if such an equation is satisfied by more than p values of x, it must be an identical equation, or one in which the coefficients of the several powers of x separately vanish. But the equation we have written down is satisfied for each of the n values $x = \alpha$, $x = \beta$, &c., no matter what the values of a and b may be. And if we substitute any other two values of x, then, by solving for a and b from the equations so obtained, we can determine a and b so that the equation may be satisfied for these two values. It is, therefore, satisfied for $n + 2$ values of x, and since it is only an equation of the $n + 1^{\text{st}}$ degree, it must be an identical equation. And the corresponding equation in general, which is of the $n + k - 1$ degree, is satisfied immediately for any of the n values $x = \alpha$, &c.; while B being of the $k - 1$ degree we can determine the k constants which occur in its general expression, so that the equation may be satisfied for k other values; the equation is, therefore, an identical equation.

48. We have now proved that the functions written in Art. 45 being of the form $AV - BU$ are either identical with Sturm's remainders, or only differ from them by constant factors. It remains to find out the value of these factors, which is an essential matter, since it is on the signs of the functions that

everything turns. Calling Sturm's remainders, as before, R_2, R_3, &c., let Mr. Sylvester's forms (Art. 45) be T_2, T_3, &c., then we have proved that the latter are of the form $T_2 = \lambda_2 R_2$, $T_3 = \lambda_3 R_3$, &c., and we want to determine λ_2, λ_3, &c. We can at once determine λ_2 by comparing the coefficients of the highest powers of x on both sides of the identity $T_2 = A_2 V - B_2 U$; for x^n does not occur in T_2, while in V the coefficient of x^{n-1} is n, and the coefficient of x is also n in A_2, which $= \Sigma (x - \alpha)$; hence $B_2 = n^2$. But the equation $T_2 = A_2 V - B_2 U$ must be identical with the equation $R_2 = Q_1 V - U$ multiplied by λ_2; we have, therefore, $\lambda_2 = n^2$.

To determine in general λ_k, it is to be observed that since any equation $T_k = A_k V - B_k U$ is λ_k times the corresponding equation for R_k, and since in the latter case it was proved (note, p. 38) that $A_k B_{k+1} - A_{k+1} B_k = 1$, the corresponding quantity for T_k, T_{k+1} must $= \lambda_k \lambda_{k+1}$. Now from the equations

$$T_k = A_k V - B_k U, \quad T_{k+1} = A_{k+1} V - B_{k+1} U,$$

we have

$$A_{k+1} T_k - A_k T_{k+1} = (A_k B_{k+1} - A_{k+1} B_k) U = \lambda_k \lambda_{k+1} U.$$

Now, comparing the coefficients of the highest powers of x on both sides of the equation, and observing that the highest power does not occur in $A_k T_{k+1}$, we have the product of the leading coefficients of A_{k+1} and $T_k = \lambda_k \lambda_{k+1}$. But if we write

$$\Sigma (\alpha - \beta)^2 = p_2, \quad \Sigma (\alpha - \beta)^2 (\alpha - \gamma)^2 (\beta - \gamma)^2 = p_3, \text{ &c.,}$$

we have, on inspection of the values in Arts. 45, 47, the leading coefficient in $T_2 = p_2$, in $T_3 = p_3$, &c., and in $A_2 = n$, in $A_3 = p_2$, in $A_4 = p_3$, &c. Hence

$$p_2^2 = \lambda_2 \lambda_3, \quad p_3^2 = \lambda_3 \lambda_4, \quad p_4^2 = \lambda_4 \lambda_5, \text{ &c., whence } \lambda_3 = \frac{p_2^2}{n^2}, \quad \lambda_4 = \frac{n^2 p_3^2}{p_2^2}, \text{ &c.}$$

The important matter then is, that these coefficients are all positive squares, and, therefore, as in using Sturm's theorem we are only concerned with the signs of the functions, we may omit them altogether.

49. When we want to know the total number of imaginary roots of an equation, it is well known that we are only con-

cerned with the coefficients of the highest powers of x in Sturm's functions, there being as many pairs of imaginary roots as there are variations in the signs of these leading terms. And since the signs of the leading terms of T'_2, T_3, &c. are the same as those of R_2, R_3, &c., it follows that an equation has as many pairs of imaginary roots as there are variations in the series of signs of 1, n, $\Sigma(\alpha-\beta)^2$, $\Sigma(\alpha-\beta)^2(\beta-\gamma)^2(\gamma-\alpha)^2$, &c. This theorem may be stated in a different form by means of Ex. 3, p. 20, and we learn that an equation has as many pairs of imaginary roots as there are variations in the signs of 1, s_0, and the series of determinants

$$\begin{vmatrix} s_0, & s_1 \\ s_1, & s_2 \end{vmatrix}, \quad \begin{vmatrix} s_0, & s_1, & s_2 \\ s_1, & s_2, & s_3 \\ s_2, & s_3, & s_4 \end{vmatrix}, \quad \begin{vmatrix} s_0, & s_1, & s_2, & s_3 \\ s_1, & s_2, & s_3, & s_4 \\ s_2, & s_3, & s_4, & s_5 \\ s_3, & s_4, & s_5, & s_6 \end{vmatrix}, \&c.,$$

the last in the series being the discriminant; and the condition that the roots of an equation should be all real is simply that every one of these determinants should be positive.

50. We return now, from this digression on Sturm's theorem, to M. Borchardt's proof, of which we commenced to give an account, Art. 45; and it is evident that in order to apply the test just obtained, to prove the reality of the roots of the equation got by expanding the determinant of Art. 44, it will be first necessary to form the sums of the powers of the roots of that equation. For the sake of brevity, we confine our proof to the determinant of the third order, it being understood that precisely the same process applies in general; and, for convenience we change the sign of λ which will not affect the question as to the *reality* of its values. Then it appears immediately, on expanding the determinant, that $s_1 = a_{11} + a_{22} + a_{33}$, since the determinant is of the form $\lambda^3 - \lambda^2(a_{11} + a_{22} + a_{33}) + \&c.$ And in the general case s_1 is equal to the sum of the leading terms. We can calculate s_2 as follows: The determinant may be supposed to have been derived by eliminating x, y, z between the equations

$$\lambda x = a_{11}x + a_{12}y + a_{13}z, \quad \lambda y = a_{21}x + a_{22}y + a_{23}z, \quad \lambda z = a_{31}x + a_{32}y + a_{33}z.$$

SYMMETRICAL DETERMINANTS. 43

Multiply all these equations by λ, and substitute on the right-hand side for λx, λy, λz their values, when we get

$\lambda^2 x = (a_{11}^2 + a_{12}^2 + a_{13}^2)x + (a_{11}a_{21} + a_{12}a_{22} + a_{13}a_{23})y + (a_{11}a_{31} + a_{12}a_{32} + a_{13}a_{33})z,$

$\lambda^2 y = (a_{21}a_{11} + a_{22}a_{12} + a_{23}a_{13})x + (a_{21}^2 + a_{22}^2 + a_{23}^2)y + (a_{21}a_{31} + a_{22}a_{32} + a_{23}a_{33})z,$

$\lambda^2 z = (a_{31}a_{11} + a_{32}a_{12} + a_{33}a_{13})x + (a_{31}a_{21} + a_{32}a_{22} + a_{33}a_{23})y + (a_{31}^2 + a_{32}^2 + a_{33}^2)z,$

from which eliminating x, y, z, we have a determinant of form exactly similar to that which we are discussing, and which may be written

$$\begin{vmatrix} b_{11} - \lambda^2, & b_{12}, & b_{13} \\ b_{21}, & b_{22} - \lambda^2, & b_{23} \\ b_{31}, & b_{32}, & b_{33} - \lambda^2 \end{vmatrix}.$$

Then, of course, in like manner,

$s_2 = b_{11} + b_{22} + b_{33} = a_{11}^2 + a_{22}^2 + a_{33}^2 + 2a_{12}^2 + 2a_{23}^2 + 2a_{31}^2.$

The same process applies in general and enables us from s_p to compute s_{p+1}. Thus suppose we have got the system of equations

$\lambda^p x = d_{11}x + d_{12}y + d_{13}z, \quad \lambda^p y = d_{21}x + d_{22}y + d_{23}z, \quad \lambda^p z = d_{31}x + d_{32}y + d_{33}z,$

from which we could deduce, as above, $s_p = d_{11} + d_{22} + d_{33}$; then multiplying both sides by λ, and substituting for λx, &c. their values, we get

$\lambda^{p+1}x = (d_{11}a_{11} + d_{12}a_{12} + d_{13}a_{13})x + (d_{11}a_{21} + d_{12}a_{22} + d_{13}a_{23})y$
$\qquad + (d_{11}a_{31} + d_{12}a_{32} + d_{13}a_{33})z,$

$\lambda^{p+1}y = (d_{21}a_{11} + d_{22}a_{12} + d_{23}a_{13})x + (d_{21}a_{21} + d_{22}a_{22} + d_{23}a_{23})y$
$\qquad + (d_{21}a_{31} + d_{22}a_{32} + d_{23}a_{33})z,$

$\lambda^{p+1}z = (d_{31}a_{11} + d_{32}a_{12} + d_{33}a_{13})x + (d_{31}a_{21} + d_{32}a_{22} + d_{33}a_{23})y$
$\qquad + (d_{31}a_{31} + d_{32}a_{32} + d_{33}a_{33})z,$

whence $s_{p+1} = d_{11}a_{11} + d_{22}a_{22} + d_{33}a_{33} + 2d_{12}a_{12} + 2d_{23}a_{23} + 2d_{31}a_{31}.$

51. We shall now show, by the help of these values for s_p, &c. and of the principle established Art. 24, that every one of the determinants at the end of Art. 49 can be expressed as the sum of a number of squares, and is therefore essentially positive.* Thus write down the set of constituents

$$\begin{Vmatrix} 1, & 1, & 1, & 0, & 0, & 0, & 0, & 0, & 0 \\ a_{11}, & a_{22}, & a_{33}, & a_{23}, & a_{31}, & a_{12}, & a_{23}, & a_{31}, & a_{12} \end{Vmatrix},$$

---

* M. Kummer first found out by actual trial that the discriminant of the cubic which determines the axes of a surface of the second degree is resolvable into a sum

then it is easy to see that $\begin{vmatrix} s_0, & s_1 \\ s_1, & s_2 \end{vmatrix}$ is the determinant formed from this by the method of Art. 24, and which expresses the sum of all possible squares of determinants which can be formed by taking any two of the nine columns written above. The determinant $\begin{vmatrix} s_0, & s_1 \\ s_1, & s_2 \end{vmatrix}$ is thus seen to be resolvable into the sum of the squares

$$(a_{11} - a_{22})^2 + (a_{22} - a_{23})^2 + (a_{33} - a_{11})^2 + 6(a_{23}^2 + a_{31}^2 + a_{12}^2),$$

and is therefore essentially positive. Again, if we write down

$$\begin{Vmatrix} 1, & 1, & 1, & 0, & 0, & 0, & 0, & 0, & 0 \\ a_{11}, & a_{22}, & a_{33}, & a_{23}, & a_{31}, & a_{12}, & a_{23}, & a_{31}, & a_{12} \\ b_{11}, & b_{22}, & b_{33}, & b_{23}, & b_{31}, & b_{12}, & b_{23}, & b_{31}, & b_{12} \end{Vmatrix},$$

where b_{11}, &c. have the meaning already explained, it will be easily seen from the values we have found that $\begin{vmatrix} s_0, & s_1, & s_2 \\ s_1, & s_2, & s_3 \\ s_2, & s_3, & s_4 \end{vmatrix}$ is the determinant which, in like manner, is equal to the sum of the squares of all possible determinants which can be formed out of the above matrix. And so in like manner in general.

LESSON VII.

SYMMETRIC FUNCTIONS.

52. WE assume the reader to be acquainted with the theory of the symmetric functions of roots of equations as usually given in works on the Theory of Equations. Thus we suppose him to be acquainted with Newton's formula for calculating the sums of the powers of the roots of the equation

$$x^n - p_1 x^{n-1} + p_2 x^{n-2} - p_3 x^{n-3} + \&c. = 0,$$

viz. $s_1 - p_1 = 0$, $s_2 - p_1 s_1 + 2p_2 = 0$, $s_3 - p_1 s_2 + p_2 s_1 - 3p_3 = 0$, &c.,

---

of squares. (*Crelle*, Vol. XXVI., p. 268). The general theory given here is due, as we have said, to M. Borchardt.

whence $s_1 = p_1$, $s_2 = p_1^2 - 2p_2$, $s_3 = p_1^3 - 3p_1 p_2 + 3p_3$, &c.;
and with the formulæ

$$\Sigma a^m \beta^p = s_m s_p - s_{m+p},$$

$$\Sigma a^m \beta^p \gamma^q = s_m s_p s_q - s_{m+p} s_q - s_{m+q} s_p - s_{p+q} s_m + 2 s_{m+p+q}, \text{ &c.}$$

If we have any homogeneous function of the coefficients p_1, p_2, &c., we shall use the word *order* of that function, in the usual sense, viz. to denote the number of factors of which each term consists. Thus, if any term were $p_1^r p_2^s p_3^t$, the order of the function would be $r + s + t$. If the function be not homogeneous, the order of the function is as usual regulated by the order of the highest term. By the *weight* of a function we shall understand the sum of the suffixes attached to each factor. Thus, if any term were $p_1^r p_2^s p_3^t$, the *weight* of the function would be $r + 2s + 3t$; or, again, if any term were $p_r p_s p_t$, this term would be of the third order, while its weight would be $r + s + t$. In the case of every function, with which we shall be concerned, the weight will be the same for every term.

53. On inspecting the expressions given above for s_1, s_2, s_3, &c. in terms of the coefficients, it is obvious that the weight of every term in s_2 is two, in s_3 is three, and it is easy to conclude by induction that the weight of every term in s_n is n. In like manner, it is evident that the weight of $\Sigma a^m \beta^p$ is $m + p$, of $\Sigma a^m \beta^p \gamma^q$ is $m + p + q$, &c.

This may be proved in general as follows: If for every root α, β, γ, &c. we substitute λ times α, λ times β, λ times γ, &c., we evidently multiply the function $\Sigma a^m \beta^p \gamma^q$ by λ^{m+p+q}. But it is known that if we multiply every root by λ, we multiply p_1 by λ, p_2 by λ^2, p_3 by λ^3, &c. It follows then that $\Sigma a^m \beta^p \gamma^q$ expressed in terms of the coefficients must be such that if we substitute for p_1, λp_1, for p_2, $\lambda^2 p_2$, and so on, we shall multiply every term by λ^{m+p+q}; and this, in other words, is saying that the weight of every term is $m + p + q$.

54. Since

$$p_1 = \alpha + \beta + \gamma + \text{&c.}, \quad p_2 = \alpha(\beta + \gamma + \text{&c.}) + \beta\gamma + \text{&c.}, \quad \text{&c.,}$$

and none of the coefficients, p_3, p_4, &c. contains any power of α beyond the first, it is plain that the order of any symmetric function $\Sigma \alpha^m \beta^p \gamma^q$ (where m is supposed to be greater than p or q) must be at least m. For of course, unless there are at least m factors, each containing α, α^m cannot appear in the product. But, conversely, any symmetric function, whose order is m, will contain some terms involving α^m. For if q_1, q_2, q_3, &c. be the sum, sum of products in pairs, in threes, &c. of β, γ, δ, &c., we have $p_1 = \alpha + q_1$, $p_2 = \alpha q_1 + q_2$, $p_3 = \alpha q_2 + q_3$, &c., and the coefficient of the highest power of α in such a term as $p_2^r p_3^s p_4^t$, will be $q_1^r q_2^s q_3^t$; and, conversely, the multiplier $q_1^r q_2^s q_3^t$ can *only* arise from the term $p_2^r p_3^s p_4^t$. It therefore cannot be made to vanish by the addition of other terms. It follows then that the order of any symmetric function $\Sigma \alpha^m \beta^p \gamma^q$ is equal to the greatest of the numbers m, p, q; for we have proved that it cannot be less than that number, and that it cannot be greater, since functions of a higher order would contain higher powers of α than α^m.

By the help of the two principles just proved we can write down the literal part of any symmetric function, and it only remains to determine the coefficients. Thus if it were required to form $\Sigma \alpha^2 (\beta - \gamma)^2$, we see on inspection that this is a function whose weight is four, and that it is of the second order; that is to say, there cannot be more than two factors in any term. The only terms then that can enter into such a function are p_4, $p_3 p_1$, p_2^2, and the calculation would be complete if we knew with what coefficients these terms are to be affected.

55. Symmetric functions of the differences of the roots of equations* being those with which we shall have most to deal, it may not be amiss to give a theorem by which the sum of any powers of the differences can be expressed in terms of the sums of the powers of the roots of the given equation. Expanding $(x - \alpha)^m$ by the binomial theorem, and adding the similar expansions for $(x - \beta)^m$, &c., we have at once

$$\Sigma (x - \alpha)^m = s_0 x^m - m s_1 x^{m-1} + \tfrac{1}{2} m (m-1) s_2 x^{m-2} - \&c.$$

Now if we substitute α for x in $\Sigma (x - \alpha)^m$ it becomes $(\alpha - \beta)^m + (\alpha - \gamma)^m + $ &c.; similarly if we substitute β for x it

---

* Such functions have been called *critical* functions.

becomes $(\beta-\alpha)^m + (\beta-\gamma)^m + \&c.$, and so on; and if we add the results of all these substitutions, if m be odd, the sum vanishes, since the terms $(\alpha-\beta)^m$, $(\beta-\alpha)^m$ cancel each other. If m be even, the result is $2\Sigma(\alpha-\beta)^m$. But when the same substitutions are made on the right-hand side of the equation last written, and the results added together, we get

$$s_0 s_m - m s_1 s_{m-1} + \tfrac{1}{2} m(m-1) s_2 s_{m-2} - \&c.$$

If m be odd, the last term will be $-s_m s_0$, which will cancel the first term, and, in like manner, all the other terms will destroy each other. But if m be even, the last term will be identical with the first, and so on, and the equation will be divisible by two. Thus then when m is even, we have

$$\Sigma(\alpha-\beta)^m = s_0 s_m - m s_1 s_{m-1} + \tfrac{1}{2} m(m-1) s_2 s_{m-2} - \&c.,$$

where the coefficients are those of the binomial until we come to the middle term with which we stop, and which must be divided by two. Thus

$$\Sigma(\alpha-\beta)^4 = s_0 s_4 - 4 s_1 s_3 + 3 s_2^2, \quad \Sigma(\alpha-\beta)^6 = s_0 s_6 - 6 s_1 s_5 + 15 s_2 s_4 - 10 s_3^2, \&c.$$

56. Any function of the differences will of course be unchanged if we increase or diminish all the roots by the same quantities, as, for instance, if we substitute $x - \lambda$ for x in the given equation. It then becomes

$$x^n - (p_1 + n\lambda) x^{n-1} + \{p_2 + (n-1)\lambda p_1 + \tfrac{1}{2} n(n-1)\lambda^2\} x^{n-2}$$
$$- \{p_3 + (n-2)\lambda p_2 + \&c.\} x^{n-3} + \&c. = 0.$$

Now any function ϕ of the coefficients p_1, p_2, &c. will, when we alter p_1 into $p_1 + \delta p_1$, p_2 into $p_2 + \delta p_2$, &c., become

$$\phi + \left\{\frac{d\phi}{dp_1}\delta p_1 + \frac{d\phi}{dp_2}\delta p_2 + \&c.\right\} + \frac{1}{1.2}\left\{\frac{d^2\phi}{dp_1^2}(\delta p_1)^2 + \&c.\right\} + \&c.$$

If then, in any function of p_1, p_2, &c., we substitute $p_1 + n\lambda$ for p_1, $p_2 + (n-1)\lambda p_1 + \tfrac{1}{2} n(n-1)\lambda^2$ for p_2, &c. and arrange the result according to the powers of λ, it becomes

$$\phi + \lambda \left\{ n\frac{d\phi}{dp_1} + (n-1) p_1 \frac{d\phi}{dp_2} + (n-2) p_2 \frac{d\phi}{dp_3} + \&c. \right\} + \lambda^2 (\&c.) = 0.$$

But since we have seen that any function of the differences is unchanged by the substitution, no matter how small λ be, it is necessary that any function of the differences, when expressed in terms of the coefficients, should satisfy the differential equation

$$n\frac{d\phi}{dp_1} + (n-1)p_1\frac{d\phi}{dp_2} + (n-2)p_2\frac{d\phi}{dp_3} + \&c. = 0.$$

Ex. 1. Let it be required to form $\Sigma(\alpha-\beta)^2$. We know that its order and weight are both $= 2$. It must therefore be of the form $Ap_2 + Bp_1^2$. Applying the differential equation, we have $\{(n-1)A + 2nB\}p_1 = 0$, whence B is proportional to $n-1$ and A to $-2n$. The function then can only differ by a factor from $(n-1)p_1^2 - 2np_2$.

The factor may be shewn to be unity by supposing $\alpha = 1$ and all the other roots $= 0$, when $p_1 = 1$, $p_2 = 0$, and the value just written reduces to $n-1$ as it ought to do.

Ex. 2. To form for a cubic the product of the squares of the differences $(\alpha - \beta)^2(\beta - \gamma^2)^2(\gamma - \alpha)^2$. This is a function of the order 4 and weight 6. It must therefore be of the form

$$Ap_3^2 + Bp_3p_2p_1 + Cp_3p_1^3 + Dp_2^3 + Ep_2^2p_1^2.$$

Operating with $3\frac{d}{dp_1} + 2p_1\frac{d}{dp_2} + p_2\frac{d}{dp_3}$ it becomes

$$(2A + 3B)p_3p_2 + (2B + 9C)p_3p_1^2 + (B + 6D + 6E)p_2^2p_1 + (C + 4E)p_2p_1^3,$$

and as this is to vanish identically, we must have $C = -4E$, $B = 18E$, $A = -27E$, $D = -4E$, or the function can only differ by a factor from

$$p_1^2p_2^2 + 18p_1p_2p_3 - 4p_2^3 - 4p_3p_1^3 - 27p_3^2.$$

The factor may be shown to be unity by supposing γ and consequently p_3 to be $= 0$.

57. We shall in future usually employ homogeneous equations. Thus writing $\dfrac{x}{y}$ for x, and clearing of fractions, the equation we have used becomes

$$x^n - p_1 x^{n-1} y + p_2 x^{n-2} y^2 \ldots \pm p_n y^n = 0.$$

We give x^n a coefficient for the sake of symmetry; and we find it convenient to give the terms the same coefficients as in the binomial theorem; and so write the equation

$$a_0 x^n + na_1 x^{n-1} y + \tfrac{1}{2}n(n-1)a_2 x^{n-2} y^2 + \ldots na_{n-1} xy^{n-1} + a_n y^n = 0.$$

One advantage of using the binomial coefficients is, that thus all functions of the differences of the roots will, when expressed in terms of the coefficients, be such that the sum of the numerical coefficients will be nothing. For we get the sum of the numerical coefficients by making $a_0 = a_1 = a_2 = \&c. = 1$; but on this supposition all the roots of the original equation become equal, and all the differences vanish.

When we speak of a symmetric function of the roots of the homogeneous equation, we understand that the equation having been divided by $a_0 y^n$, the corresponding symmetric function has been formed of the coefficients $\frac{a_1}{a_0}$, $\frac{a_2}{a_0}$, &c. of the equation in $\frac{x}{y}$, and that it has been cleared of fractions by multiplying by the highest power of a_0 in any denominator. In this way, every symmetric function will be a homogeneous function of the coefficients a_0, a_1, &c.; for before it was cleared of fractions, it was a homogeneous function of the degree 0, and it remains homogeneous when every term is multiplied by the same quantity. Or we may state the theory of the symmetric functions of the roots of the homogeneous equation, without first transforming it to an equation in $\frac{x}{y}$. If one of the roots of the latter equation be α, then it is evident that the homogeneous equation is satisfied by any system of values x', y' for which we have $x' = \alpha y'$, since it is manifest that we are only concerned with the ratio $x' : y'$. And since the equation divided by y^n is resolvable into factors, so the homogeneous equation is plainly reducible to a product of factors $(y'x - yx')(y''x - yx'')(y'''x - yx''')$, &c. Actually multiplying and comparing with the original equation, we get

$a_0 = y'y''y'''$, &c., $na_1 = -\Sigma x'y''y'''$, &c., $\frac{1}{2}n(n-1)a_2 = \Sigma x'x''y'''$, &c.

$a_n = \pm x'x''x'''$, &c., $na_{n-1} = \mp \Sigma y'x''x'''$, &c.

By making all the y's $= 1$, these expressions become the ordinary expressions for the coefficients of an equation in terms of its roots x', x'', &c. And conversely, any symmetric function expressed in the ordinary way in terms of the roots x', x'', may be reduced to the other form, by imagining each x' divided by the corresponding y', and then the whole multiplied by such a power of $y'y''$, &c. as will clear it of fractions. Thus the sum of the squares of the differences $\Sigma (x' - x'')^2$ becomes $\Sigma (x'y'' - y'x'')^2 y'''^2 y''''^2$, &c. And generally any functions of the differences will consist of the sum of products of determinants of the form $(x'y'' - y'x'')(x'y''' - y'x''')$, &c. by powers of y', y'', &c.

58. The differential equation which we have given for functions of the differences of the roots, requires to be modified when the equation has been written with binomial coefficients.

Thus, if in the equation $a_0 x^n + n a_1 x^{n-1} y + \&c. = 0$, we write $x + \lambda$ for x, the new a_1 becomes $a_1 + \lambda a_0$, a_2 becomes $a_2 + 2 a_1 \lambda + a_0 \lambda^2$, a_3 becomes $a_3 + 3 \lambda a_2 + 3 \lambda^2 a_1 + \lambda^3 a_0$, &c., and any function ϕ of the coefficients is altered by this substitution into

$$\phi + \lambda \left(a_0 \frac{d\phi}{da_1} + 2 a_1 \frac{d\phi}{da_2} + 3 a_2 \frac{d\phi}{da_3} + \&c. \right) + \&c.$$

Any function then of the differences, since it remains unaltered when for x we substitute $x + \lambda$, must satisfy the equation

$$a_0 \frac{d\phi}{da_1} + 2 a_1 \frac{d\phi}{da_2} + 3 a_2 \frac{d\phi}{da_3} + \&c. = 0.$$

In like manner any function which remains unaltered, when for y we substitute $y + \lambda$, must satisfy the equation

$$n a_1 \frac{d\phi}{da_0} + (n-1) a_2 \frac{d\phi}{da_1} + (n-2) a_3 \frac{d\phi}{da_2} + \&c. = 0.$$

Functions of the latter kind are functions of the differences of the reciprocals of the roots, and in the homogeneous notation consist of products of determinants of the form $x'y'' - y'x''$, &c. by powers of x', x'', &c. Functions of the determinants $x'y'' - y'x''$ alone, and not multiplied by any powers of the x's or the y's, will satisfy both the differential equations.

59. It is to be observed, that the condition

$$a_0 \frac{d\phi}{da_1} + 2 a_1 \frac{d\phi}{da_2} + 3 a_2 \frac{d\phi}{da_3} + \&c. = 0$$

is not only necessary but sufficient, in order that ϕ should be unaltered by the transformation $x + \lambda$ for x. We have seen that the coefficient of λ in the transformed equation then vanishes, and the coefficient of λ^2 is without difficulty found to be

$$a_0 \frac{d\phi}{da_2} + 3 a_1 \frac{d\phi}{da_3} + 6 a_2 \frac{d\phi}{da_4} + \&c. + \frac{1}{1.2} \left(a_0 \frac{d}{da_1} + 2 a_1 \frac{d}{da_2} + \&c. \right)^2 \phi,$$

where in the latter symbol the a_0, a_1, &c. which appear explicitly, are not to be differential. But it will be seen that this is precisely

$$\frac{1}{1.2} \left(a_0 \frac{d}{da_1} + 2 a_1 \frac{d}{da_2} + \&c. \right) \left(a_0 \frac{d}{da_1} + 2 a_1 \frac{d}{da_2} + \&c. \right) \phi.$$

SYMMETRIC FUNCTIONS.   51

For when we operate with the symbol on itself, the result will be the sum of the terms got by differentiating the a_1, a_2, &c. which appear explicitly, together with the result on the supposition that these a_1, a_2, &c. are constant. Thus then, the coefficient of λ^2 vanishes, since $a_0 \dfrac{d\phi}{da_1} +$ &c. is supposed to vanish identically. So in like manner, for the coefficients of the other powers of λ.

Ex. To form for the cubic $a_0 x^3 + 3a_1 x^2 y + 3a_2 x y^2 + a_3 y^3$, the function
$$\Sigma (x_1 y_2 - x_2 y_1)^2 (x_2 y_3 - x_3 y_2)^2 (x_3 y_1 - x_1 y_3)^2.$$
This can be derived from Ex. 2, p. 48, or else directly as follows. The function is to be of the order 4 and weight 6. It must therefore be of the form
$$A a_3 a_2 a_0 a_0 + B a_3 a_2 a_1 a_0 + C a_3 a_1 a_1 a_1 + D a_2 a_2 a_2 a_0 + E a_2 a_2 a_1 a_1.$$
Operate with $a_0 \dfrac{d}{da_1} + 2a_1 \dfrac{d}{da_2} + 3a_2 \dfrac{d}{da_3}$, and we get
$$(B + 6A) a_3 a_2 a_0 a_0 + (3C + 2B) a_3 a_1 a_1 a_0 + (2E + 6D + 3B) a_2 a_2 a_1 a_0 + (4E + 3C) a_2 a_1 a_1 a_1 = 0.$$
Equating separately to 0 the coefficient of each term, and taking $A = 1$, we find
$$B = -6, \ C = 4, \ D = 4, \ E = -3.$$

60. M. Serret writes the operation $a_0 \dfrac{d\phi}{da_1} +$ &c. in a compact form, which is sometimes convenient. If we imagine a fictitious variable ζ, of which the coefficients a_0, a_1, &c. are such functions, that
$$\frac{da_1}{d\zeta} = a_0, \ \frac{da_2}{d\zeta} = 2a_1, \ \frac{da_3}{d\zeta} = 3a_2, \ \&c.,$$
then evidently $\dfrac{d\phi}{d\zeta} = a_0 \dfrac{d\phi}{da_1} + 2a_1 \dfrac{d\phi}{da_2} + 3a_2 \dfrac{d\phi}{da_3} +$ &c.

In like manner $na_1 \dfrac{d\phi}{da_0} +$ &c. may be written in the compact form $\dfrac{d\phi}{d\eta}$, where η is a variable, of which a_0, a_1, &c. are supposed to be such functions that
$$\frac{da_0}{d\eta} = na_1, \ \frac{da_1}{d\eta} = (n-1) a_2, \ \&c.$$

61. It is worth while to add an observation of M. Brioschi as to the meaning of the first of these operations when expressed in terms of the roots. Let there be any function of the coeffi-

cients p_1, p_2, &c., let us express these coefficients in terms of the roots, and examine the differential of the function with respect to any root. Now if q_1, q_2, &c. denote the sum, sum of products in pairs, &c. of all the roots, omitting any one root α, then evidently

$$\frac{dp_1}{d\alpha}=1,\ \frac{dp_2}{d\alpha}=q_1,\ \frac{dp_3}{d\alpha}=q_2,\ \&c.,$$

and
$$\frac{d}{d\alpha} = \frac{d}{dp_1} + q_1 \frac{d}{dp_2} + q_2 \frac{d}{dp_3} + \&c.$$

We have of course corresponding expressions for the differentials with respect to the other roots, and if now we add all together, the coefficient of $\dfrac{d}{dp_1}$ will evidently be n. That of $\dfrac{d}{dp_2}$ will be $(n-1)p_1$; for since $q_1 = p_1 - \alpha$, when we add, the coefficient in question will be $np_1 - (\alpha + \beta + \gamma + \&c.) = (n-1)p_1$. In like manner the coefficient of $\dfrac{d}{dp_3} = (n-2)p_2$, and we have

$$\left(\frac{d}{d\alpha} + \frac{d}{d\beta} + \&c.\right) = n\frac{d}{dp_1} + (n-1)p_1\frac{d}{dp_2} + (n-2)p_2\frac{d}{dp_3} + \&c.$$

It is evident now why, when the operation on the right-hand side of this equation is applied to any function of the differences of the roots, the result vanishes, since when the equivalent operation $\dfrac{d}{d\alpha} + \&c.$ is applied to any difference, the result vanishes.

In the same manner, if the equation had been written with binomial coefficients, we should have had

$$n\frac{da_1}{d\alpha} = a_0,\ \tfrac{1}{2}n(n-1)\frac{da_2}{d\alpha} = na_1,\ \&c.,$$

whence, exactly as before,

$$\frac{d}{d\alpha} + \frac{d}{d\beta} + \&c. = a_0 \frac{d}{da_1} + 2a_1 \frac{d}{da_2} + \&c.$$

LESSON VIII.

ORDER AND WEIGHT OF ELIMINANTS.

62. WHEN we are given k homogeneous equations in k variables (or, what comes to the same thing, k non-homogeneous equations in $k-1$ variables) it is always possible so to combine the equations as to obtain from them a single equation $\Delta = 0$, in which these variables do not appear. We are then said to have eliminated the variables, and the quantity Δ is called the *Eliminant** of the system of equations. Let us take the simplest example, that which we have already considered in the first lesson, where we are given two equations of the first degree $ax + b = 0$, $a'x + b' = 0$. If we multiply the first equation by a', and the second by a, and subtract the first equation from the second, we get $ab' - a'b = 0$, and the quantity $ab' - a'b$ is the eliminant of the two equations. Now it will be observed, that we cannot draw the inference $ab' - a'b = 0$ unless the two given equations are supposed to be simultaneous, that is to say, unless it is supposed that both can be satisfied by the same value of x. For evidently when we combine two equations $\phi(x) = 0$, $\psi(x) = 0$, and draw such an inference as $l\phi(x) + m\psi(x) = 0$, it is assumed that x means the same thing in both equations. It follows then that $ab' - a'b = 0$ is the condition, that the two equations can be satisfied by the same value of x, as may also be seen immediately by solving both equations for x, and equating the resulting values. And so generally, if we are given any number of equations $U = 0$, $V = 0$, $W = 0$, &c., we may proceed to combine them, and draw an inference such as $lU + mV + nW = 0$, only if the variables have the same values in all the equations. And if by combining the equations, we arrive at a result not containing the variables, this will vanish if the equations can be satisfied by a common system of values of the variables, and not otherwise. The eliminant may then in general be defined as *that function of the coefficients of the*

---

* Eliminants are also called *resultants*.

system of equations, whose vanishing expresses that they can be satisfied by a common system of values of the variables.

63. We have now to show how elimination can be performed, and what is the nature of the results arrived at. We commence with two equations written in the non-homogeneous form

$$x^m - p_1 x^{m-1} + p_2 x^{m-2} - \&c. = 0, \text{ or } \phi(x) = 0,$$
$$x^n - q_1 x^{n-1} + q_2 x^{n-2} - \&c. = 0, \text{ or } \psi(x) = 0.$$

The eliminant of these equations is, as we have seen, the condition that they should have a common root. If this be the case, some one of the roots of the first equation must satisfy the second. Let the roots of the first equation be α, β, γ, &c.; and let us substitute these values successively in the second equation, then some one of the results $\psi(\alpha)$, $\psi(\beta)$, &c. must vanish, and therefore the product of all must be sure to vanish. But this product is a symmetric function of the roots of the first equation, and therefore can be expressed in terms of its coefficients, in which state it is the eliminant required. The rule then for elimination by this method is to take the m factors

$$\psi(\alpha) = \alpha^n - q_1 \alpha^{n-1} + q_2 \alpha^{n-2} - \&c.,$$
$$\psi(\beta) = \beta^n - q_1 \beta^{n-1} + q_2 \beta^{n-2} - \&c.,$$
$$\psi(\gamma) = \gamma^n - q_1 \gamma^{n-1} + q_2 \gamma^{n-2} - \&c., \&c.,$$

to multiply all together, and then substitute for the symmetric functions $(\alpha\beta\gamma)^n$, &c., their values in terms of the coefficients of the first equation.

Ex. To eliminate x between $x^2 - p_1 x + p_2 = 0$, $x^2 - q_1 x + q_2 = 0$. Multiplying $(\alpha^2 - q_1 \alpha + q_2)(\beta^2 - q_1 \beta + q_2)$, we get

$$\alpha^2\beta^2 - q_1 \alpha\beta(\alpha + \beta) + q_2(\alpha^2 + \beta^2) + q_1^2 \alpha\beta - q_1 q_2(\alpha + \beta) + q_2^2;$$

and then substituting $\alpha + \beta = p_1$, $\alpha\beta = p_2$, $\alpha^2 + \beta^2 = p_1^2 - 2p_2$, we have

$$p_2^2 - p_1 p_2 q_1 + q_2(p_1^2 - 2p_2) + p_2 q_1^2 - q_1 q_2 p_1 + q_2^2,$$

or $\qquad (p_2 - q_2)^2 + (p_1 - q_1)(p_1 q_2 - p_2 q_1),$

which is the eliminant required.

64. We obtain in this way the same result (or at least results differing only in sign), whether we substitute the roots of the first equation in the second, or those of the second in the first. In other words, if α', β', γ', &c. be the roots of the

ORDER AND WEIGHT OF ELIMINANTS.   55

second equation, the eliminant may be written at pleasure as the continued product of $\phi(\alpha')$, $\phi(\beta')$, $\phi(\gamma')$, &c., or as the product of $\psi(\alpha)$, $\psi(\beta)$, $\psi(\gamma)$, &c. For remembering that $\phi(x) = (x-\alpha)(x-\beta)(x-\gamma)$, &c., the first form is

$$(\alpha'-\alpha)(\alpha'-\beta)(\alpha'-\gamma), \&c.\ (\beta'-\alpha)(\beta'-\beta)(\beta'-\gamma), \&c.,$$

and the second is

$$(\alpha-\alpha')(\alpha-\beta')(\alpha-\gamma'), \&c.\ (\beta-\alpha')(\beta-\beta')(\beta-\gamma'), \&c.$$

In either case we get the product of all possible differences between a root of the first equation and a root of the second; and the two products can at most differ in sign.

65. If the equations had been given in the homogeneous form, with or without binomial coefficients,

$$a_0 x^m + m a_1 x^{m-1} y + \tfrac{1}{2} m (m-1) a_2 x^{m-2} y^2 + \&c. = 0,$$
$$b_0 x^n + n b_1 x^{n-1} y + \tfrac{1}{2} n (n-1) \ x^{n-2} y^2 + \&c. = 0,$$

we can reduce them to the preceding form by dividing them respectively by $a_0 y^m$, $b_0 y^n$, when we have $p_1 = -\dfrac{m a_1}{a_0}$, $q_1 = -\dfrac{n b_1}{b_0}$, &c. We substitute then these values for p_1, q_1, &c. in the result obtained by the method of the last article, and then clear of fractions by multiplying by the highest power of a_0, or b_0 in any denominator. Thus the eliminant of $a_0 x^2 + 2 a_1 x y + a_2 y^2$, $b_0 x^2 + 2 b_1 x y + b_2 y^2$, obtained in this manner from the result of Ex., Art. 63, is

$$(a_0 b_2 - a_2 b_0)^2 + 4(a_1 b_0 - a_0 b_1)(a_1 b_2 - a_2 b_1).$$

It is evident thus that *the eliminant is always a homogeneous function of the coefficients of either equation.* For before we cleared of fractions, it was evidently a homogeneous function of the degree 0, and it remains homogeneous when every term is multiplied by the same quantity.

The same thing may be seen by applying to the equations directly the process of Art. 63. Let the values which satisfy the first equation be $x'y'$, $x''y''$, &c.; then, if the equations have a common factor, some one of these values must satisfy the second equation. We must then multiply together

$$(b_0 x'^n + n b_1 x'^{n-1} y' + \&c.)(b_0 x''^n + n b_1 x''^{n-1} y'' + \&c.)(\&c.),$$

and then substitute for the symmetric functions $(x'x'', \&c.)^n$, &c. their values in terms of the coefficients of the first equation, as found by Art. 57.

66. *The eliminant of two equations of the mth and n^{th} orders respectively, is of the nth order in the coefficients of the first equation, and of the mth in the coefficients of the second.*

For it may be written either as the product of m factors $\psi(\alpha)$, $\psi(\beta)$, &c., each containing the coefficients of the second equation in the first degree, or else as the product of n factors $\phi(\alpha')$, $\phi(\beta')$, &c., each containing in the first degree the coefficients of the first equation. Or confining our attention to the form $(\psi\alpha)(\psi\beta)$, &c. we can see that this form, which obviously contains the coefficients of the second equation in the degree m, contains those of the first in the degree n, since the symmetric functions which occur in it may contain the nth, and no higher, power of any root (Art. 54).

67. *The weight of the eliminant is mn;* that is to say, the sum of the suffixes in every term is constant and $= mn$. For if each of the roots $\alpha, \beta; \alpha', \beta'$, &c. be multiplied by the same factor λ, then since each of the mn differences $\alpha - \alpha'$ (see Art. 64) is multiplied by this factor λ, the eliminant will be multiplied by λ^{mn}. But the roots of the two equations will be multiplied by λ if for p_1, q_1 we substitute $\lambda p_1, \lambda q_1$; for p_2, q_2; $\lambda^2 p_2, \lambda^2 q_2$; &c. We see then that if we make this substitution in the eliminant, the effect will be that every term will be multiplied by λ^{mn}; or, in other words, the sum of the suffixes in every term will be mn. The same thing may also be seen to follow from the principle of Art. 53. In $\psi(x)$ the sum of the index of every term and the suffix of the corresponding coefficient is n; that is to say, $\psi(x)$ consists of the sum of a number of terms, each of the form $q_{n-i}x^i$. If then we take any terms at random in each of the factors $\psi(\alpha), \psi(\beta)$, &c., the corresponding term in the product will be $q_{n-i}q_{n-j}q_{n-k}\alpha^i\beta^j\gamma^k$, &c., and if we combine with this all other terms in which the same coefficients of the second equation occur, we get $q_{n-i}q_{n-j}q_{n-k}\Sigma\alpha^i\beta^j\gamma^k$, &c. The sum of the suffixes of the q's is $n-i+n-j+n-k+\&c.$, or since there are m factors, the sum is $mn - (i+j+k+\&c.)$.

But, by Art. 53, the sum of the suffixes of the p's in the expression for $\Sigma a^i \beta^j \gamma^k$, &c. is $i+j+k+$&c. Therefore the sum of both sets of suffixes is mn, which was to be proved.

The result at which we have arrived may be otherwise stated thus:* *If p_1, q_1 contain any new variable z in the first degree; if p_2, q_2 contain it in the second and lower degrees; if p_3, q_3 in the third, and so on; then the eliminant will in general contain this variable in the mn^{th} degree.*

It is evident that the results of this and of the last article are equally true if the equations had been written in the homogeneous form $a_0 x^m +$&c., because the suffixes in the two forms mutually correspond. And again, from symmetry it follows that the result of this article would be equally true if the equations had been written in the form $a_m x^m + m a_{m-1} x^{m-1} y +$&c., where the suffix of any coefficient corresponds to the power of x which it multiplies, instead of to the power of y.

68. Since the eliminant is a function of the differences between a root of one equation and a root of the other, it will be unaltered if the roots of each equation be increased by the same quantity; that is to say, if we substitute $x + \lambda$ for x in each equation. It follows then, as in Art. 56, that the eliminant must satisfy the differential equation

$$m \frac{d\Delta}{dp_1} + (m-1) p_2 \frac{d\Delta}{dp_1} + (m-2) p_3 \frac{d\Delta}{dp_2} + \&c.$$
$$+ n \frac{d\Delta}{dq_1} + (n-1) q_2 \frac{d\Delta}{dq_1} + \&c. = 0,$$

or, as in Art. 58, if the equations had been written with binomial coefficients, we have

$$a_0 \frac{d\Delta}{da_1} + 2a_1 \frac{d\Delta}{da_2} + \&c. + b_0 \frac{d\Delta}{db_1} + 2b_1 \frac{d\Delta}{db_2} + \&c. = 0.$$

69. *Given two homogeneous equations between three variables, of the mth and n^{th} degrees respectively, the number of systems of*

---

* Or again thus: if in the eliminant we substitute for each coefficient p_a, the term x^a which it multiplies in the original equation, every term of the eliminant will be divisible by x^{mn}. Or, in the homogeneous form, if we substitute for each coefficient a_a the term $x^a y^{m-a}$, which it multiplies, every term of the eliminant will be divisible by $x^{mn} y^{mn}$.

values of the variables which can be found to satisfy simultaneously the two equations is mn.*

Let the two equations, arranged according to powers of x, be

$$ax^m + (by + cz)\, x^{m-1} + (dy^2 + eyz + fz^2)\, x^{m-2} + \&c. = 0,$$
$$a'x^n + (b'y + c'z)\, x^{n-1} + (d'y^2 + e'yz + f'z^2)\, x^{n-2} + \&c. = 0.$$

If now we eliminate x between these equations, since the coefficient of x^{m-1} is a homogeneous function of y and z of the first degree, that of x^{m-2} is a similar function of the second degree, and so on,—it follows from the last Article that the eliminant will be a homogeneous function of y and z of the mn^{th} degree. It follows then that mn values of y and z† can be found which will make the eliminant $= 0$. If we substitute any one of these in the given equations, they will now have a common root when solved for x (since their eliminant vanishes); and this value of x, combined with the values of y and z already found, gives one system of values satisfying the given equations. So we plainly have in all mn such systems of values. We shall, in Lesson X., give a method by which, when two equations have a common root, that common root can immediately be found.

Ex. To find the co-ordinates of the four points of intersection of the two conics
$ax^2 + by^2 + cz^2 + 2fyz + 2gzx + 2hxy = 0$, $\quad a'x^2 + b'y^2 + c'z^2 + 2f'yz + 2g'zx + 2h'xy = 0$.
Arrange the equations according to the powers of x, and eliminate that variable; then, by Art. 32, the result is
$\{(ab')\, y^2 + 2\, (af')\, yz + (ac')\, z^2\}^2$
$+ 4\,[(ah')\, y + (ag')\, z]\, [(bh')\, y^3 + \{(bg') + 2\,(fh')\}\, y^2z + \{(ch') + 2\,(fg')\}\, z^2y + (cg')\, z^3] = 0$,
where, as in Lesson I., we have written (ab') for $ab' - a'b$. This equation, solved for $y : z$, determines the values corresponding to the four points of intersection. Having found these, by substituting any one of them in both equations, and finding their common root, we obtain the corresponding value of $x : z$. We might have at once got the four values of $x : z$ by eliminating y between the equations, but substitution in the equations is necessary in order to find which value of y corresponds to each value of x. By making $z = 1$, what has been said is translated into the language of ordinary Cartesian co-ordinates.

---

* These equations may be considered as representing two curves of the mth and n^{th} degrees respectively; the geometrical interpretation of the proposition of this Article being, that two such curves intersect in mn points. The equations are reduced to ordinary Cartesian equations by making $z = 1$.

† The reader will remember that when we use homogeneous equations, the *ratio* of the variables is all with which we are concerned. Thus here, z' may be taken arbitrarily, the corresponding value of y being determined by the equation in $y : z$.

70. *Any symmetric functions of the mn values which simultaneously satisfy the two equations can be expressed in terms of the coefficients of those equations.*

In order to be more easily understood, we first consider non-homogeneous equations in two variables. Then it is plain enough that we can so express symmetric functions involving either variable alone. For eliminating y, we have an equation in x, in terms of whose coefficients can be expressed all symmetric functions of the mn values of x which satisfy both equations. Similarly for y. Thus, for example, in the case of two conics; $x_i y_i$, &c., being the co-ordinates of their points of intersection, we see at once how to express such symmetric functions as

$$x_i + x_{ii} + x_{iii} + x_{iiii}, \quad y^2_i + y^2_{ii} + y^2_{iii} + y^2_{iiii}, \text{ &c.,}$$

and the only thing requiring explanation is how to express symmetric functions into which both variables enter, such as

$$x_i y_i + x_{ii} y_{ii} + x_{iii} y_{iii} + x_{iiii} y_{iiii}.$$

To do this, we introduce a new variable, $t = \lambda x + \mu y$, and by the help of this assumed equation eliminate both x and y from the given equations. Thus y is immediately eliminated by substituting in both its value derived from $t = \lambda x + \mu y$, and then we have two equations of the mth and n^{th} degrees in x, the eliminant of which will be of the mn^{th} degree in t, and its roots will be obviously $\lambda x_i + \mu y_i$, $\lambda x_{ii} + \mu y_{ii}$, &c., where $x_i y_i$, $x_{ii} y_{ii}$ are the values of x and y common to the two equations. The coefficients of this equation in t will of course involve λ and μ. We next form the sum of the kth powers of the roots of this equation in t, which must plainly be $= (\lambda x_i + \mu y_i)^k + (\lambda x_{ii} + \mu y_{ii})^k + $&c. The coefficient, then, of λ^k in this sum will be Σx_i^k: the coefficient of $\lambda^{k-1} \mu$ gives us $\Sigma x_i^{k-1} y_i$, and so on.

Little need be said in order to translate the above into the language of homogeneous equations. We see at once how to form symmetric functions involving two variables only, such as $\Sigma y_i z_{ii} z_{iii} z_{iiii}$, for these are found, as explained, Art. 57, from the homogeneous equation obtained on eliminating the remaining variable; the only thing requiring explanation is how to form symmetric functions involving all these variables, and this is done precisely as above, by substituting $t = \lambda x + \mu y$.

Ex. To form the symmetric functions of the co-ordinates of the four points common to two conics. The equation in the last Example gives at once

$$y_{,}y_{,,}y_{,,,}y_{,,,,} = (ac')^2 + 4(ag')(cg'); \quad z_{,}z_{,,}z_{,,,}z_{,,,,} = (ab')^2 + 4(ah')(bh');$$

and from symmetry, $\quad x_{,}x_{,,}x_{,,,}x_{,,,,} = (bc')^2 + 4(bf')(cf'),$

$$-\Sigma(y_{,}y_{,,}z_{,,,}z_{,,,,}) = 4\{(ac')(af') + (ah')(cg') + (ag')(ch') + 2(ag')(fg')\}, \&c.$$

To take an Example of a function involving three variables, let us form

$$\Sigma(x_{,}y_{,}z^2_{,,}z^2_{,,,}z^2_{,,,,}),$$

which corresponds to $\Sigma(x'y')$ when the equations are written in the non-homogeneous form.

By the preceding theory we are to eliminate between the given equations, and $t = \lambda x + \mu y$; and the required function will be half the coefficient of $\lambda\mu$ in $\Sigma(t^2 z^2_{,} z^2_{,,} z^2_{,,,})$. If the result of elimination be

$$At^4 + (B\lambda + C\mu) t^3 z + (D\lambda^2 + E\lambda\mu + F\mu^2) t^2 z^2 + \&c.$$

$$\Sigma(t^2 z^2_{,} z^2_{,,} z^2_{,,,}) = (B\lambda + C\mu)^2 - 2A(D\lambda^2 + E\lambda\mu + F\mu^2),$$

and $\qquad \Sigma(x_{,}y_{,}z^2_{,,}z^2_{,,,}z^2_{,,,,}) = BC - AE.$

By actual elimination

$$A = (ab')^2 + 4(ah')(bh'), \quad B = 4\{(ba')(bg') + (bf')(ah') + (bh')(af') + 2(bh')(gh')\},$$

$$C = 4\{(ab')(af') + (ag')(bh') + (ah')(bg') + 2(ah')(fh')\},$$

$$E = 4\{(ac')(bh') + (bc')(ah') - 2(af')(hf') - 2(bg')(hg') + 4(hf')(hg')\}.$$

71. *To form the eliminant of three homogeneous equations in three variables, of the mth, n^{th}, and p^{th} degrees respectively.* The vanishing of the eliminant is the condition that a system of values of x, y, z can be found to satisfy all three equations.* When this, then, is the case, if we solve for any two of the equations, and substitute successively in the remaining one the values so found for x, y, z, some one of these sets of values must satisfy that equation, and therefore the product of all the results of substitution must vanish. Let, then, x', y', z'; x'', y'', z'', &c. be the system of values which satisfy the last two equations, and which (Art. 69) are np in number: we substitute these values in the first, and multiply together the np results $\phi(x', y', z')$, $\phi(x'', y'', z'')$, &c. The product will plainly involve only symmetric functions of x', y', z', &c. which (Art. 70) can all be expressed in terms of the coefficients of the last two equations; and, when they are so expressed, it is the eliminant required.

---

* If the three equations represent curves, the vanishing of the eliminant is the condition that all three curves should pass through a common point.

ORDER AND WEIGHT OF ELIMINANTS.   61

72. *The eliminant is a homogeneous function of the np^{th} order in the coefficients of the first equation; of the mp^{th} in those of the second; and of the mn^{th} in those of the third.*

For each of the np factors $\phi(x', y', z')$ is a homogeneous function of the first degree in the coefficients of the first equation; and the expression of the symmetric functions in terms of the coefficients only involves coefficients of the last two equations, from solving which x', y', z', &c. were obtained. The eliminant is therefore of the np^{th} degree in the coefficients of the first equation; and in like manner its degree in the coefficients of the others may be inferred.

73. *The weight of the eliminant will be mnp;* that is to say, *If all the coefficients in the equations which multiply the first power of one of the variables, z, be affected with a suffix 1, those which multiply z^2 with a suffix 2, and so on; the sum of all the suffixes in each term of the eliminant will be equal to mnp.* In other words: *If all the coefficients which multiply z contain a new variable in the first degree;—if those which multiply z^2 contain it in the second and lower degrees, and so on; then the eliminant will contain this variable in the degree mnp.*

This is proved as in Art. 67. In the first place, it is evident that if a homogeneous equation of the mth degree be satisfied by values x', y', z'; and if the equation be altered by multiplying each coefficient by a power of λ, equal to the power of z, which the coefficient multiplies, then the equation so transformed will be satisfied by the values $\lambda x'$, $\lambda y'$, z'; or, in general, that the result of substituting $\lambda x'$, $\lambda y'$, z' in the transformed equation is λ^m times the result of substituting x', y', z' in the untransformed. Thus, take the equation $x^3 + y^3 - z^3 - z^2x - zy^2$, the transformed is $x^3 + y^3 - \lambda^3 z^3 - \lambda^2 z^2 x - \lambda z y^2$; and obviously the result of substituting $\lambda x'$, $\lambda y'$, z' in the second is λ^3 times the result of substituting x', y', z' in the given equation. If, then, the three given equations be all transformed by multiplying each coefficient by a power of λ equal to the power of z, which the coefficient multiplies, then it follows, that if x', y', z' be one of the system of values which satisfy the two last of the original equations, then the transformed equations will be satisfied by $(\lambda x', \lambda y', z')$, and the result of substituting these values in the first will be $\lambda^m \phi(x', y', z')$. The eliminant, then, which is the product of np factors of the form $\phi(x', y', z')$ will be multiplied by λ^{mnp}. If then any term in the eliminant be $a_i b_l c_m$, &c., where the suffix corresponds to the power of z, which the coefficient multiplies, since the alteration of a_k into $\lambda^k a_k$, b_l into $\lambda^l b_l$, &c., multiplies the term by λ^{mnp}, we must have $k + l + $ &c. $= mnp$. Q.E.D.

74. It is proved in like manner that three equations are in general satisfied by mnp common values; that any symmetric function of these values can be expressed in terms of their coefficients; and that we can form the eliminant of four equations by solving from any three of them, substituting successively in the fourth each of the systems of values so found, forming the product of the results of substitution, and then, by the method of symmetric functions, expressing the product in terms of the coefficients of the equations. In this way we can form the eliminant of any number of equations; and we have the following general theorems: *The eliminant of k equations in $k-1$ independent variables is a homogeneous function of the coefficients of each equation, whose order is equal to the product of the degrees of all the remaining equations. If each coefficient in all the equations be affected with a suffix equal to the power of any one variable which it multiplies, then the sum of the suffixes in every term of the eliminant will be equal to the product of the degrees of all the equations. And, again, if we are given k equations in k variables, the number of systems of common values of the variables, which can be found to satisfy all the equations, will be equal to the product of the orders of the equations.*

LESSON IX.

EXPRESSION OF ELIMINANTS AS DETERMINANTS.

75. THE method of elimination by symmetric functions is, in a theoretical point of view, perhaps preferable to any other, it being universally applicable to equations in any number of

variables; yet as it is not very expeditious in practice, and does not yield its results in the most convenient form, we shall in this Lesson give an account of some other methods of elimination. The following is the method which most obviously presents itself. It is in substance identical with what is called elimination by the process of finding the greatest common measure. We have already seen that the eliminant of two linear equations $ax + b = 0$, $a'x + b' = 0$ is the determinant $ab' - ba' = 0$. If now we have two quadratic equations
$$ax^2 + bx + c = 0, \quad a'x^2 + b'x + c' = 0,$$
multiplying the first by a', the second by a, and subtracting, we get
$$(ab')\, x + (ac') = 0;$$
and, again, multiplying the first by c', the second by c, subtracting, and dividing by x, we get
$$(ac')\, x + (bc') = 0.$$
The problem is now reduced to elimination between two linear equations, and the result is
$$(ac')^2 + (ba')(bc') = 0.$$

76. So, again, if we have two cubic equations
$$ax^3 + bx^2 + cx + d = 0, \quad a'x^3 + b'x^2 + c'x + d' = 0,$$
we multiply the first by a', the second by a, and subtract; and also multiply the first by d', the second by d, subtract and divide by x. The problem is thus reduced to elimination between the two quadratics
$$(ab')\, x^2 + (ac')\, x + (ad') = 0, \quad (ad')\, x^2 + (bd')\, x + (cd') = 0.$$
By the last article, the result is
$$\{(ad')^2 - (ab')(cd')\}^2 + \{(ad')(ac') - (ab')(bd')\}\{(ad')(db') - (ac')(dc')\} = 0.$$
Now it is to be observed that the equation
$$(ab')(cd') + (ac')(db') + (ad')(bc') = 0$$
is identically true. Consequently when we multiply out, the preceding result becomes divisible by (ad'), and the reduced result is
$$(ad')^3 - 2\,(ad')(ab')(cd') - (ad')(ac')(bd')$$
$$+ (ac')^2 (cd') + (bd')^2 (ab') - (ab')(bc')(cd') = 0.$$

The reason that in this process the irrelevant factor (ad') is introduced is that, if $ad' = a'd$, and therefore a to a' in the same ratio as d to d', we must get results differing only by a factor, if from the first equation multiplied by a' we subtract the second equation multiplied by a, or if from the first equation multiplied by d', we subtract the second equation multiplied by d. Thus, on the supposition $(ad') = 0$, even though the original two cubics have not a common factor, the two quadratics to which we reduced them would have a common factor. In general then, when we eliminate by this process, irrelevant factors are introduced, and therefore other methods are preferable.

77. *Euler's Method.* If two equations of the mth and n^{th} degrees respectively have a common factor of the first degree, we must obtain identical results, whether we multiply the first equation by the remaining $n-1$ factors of the second, or the second by the remaining $m-1$ factors of the first. If then we multiply the first by an arbitrary function of the $(n-1)^{th}$ degree, which, of course, introduces n arbitrary constants; if we multiply the second by an arbitrary function of the $(m-1)^{th}$ degree, introducing thus m constants; and if we then equate, term by term, the two equations of the $(m+n-1)^{th}$ degree so formed, we shall have $m+n$ equations, from which we can eliminate the $m+n$ introduced constants, which all enter into those equations only in the first degree; and we shall thus obtain, in the form of a determinant, the eliminant of the two given equations.

Ex. To eliminate between $ax^2 + bxy + cy^2 = 0$, $a'x^2 + b'xy + c'y^2 = 0$.
We are to equate, term by term,
$$(Ax + By)(ax^2 + bxy + cy^2) \text{ and } (A'x + B'y)(a'x^2 + b'xy + c'y^2).$$
The four resulting equations are
$$Aa \ \ldots\ldots\ - A'a'\ \ldots\ldots = 0,$$
$$Ab + Ba - A'b' - B'a' = 0,$$
$$Ac + Bb - A'c' - B'b' = 0,$$
$$Bc \ \ldots\ldots\ - B'c' = 0,$$
from which eliminating A, B, A', B', the result is the determinant
$$\begin{vmatrix} a, & 0, & a', & 0 \\ b, & a, & b', & a' \\ c, & b, & c', & b' \\ 0, & c, & 0, & c' \end{vmatrix}.$$

78. This method may be extended to find the conditions that the equations should have two common factors. In this case it is evident, in like manner, that we shall obtain the same result whether we multiply the first by the remaining $n-2$ factors of the second, or the second by the remaining $m-2$ factors of the first. As before, then, we multiply the first by an arbitrary function of the $n-2$ degree (introducing $n-1$ constants), and the second by an arbitrary function of the $m-2$ degree; and equating, term by term, the two equations of the $m+n-2$ degree so found, we have $m+n-1$ equations, from any $m+n-2$ of which, eliminating the $m+n-2$ introduced constants, we obtain $m+n-1$ conditions, equivalent, of course, to only two independent conditions.

Ex. To find the conditions that
$$ax^3 + bx^2y + cxy^2 + dy^3 = 0, \quad a'x^3 + b'x^2y + c'xy^2 + d'y^3 = 0,$$
should have two common factors. Equating
$$(Ax + By)(ax^3 + bx^2y + cxy^2 + dy^3) = (A'x + B'y)(a'x^3 + b'x^2y + c'xy^2 + d'y^3),$$
we have
$$Aa \quad \ldots\ldots \quad - A'a' \quad \ldots\ldots = 0,$$
$$Ab + Ba - A'b' - B'a' = 0,$$
$$Ac + Bb - A'c' - B'b' = 0,$$
$$Ad + Bc - A'd' - B'c' = 0,$$
$$Bd \quad \ldots\ldots \quad - B'd' = 0,$$
from which, eliminating A, B, A', B', we have the system of determinants [for the notation used, see Art. 3],
$$\begin{Vmatrix} a, & b, & c, & d, & 0 \\ 0, & a, & b, & c, & d \\ a', & b', & c', & d', & 0 \\ 0, & a', & b', & c', & d' \end{Vmatrix} = 0.$$

79. *Sylvester's dialytic method.* This method is identical in its results with Euler's, but simpler in its application, and more easily capable of being extended. Multiply the equation of the mth degree by x^{n-1}, $x^{n-2}y$, $x^{n-3}y^2$, &c.; and the second equation by x^{m-1}, $x^{m-2}y$, $x^{m-3}y^2$, &c., and we thus get $m+n$ equations, from which we can eliminate linearly the $m+n$ quantities x^{m+n-1}, $x^{m+n-2}y$, $x^{m+n-3}y^2$, &c., considered as independent unknowns. Thus in the case of two quadratics, multiply both by x and by y, and we get the equations
$$ax^3 + bx^2y + cxy^2 \qquad = 0,$$
$$ax^2y + bxy^2 + cy^3 = 0,$$
$$a'x^3 + b'x^2y + c'xy^2 \qquad = 0,$$
$$a'x^2y + b'xy^2 + c'y^3 = 0,$$

from which, eliminating x^3, x^2y, xy^2, y^3, we get the same determinant as before

$$\begin{vmatrix} a, & b, & c, & \\ & a, & b, & c \\ a', & b', & c', & \\ & a', & b', & c' \end{vmatrix}.$$

In general, it is evident by this method, that the eliminant is expressed as a determinant of which n rows contain the coefficients of the first equation, and m rows contain the coefficients of the second. Thus we obtain the rule already stated for the order of the eliminant in the coefficients of each equation.

80. *Bezout's method.* This process also expresses the eliminant in the form of a determinant; but one which can be more rapidly calculated than the preceding. The general method will, perhaps, be better understood if we apply it first to the particular case of the two equations of the fourth degree

$$ax^4 + bx^3y + cx^2y^2 + dxy^3 + ey^4 = 0, \quad a'x^4 + b'x^3y + c'x^2y^2 + d'xy^3 + e'y^4 = 0.$$

Multiplying the first by a', the second by a, and subtracting, the first term in each is eliminated, and the result, being divisible by y, gives

$$(ab')\, x^3 + (ac')\, x^2y + (ad')\, xy^2 + (ae')\, y^3 = 0.$$

Again, multiply the first by $a'x + b'y$, and the second by $ax + by$, and the two first terms in each are eliminated, and the result, being divided by y^2, gives

$$(ac')\, x^3 + \{(ad') + (bc')\}\, x^2y + \{(ae') + (bd')\}\, xy^2 + (be')\, y^3 = 0.$$

Next, multiply the first by $a'x^2 + b'xy + c'y^2$; and the second by $ax^2 + bxy + cy^2$; subtract, and divide by y^3; when we get

$$(ad')\, x^3 + \{(ae') + (bd')\}\, x^2y + \{(be') + (cd')\}\, xy^2 + (ce')\, y^3 = 0.$$

Lastly, multiply the first by $a'x^3 + b'x^2y + c'xy^2 + d'y^3$; the second by $ax^3 + bx^2y + cxy^2 + dy^3$; subtract, and divide by y^4; when we get

$$(ae')\, x^3 + (be')\, x^2y + (ce')\, xy^2 + (de')\, y^3 = 0.$$

From the four equations thus formed we can eliminate linearly the four quantities, x^3, x^2y, xy^2, y^3, and obtain for our result the determinant

$$\begin{vmatrix} (ab'), & (ac') & , (ad') & , (ae') \\ (ac'), & (ad')+(bc'), & (ae')+(bd') & , (be') \\ (ad'), & (ae')+(bd'), & +(be')+(cd'), & (ce') \\ (ae'), & (be'), & (ce'), & (de') \end{vmatrix}.$$

81. The process here employed is so evidently applicable to any two equations, both of the nth degree, that it is unnecessary to make a formal statement of the general proof. On inspection of the determinant obtained in the last article, the law of its formation is apparent, and we can at once write down the determinant which is the eliminant between two equations of the fifth degree by simply continuing the series of terms, writing an (af') after every (ae'), &c. Thus the eliminant is

$$\begin{vmatrix} (ab'),(ac') & ,(ad') & ,(ae') & ,(af') \\ (ac'),(ad')+(bc'),(ae')+(bd') & ,(af')+(be') & ,(bf') \\ (ad'),(ae')+(bd'),(af')+(be')+(cd'), & +(bf')+(ce') & ,(cf') \\ (ae'),(af')+(be') , & +(bf')+(ce'), & +(cf')+(de'),(df') \\ (af'), & +(bf'), & +(cf'), & +(df'),(ef') \end{vmatrix}$$

It appears hence that in the eliminant every term must contain a or a'; as was evident beforehand, since if both of these were $=0$, the equations would evidently have the common factor $y=0$.

It appears also that those terms which contain a or a' only in the first degree are (ab') multiplied by the eliminant of the equations got by making a and $a'=0$ in the given equations. For every element in the determinant written above must contain a constituent from the first row, and also one from the first column; but as all the constituents of the first row or column contain a or a', the only terms which contain a and a' in only the first degree, are (ab') multiplied by the corresponding minor; and this, when a and a' are made $=0$, is the next lower eliminant.

82. It only remains to show that the process here employed is applicable when the equations are of different dimensions;

and, as before, we commence with a particular example, viz., the equations

$$ax^4 + bx^3y + cx^2y^2 + dxy^3 + ey^4 = 0, \quad a'x^2 + b'xy + c'y^2 = 0.$$

Multiply the first by a', the second by ax^2, and subtract, when we have

$$(ba') x^3 + (ca') x^2y + (da') xy^2 + (ea') y^3 = 0.$$

In like manner, multiply the first by $a'x + b'y$, and the second by $(ax + by) x^2$, and we get

$$(ca') x^3 + \{(cb') + (da')\} x^2y + \{(db') + (ea')\} xy^2 + (eb') y^3 = 0.$$

This process can be carried no further; but if we join to the two equations just obtained the two equations got by multiplying the second of the original equations by x and by y, we have four equations from which to eliminate x^3, x^2y, xy^2, y^3.

And in general, when the degrees of the equations are unequal, m being the greater, it will be found that the process of Art. 80 gives us n equations of the $(m-1)^{\text{th}}$ degree, each of these equations being of the first order in the coefficients of *each* equation: to which we are to add the $m - n$ equations found by multiplying the second equation by x^{m-n-1}, $x^{m-n-2}y$, &c., and we can then eliminate the m quantities x^{m-1}, $x^{m-2}y$, &c., from the m equations we have formed. Every row of the determinant contains the coefficients of the second equation, but only n rows contain the coefficients of the first. The eliminant is, therefore, as it ought to be, of the nth degree in the coefficients of the first, and of the mth in those of the second equation.

83. *Cayley's statement of Bezout's method.* If two equations $\phi(x, y)$, $\psi(x, y)$, have a common root, then it must be possible to satisfy any equation of the form $\phi + \lambda \psi = 0$, independently of any particular value of λ. Take then the equation

$$\phi(x, y) \psi(x', y') - \phi(x', y') \psi(x, y) = 0;$$

which, if ϕ and ψ have a common factor, can be satisfied independently of any particular values of x' and y'. We may in the first place divide it by $xy' - yx'$, which is obviously a factor: then equate to 0 the coefficients of the several powers of x', y'; and then eliminate the powers of x and y as if they were independent

EXPRESSION OF ELIMINANTS AS DETERMINANTS.   69

variables, when the result comes out in precisely the same form as by the method of Art. 80.

Ex. To eliminate between $ax^2 + bxy + cy^2 = 0$, $a'x^2 + b'xy + c'y^2 = 0$,
$$(ax^2 + bxy + cy^2)(a'x'^2 + b'x'y' + c'y'^2) - (a'x^2 + b'xy + c'y^2)(ax'^2 + bx'y' + cy'^2),$$
when divided by $xy' - yx'$, gives
$$\{(ab')\,x + (ac')\,y\}\,x' + \{(ac')\,x + (bc')\,y\}\,y' = 0\,;$$
and eliminating x, y between the coefficients of x' and y', separately equated to 0, we get the eliminant
$$(ac')^2 + (ba')(bc') = 0.$$

84. We proceed now to the theory of functions of three variables, the eliminant of which, however, except in particular cases, has not been expressed as a determinant, though it can always be expressed as the quotient of one determinant divided by another. We shall show, in the first place, how to form a function of great importance in the theory of elimination. Given k equations in k variables, $u = 0$, $v = 0$, $w = 0$, &c., and if we write u_1, u_2, u_3, &c. for $\dfrac{du}{dx}, \dfrac{du}{dy}, \dfrac{du}{dz}$, &c., then the determinant

$$\begin{vmatrix} u_1, & u_2, & u_3, & \&\text{c.} \\ v_1, & v_2, & v_3, & \&\text{c.} \\ w_1, & w_2, & w_3, & \&\text{c.} \\ \&\text{c.} & & & \end{vmatrix}$$

is called Jacobi's determinant, or simply the *Jacobian* of the given equations, and will be denoted in what follows by the letter J.

85. *If any number of equations are satisfied by a common system of values, that system will satisfy the Jacobian; and when the equations are of the same degree, it will also satisfy the differentials of the Jacobian with regard to each of the variables.*

The proof of this for three variables applies in general. By the theorem of homogeneous functions, we have
$$xu_1 + yu_2 + zu_3 = nu,$$
$$xv_1 + yv_2 + zv_3 = nv,$$
$$xw_1 + yw_2 + zw_3 = nw.$$

Now if, as in Lesson IV., we write the minors of the Jacobian, got by suppressing the row and column containing u_1, v_1, &c., U_1, V_1, &c., then if we solve these equations, we find (Art. 28)

$Jx = U_1 nu + V_1 nv + W_1 nw$, from which it appears at once that if u, v, w vanish, J will vanish too. Again, differentiating the equation just found, we have

$$J + x \frac{dJ}{dx} = nu \frac{dU_1}{dx} + nv \frac{dV_1}{dx} + nw \frac{dW_1}{dx} + n(u_1 U_1 + v_1 V_1 + w_1 W_1),$$

$$x \frac{dJ}{dy} = nu \frac{dU_1}{dy} + nv \frac{dV_1}{dy} + nw \frac{dW_1}{dy} + n(u_2 U_1 + v_2 V_1 + w_2 W_1).$$

But, remembering (Art. 27) that

$$u_1 U_1 + v_1 V_1 + w_1 W_1 = J; \quad u_2 U_1 + v_2 V_1 + w_2 W_1 = 0;$$

we see that the supposition $u = 0$, $v = 0$, $w = 0$ (in consequence of which J is also $= 0$) makes $\frac{dJ}{dx}$, $\frac{dJ}{dy}$ also to vanish.

86. We can now express as a determinant the eliminant of three equations, each of the second degree. For their Jacobian is of the third degree, and therefore its differentials are of the second. We have thus three new equations of the second degree, which will be also satisfied by any system of values common to the given equations. From the six equations, then, u, v, w, $\frac{dJ}{dx}$, $\frac{dJ}{dy}$, $\frac{dJ}{dz}$, we can eliminate the six quantities x^2, y^2, z^2, yx, zx, xy, and so form the determinant required.

Again, if the equations are all of the third degree, J is of the sixth, and its differentials of the fifth, and if we multiply each of the three given equations by x^2, y^2, z^2, yz, zx, xy, we obtain eighteen equations, which, combined with the three differentials of the Jacobian, enable us to eliminate dialytically the twenty-one quantities, x^5, $x^4 y$, &c., which enter into an equation of the fifth degree. This process, however, cannot, without modification, be extended further.

87.* Mr. Sylvester has shown that the eliminant can always be expressed as a determinant when the three equations are of the same degree. Let us take, for an example, three equations of the fourth degree. Multiply each by the six terms $(x^2, xy, y^2,$ &c.) of an equation of the second degree [or gene-

---

* The beginner may omit the rest of this Lesson.

rally by the $\frac{1}{2}n(n-1)$ terms of an equation of the degree $(n-2)$]. We thus form eighteen [$\frac{3}{2}n(n-1)$] equations. But since these equations, being now of the sixth [$2n-2$] degree, consist of twenty-eight [$n(2n-1)$] terms; we require ten [$\frac{1}{2}n(n+1)$] additional equations to enable us to eliminate dialytically all the powers of the variables. These equations are formed as follows: The first of the three given equations can be written in the form $Ax^4 + By + Cz$; the second and third, in the form
$$A'x^4 + B'y + C'z, \quad A''x^4 + B''y + C''z;$$
and the determinant $(AB'C'')$ which is of the sixth degree in the variables must obviously be satisfied by any values which satisfy all the given equations. We should form two similar determinants by decomposing the equations into the form $Ay^4 + Bx + Cz$, $Az^4 + Bx + Cy$. So again, we might decompose the equations into the forms $Ax^3 + By^2 + Cz$, $A'x^3 + B'y^2 + C'z$, $A''x^3 + B''y^2 + C''z$ (for every term not divisible by x^3 or y^2 must be divisible by z); and then we obtain another determinant $(AB'C'')$ which will be satisfied when the equations vanish together. There are six determinants of this form got by interchanging x, y, and z in the rule for decomposing the equations. Lastly, decomposing into the form $Ax^2 + By^2 + Cz^2$, &c. we get a single determinant, which, added to the nine equations already found, makes the ten required. In general, we decompose the equations into the form $Ax^\alpha + By^\beta + Cz^\gamma$, such that $\alpha + \beta + \gamma = n + 2$, and form the determinant $AB'C''$; and it can be very easily proved that the number of integer solutions of the equation $\alpha + \beta + \gamma = n + 2$ is $\frac{1}{2}n(n+1)$, exactly the number required.

88. When the degrees of the equations are different, it is not possible to form a determinant in this way, which shall give the eliminant clear of extraneous factors. The reason why such factors are introduced, and the method by which they are to be got rid of, will be understood from the following theory, due to Mr. Cayley: Let us take for simplicity three equations, u, v, w, all of the second degree. If we attempt to eliminate dialytically by multiplying each by x, y, z, we get nine equations, which are not sufficient to eliminate the ten quantities x^3, x^2y, &c. Again, if we multiply each equation by the six

quantities, x^2, xy, y^2, &c., we have eighteen equations, which are *more than* sufficient to eliminate the fifteen quantities x^4, x^3y, &c. If we take at pleasure any fifteen of these equations, and form their determinant, we shall indeed have the eliminant, but it will be multiplied by an extraneous factor; since the determinant is of the fifteenth degree in the coefficients, while the eliminant is only of the twelfth (Art. 72, $mn + np + pm = 12$, when $m = n = p = 2$). The reason of this is, that the eighteen equations we have formed are not independent, but are connected by three linear relations. In fact, if we write the identity $uv = vu$, and then replace the first u by its value, $ax^2 + by^2 + $ &c., and in like manner, with the v on the right-hand side of the equation, we get

$$ax^2v + by^2v + cz^2v + 2fyzv + \&c. = a'x^2u + b'y^2u + \&c.$$

In like manner, from the identities $vw = wv$, $wu = uw$, we get two other identical relations connecting the quantities x^2u, y^2u, x^2v, x^2w, &c. The question then comes to this: "If there be $m + p$ linear equations in m variables, but these equations connected by p linear relations so as to be equivalent only to m independent equations, how to express most simply the condition that all the equations can be made to vanish together." In the present case $m = 15$, $p = 3$.

89. Let us, for simplicity, take an example with numbers not quite so large, for instance, $m = 3$, $p = 1$. That is to say, let us consider four equations, s, t, u, v, where $s = a_1x + b_1y + c_1z$, $t = a_2x + b_2y + c_2z$, &c., these equations not being independent, but satisfying the relation, $D_1s + D_2t + D_3u + D_4v = 0$. Now I say, in the first place, that if we form the determinant $(a_1b_2c_3)$ of any three of these equations, s, t, u, this must contain D_4 as a factor. For if $D_4 = 0$, we shall have s, t, u connected by a linear relation, so that any values which satisfied both s and t should satisfy u also; and therefore the supposition $D_4 = 0$ would cause the determinant $(a_1b_2c_3)$ to vanish. And in the second place, I say that we get the same result (or, at least, one differing only in sign) whether we divide $(a_1b_2c_3)$ by D_4 or $(a_1b_2c_4)$ by D_3. For (Art. 15) $D_4(a_1b_2c_4)$ is the same as the determinant whose first row is a_1, b_1, c_1, the second, a_2, b_2, c_2, and the third, D_4a_4, D_4b_4, D_4c_4: but we may substitute for D_4a_4 its value

EXPRESSION OF ELIMINANTS AS DETERMINANTS. 73

$-D_1a_1-D_2a_2-D_3a_3$, and in like manner for D_4b_4, D_4c_4. The determinant would then (Art. 18) be resolvable into the sum of three others; but two of these would vanish, having two rows the same, and there would remain $D_4(a_1b_2c_4) = -D_8(a_1b_2c_3)$. It follows then that the eliminant of the system may be expressed in any of the equivalent forms obtained by forming the determinant $(a_1b_2c_3)$ of any three of the equations, and dividing by the remaining constant D_4.

Suppose now that we had five equations s, t, u, v, w, connected by two linear relations $D_1s + D_2t + D_3u + D_4v + D_5w = 0$, $E_1s + E_2t + E_3u + E_4v + E_5w = 0$. Eliminating w from these relations, we have $(D_1E_5)s + (D_2E_5)t + (D_3E_5)u + (D_4E_5)v = 0$, and we see, precisely as before, that the supposition $(D_4E_5) = 0$ would cause the determinant $(a_1b_2c_3)$ to vanish; and that we get the same result whether we divide $(a_1b_2c_3)$ by (D_4E_5) or divide the determinant of any other three of the equations by the complemental determinant answering to (D_4E_5). This reasoning may be extended to any number of equations connected by any number of relations, and we are led to the following general rule for finding the eliminant of the system in its simplest form. Write down the constants in the $m+p$ equations, and complete them into a square form by adding the constants in the p relations: thus

$$\begin{array}{c|cc} s; \quad a_1, b_1, c_1 & D_1, E_1 \\ t; \quad a_2, b_2, c_2 & D_2, E_2 \\ u; \quad a_3, b_3, c_3 & D_3, E_3 \\ v; \quad a_4, b_4, c_4 & D_4, E_4 \\ w; \quad a_5, b_5, c_5 & D_5, E_5 \end{array}$$

then the eliminant in its most reduced form is the determinant of any m rows of the left-hand or equation columns, divided by the determinant got by erasing these rows in the right-hand columns.

Thus, then, in the example of the last Article, we take the determinant of any fifteen of the equations, and, dividing it by a determinant formed with three of the relation rows, obtain the eliminant; which is of the twelfth degree, as it ought to be.

L

90. And, in general, given three equations of the mth, n^{th}, and p^{th} degrees, we form a number of equations of the degree $m+n+p-2$, by multiplying the first equation by all the terms x^{n+p-2}, $x^{n+p-3}y$, and so on. We should in this manner have

$$\tfrac{1}{2}(n+p-1)(n+p) + \tfrac{1}{2}(p+m-1)(p+m) + \tfrac{1}{2}(m+n-1)(m+n)$$

equations. But the number of terms, $x^{m+n+p-2}$, &c., to be eliminated from the equations formed, is $\tfrac{1}{2}(m+n+p-1)(m+n+p)$, or, in general, less than the number of equations. But again, if we consider the identity $uv = vu$, which is of the degree $m+n$, and multiply it by the several terms x^{p-2}, &c., we get $\tfrac{1}{2}(p-1)p$ identical relations between the system of equations we have formed; and in like manner $\tfrac{1}{2}(n-1)n + \tfrac{1}{2}(m-1)m$ other identities; and the number of identities subtracted from the number of equations leaves exactly the number of variables to be eliminated, and gives the eliminant in the right degree.

91. If we had four equations in four variables, we should proceed in like manner, and it would be found then that the case would arise of our having $m+n$ linear equations in m variables, these equations not being independent, but connected by $n+p$ relations; these latter relations again not being independent, but connected by p other relations. And in order to find the reduced eliminant of such a system, we should divide the determinant of any m of the equations by a quantity which is itself the quotient of two determinants. I think it needless to go into further details, but I thought it necessary to explain so much of the theory, the above being, as far as I know, the only general theory of the expression of eliminants as determinants; since whenever, in the application of the dialytic method, any of the equations is multiplied by terms exceeding its own degree, we shall be sure to have a number of equations greater than the number of quantities which we want to eliminate.

LESSON X.

DETERMINATION OF COMMON ROOTS.

92. When the eliminant of any number of equations vanishes, these equations can be satisfied by a common system of values, and we purpose in this Lesson to show how that system of values can be found without actually solving the equations. The method is the same whatever be the number of the variables; but for greater simplicity we commence with the system of two equations, $\phi = 0$, $\psi = 0$, where

$$\phi = a_m x^m + a_{m-1} x^{m-1} + a_{m-2} x^{m-2} + \&\text{c.} = 0,$$

$$\psi = b_n x^n + b_{n-1} x^{n-1} + b_{n-2} x^{n-2} + \&\text{c.} = 0.$$

Let us suppose that some root of the second equation, $x = \alpha$ satisfies the first, and therefore that R the eliminant of the system vanishes. Now in ϕ we may alter the coefficients (a_m into $a_m + A_m$, a_{m-1} into $a_{m-1} + A_{m-1}$, &c.); and the transformed equation

$$a_m x^m + a_{m-1} x^{m-1} + \&\text{c.} + A_m x^m + A_{m-1} x^{m-1} + \&\text{c.} = 0$$

will obviously still be satisfied by the value $x = \alpha$, provided only that the increments A_m, A_{m-1}, &c. are connected by the single relation

$$A_m \alpha^m + A_{m-1} \alpha^{m-1} + \&\text{c.} = 0,$$

since the remaining part of the equation, by hypothesis, vanishes for $x = \alpha$. The transformed equation then has a root common with ψ, and therefore the eliminant between ψ and that transformed equation vanishes. But this eliminant is obtained from R, the eliminant of ϕ and ψ, by altering in it a_m into $a_m + A_m$, &c. The eliminant so transformed is

$$R + \left\{ A_m \frac{dR}{da_m} + A_{m-1} \frac{dR}{da_{m-1}} + \&\text{c.} \right\} + \&\text{c.} = 0.$$

We have $R = 0$ by hypothesis; and since the increments A_m, &c. may be as small as we please, the terms containing the first powers of these increments must vanish separately. We have then $A_m \dfrac{dR}{da_m} + A_{m-1} \dfrac{dR}{da_{m-1}} + \&c. = 0$. This relation must be identical with the relation $A_m \alpha'_m + A_{m-1} \alpha^{m-1} + \&c. = 0$, which we have seen is the only relation that the increments need satisfy. It follows then that the several differential coefficients are proportional to α^m, α^{m-1}, &c., and therefore that α can be found by taking the quotient of any two consecutive differential coefficients.

Cor. 1. If a_p, a_q be any two coefficients in ϕ, we must have, when $R=0$, $\dfrac{dR}{da_p} : \dfrac{dR}{da_{p-k}} :: \dfrac{dR}{da_q} : \dfrac{dR}{da_{q-k}}$; since the quotient either of the first by the second, or of the third by the fourth will $= \alpha^k$. It follows that $\dfrac{dR}{da_p}\dfrac{dR}{da_{q-k}} - \dfrac{dR}{da_q}\dfrac{dR}{da_{p-k}}$ vanishes when $R=0$, and therefore must contain R as a factor; or, in other words, $\dfrac{dR}{da_p}\dfrac{dR}{da_q} - \dfrac{dR}{da_r}\dfrac{dR}{da_s}$ contains R as a factor if we have $p+q=r+s$.

Cor. 2. It is evident, by parity of reasoning, that the differential coefficients of the eliminant, with regard to the several coefficients in ψ, are proportional to α^n, α^{n-1}, &c.; and hence, as in the last corollary, that, when $R=0$, $\dfrac{dR}{da_p} : \dfrac{dR}{da_{p-k}} :: \dfrac{dR}{db_q} : \dfrac{dR}{db_{q-k}}$; or that $\dfrac{dR}{da_p}\dfrac{dR}{db_q} - \dfrac{dR}{da_r}\dfrac{dR}{db_s}$ contains R as a factor when we have $p+q=r+s$.

Cor. 3. Or, again, if we substitute in the second equation the values of α^n, α^{n-1}, &c. given above, we have

$$b_n \dfrac{dR}{da_n} + b_{n-1} \dfrac{dR}{da_{n-1}} + \&c. = 0,$$

when $R=0$. But the left-hand side of this equation cannot contain R as a factor, for it obviously contains the coefficients of ϕ in a degree less by one than that in which R contains them. It must therefore vanish identically.

93. The results of the preceding article may be confirmed by calculating the actual values of the differential coefficients of R. We know (Art. 63) that $R = \phi(\alpha) \phi(\beta) \phi(\gamma)$ &c. But

since $\phi(\alpha) = a_m \alpha^m + a_{m-1} \alpha^{m-1} + \&c.$, we have $\dfrac{d\phi(\alpha)}{da_p} = \alpha^p$; and therefore
$$\frac{dR}{da_p} = \alpha^p \phi(\beta) \phi(\gamma) + \beta^p \phi(\alpha) \phi(\gamma) + \&c.$$
If then α satisfies ϕ, we have $\phi(\alpha)=0$, and $\dfrac{dR}{da_p} = \alpha^p \phi(\beta) \phi(\gamma)$ &c.

In like manner $\dfrac{dR}{da_q} = \alpha^q \phi(\beta) \phi(\gamma)$; and therefore, as before,
$$\frac{dR}{da_p} : \frac{dR}{da_q} :: \alpha^p : \alpha^q.$$

Again, if we multiply together,
$$\frac{dR}{da_p}\frac{dR}{da_q} = \alpha^{p+q}\{\phi(\beta)\}^2\{\phi(\gamma)\}^2 \&c. + R(\alpha^p\beta^q + \alpha^q\beta^p)\phi(\gamma)\&c.+\&c.;$$
and it can easily be seen that the series multiplying R is $\dfrac{d^2R}{da_p da_q}$. If now we subtract $\dfrac{dR}{da_r}\dfrac{dR}{da_s}$, the terms not multiplied by R will destroy each other if we have $p+q = r+s$, and there will remain
$$\frac{dR}{da_p}\frac{dR}{da_q} - \frac{dR}{da_r}\frac{dR}{da_s} = R\left(\frac{d^2R}{da_p da_q} - \frac{d^2R}{da_r da_s}\right).$$

By a similar process we can show that $\dfrac{dR}{da_p}\dfrac{dR}{db_q} - \dfrac{dR}{da_q}\dfrac{dR}{db_p}$ is divisible by R, but the quotient is not $\dfrac{d^2R}{da_p db_q} - \dfrac{d^2R}{da_q db_p}$.

94. What has been said is applicable, as we shall presently see, to a system of equations in any number of variables. The following simpler method only applies to a system of two equations. In that case we have seen (Art. 80) that the eliminant can be expressed in the form of the determinant resulting from the elimination of x^{m-1}, x^{m-2}, &c. from a system of equations linear in these quantities. When this determinant vanishes, the equations are consistent with each other, and if we leave out any one of them, the remainder will suffice to determine x. Hence if β_{11}, β_{12}, &c. represent the minors of the determinant in question, we have x^{m-1}, x^{m-2}, &c. severally proportional to β_{11}, β_{12}, β_{13} &c., or to β_{21}, β_{22}, β_{23}, &c., &c. These values are simpler than those found by the preceding method; since they

are a degree lower than the discriminant in the coefficients of *each* equation; whereas the values found by differentiating the discriminant are a degree lower than it only in the coefficients of one of the equations. For example, the common value which satisfies the pair of equations

$$ax^2 + bx + c = 0, \quad a'x^2 + b'x + c' = 0 \ .$$

is by this method found to be $-\dfrac{(bc')}{(ac')} = -\dfrac{(ac')}{(ab')}$; whereas by the preceding method it is given in the less simple form

$$\frac{2c'(ac') - b'(bc')}{a'(bc') - c'(ab')} = \frac{a'(bc') - c'(ab')}{-2a'(ac') + b'(ab')} \ .$$

All these values are equal in virtue of the relation which is supposed to be satisfied $(ac')^2 = (ab')(bc')$.

95. If we substitute in any of the equations used in the last article, the values $\dfrac{dR}{da_{m-1}}$ for x^{m-1}, &c., this equation must be satisfied when $R = 0$, and therefore the result of substitution must be divisible by R. In other words, if α_{r1}, α_{r2}, &c. be the constituents of any of the lines of the determinant of Art. 80, we must have $\alpha_{r1}\dfrac{dR}{da_{m-1}} + \alpha_{r2}\dfrac{dR}{da_{m-2}} + $ &c. divisible by R. But if we examine what α_{r1}, &c. are, we see that α_{r1} is the determinant $(a_m b_{n-r})$, &c., and thus that the function $\alpha_{r1}\dfrac{dR}{da_{m-1}} + $ &c. contains the b coefficients in a degree one higher than R while its weight exceeds that of R by $n - r + 1$. Consequently the remaining factor must be b_{n-r+1} multiplied by a numerical coefficient. To determine this coefficient we suppose all the terms of ψ to vanish except b_{n-r+1}. Now it follows at once from the method of elimination by symmetric functions that if ψ consist of factors V, W, &c., the eliminant of ϕ and ψ is the product of the eliminants of $\phi, V; \phi, W;$ &c. For if V be $(x-\alpha)(x-\beta)$ &c., and W be $(x-\alpha')(x-\beta')$ &c., the eliminant of ϕ and V is $\phi(\alpha)\phi(\beta)$ &c., that of ϕ and W is $\phi(\alpha')\phi(\beta')$ &c., and the product of all these is the eliminant of ϕ and ψ.

Again, if ψ reduce to the single term $b_a x^a y^\beta$, since the eliminant of ϕ and x is a_0 and of ϕ and y is a_m, the eliminant of

DETERMINATION OF COMMON ROOTS. 79

ϕ and ψ will be $b_a^m a_0^s a_m^\beta$. The only one then of the series of terms $\dfrac{dR}{da_{m-1}}$, &c., which will not vanish when all the coefficients of ψ, except b_a are made to vanish, will be $\dfrac{dR}{da_0}$, and this will be $\alpha b_a^m a_0^{s-1} a_m^\beta$. But in the case we are considering, it will be found that the term by which $\dfrac{dR}{da_0}$ is multiplied will be $b_a a_0$, and hence that in general, when $\alpha = n - r + 1$,

$$a_{r_1} \frac{dR}{da_{m-1}} + a_{r_2} \frac{dR}{da_{m-2}} + \&c. = (n - r + 1) R b_{n-r+1}.$$

Ex. In order to make what has been said more intelligible, we repeat the proof for the particular case of the two cubics $a_3 x^3 + a_2 x^2 + a_1 x + a_0$, $b_3 x^3 + b_2 x^2 + b_1 x + b_0$, then we have the system of equations (Art. 80)

$$(a_3 b_2) x^2 + \qquad (a_3 b_1) \qquad x + (a_3 b_0) = 0,$$
$$(a_3 b_1) x^2 + \{(a_3 b_0) + (a_2 b_1)\} x + (a_2 b_0) = 0,$$
$$(a_3 b_0) x^2 + \qquad (a_2 b_0) \qquad x + (a_1 b_0) = 0.$$

Substituting then, suppose in the second equation; the following quantity must be divisible by R,

$$(a_3 b_1) \frac{dR}{da_2} + \{(a_3 b_0) + (a_2 b_1)\} \frac{dR}{da_1} + (a_2 b_0) \frac{dR}{da_0}.$$

But, considering the order and weight of the function in question, it is seen at once that the remaining factor must be b_2 multiplied by a numerical coefficient. To determine that coefficient, let b_0, b_1, b_3 all vanish, then the quantity we are discussing reduces to $-b_2 \left(a_1 \dfrac{dR}{da_1} + a_0 \dfrac{dR}{da_0}\right)$. But R, on the same supposition, reduces to $b_2^2 a_3 a_0^2$; and therefore the function we are calculating at most differs in sign from $2 b_2 R$.

96. There is no difficulty in applying the method of Arts. 92, 93 to the case of any number of variables. For greater clearness we confine ourselves to three variables, but the same proof applies word for word to any number of variables. Let there be three equations $\phi = 0$, $\psi = 0$, $\chi = 0$, where $\phi = a_{m, 0, 0} x^m + \ldots + a_{\alpha, \beta, \gamma} x^\alpha y^\beta z^\gamma + \&c.$, and let the values $x' y' z'$ satisfy all the equations; then they will still satisfy them if in ϕ we alter $a_{m, 0, 0}$, $a_{\alpha, \beta, \gamma}$ into $a_{m, 0, 0} + A_{m, 0, 0}$, $a_{\alpha, \beta, \gamma} + A_{\alpha, \beta, \gamma}$, &c. provided only that $A_{m, 0, 0} x'^m + \&c. + A_{\alpha, \beta, \gamma} x'^\alpha y'^\beta z'^\gamma + \&c. = 0$. But, as in Art. 92, the equation must also be satisfied

$$A_{m, 0, 0} \frac{dR}{dA_{m, 0, 0}} + A_{\alpha, \beta, \gamma} \frac{dR}{dA_{\alpha, \beta, \gamma}} + \&c. = 0;$$

and comparing these two equations, we see that the value of

each term $x'^\alpha y'^\beta z'^\gamma$ must be proportional to the differential of the eliminant with respect to the coefficient which multiplies it. We obtain the values of x', y', z', by taking the ratios of the differentials of R with respect to the coefficients of any terms which are in the ratio of x, y, z. And this may be verified as in Art. 93. For let the common roots of ψ, χ substituted in ϕ, give results ϕ', ϕ'', &c. Then $R = \phi'\phi''\phi''' $ &c. And

$$\frac{dR}{da_{\alpha,\beta,\gamma}} = x'^\alpha y'^\beta z'^\gamma \phi''\phi''' \text{ &c.} + x''^\alpha y''^\beta z''^\gamma \phi'\phi'' \text{ &c.} + \text{&c.} ;$$

and if we suppose ϕ' to vanish, the value of this differential coefficient reduces to its first term, and it is seen as before, that the differential with regard to each coefficient is proportional to the term which that coefficient multiplies. The same corollaries may be drawn as in Art. 92.

97. And more generally, in like manner, if the coefficients of ϕ be functions of any quantities a, b, c, &c. which do not enter into ψ, χ, it is proved by the same method, that $\frac{dR}{da} : \frac{dR}{db} :: \frac{d\phi}{da} : \frac{d\phi}{db}$ where in the latter differentials x, y, z, are supposed to have the values x', y', z', which satisfy all the equations. For either, as in Art. 93, we have when $\phi' = 0$, $\frac{dR}{da} = \frac{d\phi}{da}\phi''\phi'''$ &c. $\frac{dR}{db} = \frac{d\phi}{db}\phi''\phi'''$ &c.; or again, as in Art. 92, if a, b, c be varied, so that the same system of values continues to satisfy ϕ, we have

$$\frac{d\phi}{da}\delta a + \frac{d\phi}{db}\delta b + \frac{d\phi}{dc}\delta c + \text{&c.} = 0;$$

while, because in this case, the eliminant of the transformed ϕ and of the other equations continues to vanish, we have

$$\frac{dR}{da}\delta a + \frac{dR}{db}\delta b + \frac{dR}{dc}\delta c + \text{&c.} = 0,$$

and these two equations must be identical.

98. The formulæ become more complicated if we take the differentials of the eliminant with respect to quantities a, b, &c. which enter into all the equations. As before, if we give these

DETERMINATION OF COMMON ROOTS. 81

quantities variations, consistent with the supposition that the eliminant still vanishes, we have

$$\frac{dR}{da}\delta a + \frac{dR}{db}\delta b + \frac{dR}{dc}\delta c + \&c. = 0.$$

Now in the former case, where a, b, c, &c. only entered into one of the equations, a change in these quantities produced no change in the value of the common roots, since the coefficients remained constant in the other equations, whose system of common roots was therefore fixed and determinate. But this will now no longer be the case, and the common roots of the transformed equations may be different from those of the original system. Let the new system of common roots be $x' + \delta x'$, $y' + \delta y'$, $z' + \delta z'$, &c., then the variations are connected by the relations

$$\frac{d\phi}{da}\delta a + \frac{d\phi}{db}\delta b + \&c. + \frac{d\phi}{dx}\delta x' + \frac{d\phi}{dy}\delta y' + \&c. = 0,$$

$$\frac{d\psi}{da}\delta a + \frac{d\psi}{db}\delta b + \&c. + \frac{d\psi}{dx}\delta x' + \frac{d\psi}{dy}\delta y' + \&c. = 0, \&c.$$

If there are k such equations, there will be $k-1$ independent variables;* we may therefore, between these k relations eliminate the $k-1$ variations $\delta x'$, $\delta y'$, &c., and so arrive at a relation between the variations δa, δb, &c. only; the coefficients of which must be severally proportional to $\frac{dR}{da}$, $\frac{dR}{db}$, &c.

Ex. 1. Let there be two equations and one variable. The final relation then is

$$\left(\frac{d\phi}{da}\frac{d\psi}{dx} - \frac{d\psi}{da}\frac{d\phi}{dx}\right)\delta a + \left(\frac{d\phi}{db}\frac{d\psi}{dx} - \frac{d\psi}{db}\frac{d\phi}{dx}\right)\delta b + \&c. = 0,$$

and the several coefficients are proportional to $\frac{dR}{da}$, $\frac{dR}{db}$, &c. If the equations had been given in the homogeneous form, we might have taken x as constant, and substituted $\frac{d\phi}{dy}$, $\frac{d\psi}{dy}$ for $\frac{d\phi}{dx}$, $\frac{d\psi}{dx}$ in the preceding formula. This makes no change, because it was proved, Art. 85, that the common root satisfies the Jacobian, or makes

$$\frac{d\phi}{dx}\frac{d\psi}{dy} = \frac{d\phi}{dy}\frac{d\psi}{dx}.$$

---

* If the equations had been given as homogeneous functions of k variables, still since their ratio is all we are concerned with, we may suppose any of the variables z to be the same in all the equations and may suppose $\delta z' = 0$.

Ex. 2. If there are three equations, the coefficient of δa is

$$\begin{vmatrix} \dfrac{d\phi}{da}, & \dfrac{d\psi}{da}, & \dfrac{d\chi}{da} \\ \phi_1, & \psi_1, & \chi_1 \\ \phi_2, & \psi_2, & \chi_2 \end{vmatrix},$$

where ϕ_1, ϕ_2 denote the differential coefficients of ϕ with respect to x and y, &c.

99. If a system of equations is satisfied by *two* common systems of values, not only will the eliminant R vanish, but also the differential of R with respect to every coefficient in either equation. For evidently the values of the differentials, given Art. 93, all vanish if both $\phi(\alpha)$ and $\phi(\beta) = 0$, or, in Art. 96, if ϕ', ϕ'' both $= 0$. In this case the actual values of the two common roots can be expressed by a quadratic equation in terms of the second differentials of R. The following, though for brevity, stated only for the case of two equations, applies word for word in general. We have (see Art. 93)

$$\frac{d^2 R}{da_p^2} = \alpha^p \beta^p \phi(\gamma) \phi(\delta) \text{ &c.} + \beta^p \gamma^p \phi(\alpha) \phi(\delta) \text{ &c.} + \text{&c.},$$

which, when $\phi(\alpha)$, $\phi(\beta) = 0$, reduces to the single term

$$\alpha^p \beta^p \phi(\gamma) \phi(\delta) \text{ &c.}$$

In like manner, in the same case,

$$\frac{d^2 R}{da_p da_q} = (\alpha^p \beta^q + \alpha^q \beta^p) \phi(\gamma) \phi(\delta) \text{ &c.}, \quad \frac{d^2 R}{da_q^2} = \alpha^q \beta^q \phi(\gamma) \phi(\delta) \text{ &c.}$$

If then we solve the quadratic in $\lambda : \mu$,

$$\lambda^2 \frac{d^2 R}{da_q^2} - \lambda \mu \frac{d^2 R}{da_p da_q} + \mu^2 \frac{d^2 R}{da_p^2} = 0,$$

the roots will give the ratios $\alpha^p : \alpha^q$, $\beta^p : \beta^q$.

If the equations have three common systems of values, all the second differentials of R vanish, and the common roots are found by proceeding to the third differential coefficients and solving a cubic equation.

LESSON XI.

DISCRIMINANTS.

100. BEFORE entering on the subject of discriminants, we shall explain some terms and symbols which we shall frequently find it convenient to employ. In ordinary algebra we are wholly concerned with *equations*, the object usually being to find the values of x which will make a given function $= 0$. In what follows we have little to do with equations, the most frequent subject of investigation being that on which we enter in the next Lesson: namely, the discovery of those properties of a function which are unaltered by linear transformations. It is convenient, then, to have a word to denote the function itself, without being obliged to speak of the equation got by putting the function $=0$: a word, for example, to denote $ax^2 + bxy + cy^2$ without being obliged to speak of the quadratic equation $ax^2 + bxy + cy^2 = 0$. We shall, after Mr. Cayley, use the term *quantic* to denote a homogeneous function in general; using the words quadric, cubic, quartic, quintic, n^{ic}, to denote quantics of the 2nd, 3rd, 4th, 5th, n^{th} degrees. And we distinguish quantics into binary, ternary, quaternary, n-ary, according as they contain 2, 3, 4, n variables. Thus, by a binary cubic, we mean a function such as $ax^3 + bx^2y + cxy^2 + dy^3$; by a ternary quadric, such as $ax^2 + by^2 + cz^2 + 2fyz + 2gzx + 2hxy$, &c. Mr. Cayley uses the abbreviation $(a, b, c, d\!\!\!\!\searrow\!\! x, y)^3$ to denote the quantic $ax^3 + 3bx^2y + 3cxy^2 + dy^3$, in which, as is usually most convenient, the terms are affected with the same numerical coefficients as in the expansion of $(x + y)^3$. So the ternary quadric written above would be expressed $(a, b, c, f, g, h\!\!\!\!\searrow\!\! x, y, z)^2$. When the terms are not thus affected with numerical coefficients, he puts an arrow-head on the parenthesis, writing, for instance $(a, b, c, d\!\!\!\!\searrow\!\! x, y)^3$ to denote $ax^3 + bx^2y + cxy^2 + dy^3$. When it is not necessary to mention the coefficients, the quantic of the nth degree is written $(x, y)^n$, $(x, y, z)^n$, &c.

101. If a quantic in k variables be differentiated with respect to each of the variables, the eliminant of the k differentials is called the *discriminant* of the given quantic.

If n be the degree of the quantic, its discriminant is a homogeneous function of its coefficients, and of the order $k(n-1)^{k-1}$. For the discriminant is the eliminant of k equations of the $(n-1)^{\text{th}}$ degree, and (Art. 74) must contain the coefficients of *each* of these equations in a degree equal to the product of the degrees of all the rest, that is $(n-1)^{k-1}$. And since each of these equations contains the coefficients of the original quantic in the first degree, the discriminant contains them in the $k(n-1)^{k-1}$ degree. Thus, then, the discriminant of a binary quantic is of the degree $2(n-1)$; of a ternary, is of the degree $3(n-1)^2$; &c.

102. *If in the original quantic every coefficient multiplying the first power of one of the variables x, be affected with a suffix 1, every term multiplying the second power by a suffix 2, and so on; then the sum of the suffixes in each term of the discriminant is constant and* $= n(n-1)^{k-1}$. It was proved (Art. 74) that if every coefficient in a system of equations were affected with a suffix corresponding to the power of x which it multiplies, then the sum of the suffixes in every term of their eliminant would be equal to the product of the degrees of those equations, viz., $= mnp$ &c. Now suppose, that in the first of these equations the suffix of x^0, instead of being 0, was l; that of x^1 was $l+1$, and so on; it is evident that the effect would be to increase the sum of the suffixes by l for every coefficient of the first equation which enters into the eliminant; and since (Art. 74) every term contains np &c. coefficients of the first equation; the total sum of suffixes is mnp &c. $+ lnp$ &c. $= (m+l)np$ &c. Now, in the present example, it is evident that every coefficient in the $k-1$ differentials U_2, U_3, &c.,* multiplies the same power of x as it did in the original quantic U. But in the remaining differential, U_1, every coefficient multiplies a power of x one less than in U, and the coefficient multiplying any term x^l in this

---

* We write, as before, U_1, U_2, U_3, &c. to denote the differential coefficients of U with respect to x, y, z, &c.

differential will be marked with the suffix $l+1$, since it arose from differentiating a term x^{l+1} in the original quantic. It follows, then, that the sum of suffixes in the discriminant must $= (n-1)^k + (n-1)^{k-1} = n(n-1)^{k-1}$.

We shall briefly express the results of this and of the last article by saying that the *order* of the discriminant is $k(n-1)^{k-1}$; and its *weight*, $n(n-1)^{k-1}$. Thus for a binary quantic the weight of the discriminant is $n(n-1)$.

103. If a binary quantic contain a square factor, then, as is well known, the discriminant vanishes identically. For the two differentials must each contain that factor in the first degree, and therefore, since they have a common factor, their eliminant vanishes. In like manner, if a ternary quantic be of the form $X^2\phi + XY\psi + Y^2\chi$, where $X = ax + by + cz$, $Y = a'x + b'y + c'z$, then the discriminant must vanish, since every term in any of the differentials must contain either X or Y, and therefore the differentials have common the system of roots derived from the equations $X = 0$, $Y = 0$. In like manner, the discriminant of a quaternary quantic vanishes, if the quantic can be expressed as a function in the second degree of X, Y, Z, these being any linear functions of the variables.* We shall call those values which make all the differentials vanish, the *singular roots* of the quantic.

104. We shall now discuss the properties of the discriminant of the binary quantic $U = a_0 x^n + n a_1 x^{n-1} y + \tfrac{1}{2} n(n-1) a_2 x^{n-2} y^2 + \&c.$

The eliminant of U and U_1 is a_0 times the discriminant; and the eliminant of U and U_2 is a_n times the discriminant.† For since $nU = xU_1 + yU_2$, the result of substituting in nU any root of U_1 is $y'U_2'$; and when all the results of substitution are multiplied together, the product will be $y'y''y'''$ &c. (which is $= a_0$, see Art. 57), multiplied by the product of the results of substituting the same roots in U_2, which is the discriminant.

---

* In other words, the vanishing of the discriminant of an algebraical equation expresses the condition that the equation shall have equal roots; and the vanishing of the discriminant of the equation of a curve or surface expresses the condition that the curve or surface shall have a double point.

† We do not take account of mere numerical factors.

105. *To express the discriminant in terms of the values* $x_1 y_1$, $x_2 y_2$, *&c., which make the quantic vanish.*

Let $U = (xy_1 - yx_1)(xy_2 - yx_2)(xy_3 - yx_3)$ &c. (see Art. 57); then

$$U_1 = y_1(xy_2 - yx_2)(xy_3 - yx_3) \,\&c. + y_2(xy_1 - yx_1)(xy_3 - yx_3) \,\&c. + \&c.;$$

and the result of the substitution in U_1 of any root $x_1 y_1$ of U is $y_1(x_1 y_2 - y_1 x_2)(x_1 y_3 - y_1 x_3)$ &c. Similarly, the result of substituting $x_2 y_2$ is $y_2(x_2 y_1 - x_1 y_2)(x_2 y_3 - y_2 x_3)$ &c. If, then, all the results of substitution are multiplied together, the product is

$$\pm y_1 y_2 y_3 \,\&c. \, (x_1 y_2 - y_1 x_2)^2 (x_1 y_3 - y_1 x_3)^2 (x_2 y_3 - y_2 x_3)^2 \,\&c.$$

This, then, is the eliminant of U and U_1, and if we divide it by a_0, which is $= y_1 y_2 y_3$ &c., we shall have the discriminant $= (x_1 y_2 - y_1 x_2)^2 (x_1 y_3 - y_1 x_3)^2$ &c. If we make in it all the y's $= 1$, we get the theorem in the well-known form that the discriminant is equal to the product of the squares of all the differences of any two roots of the equation. We shall, for simplicity, refer to the theorem in the latter form.

106. *The discriminant of the product of two quantics is equal to the product of their discriminants multiplied by the square of their eliminant.* For the product of the squares of differences of all the roots evidently consists of the product of the squares of differences of two roots both belonging to the same quantic, multiplied by the square of the product of all differences between a root of one and a root of the other, and this latter product is the eliminant (Art. 64). As a particular case of this, the discriminant of $(x - \alpha) \phi(x)$ is the discriminant of $\phi(x)$ multiplied by the square of $\phi(\alpha)$. For if β, γ, &c. be the roots of $\phi(x)$, then $(\alpha - \beta)^2 (\alpha - \gamma)^2 (\beta - \gamma)^2$ &c. is equal to the square of $(\alpha - \beta)(\alpha - \gamma)$ &c. which is $\phi(\alpha)$, multiplied by the product of the squares of all differences not containing α.

107. *The discriminant of* $(a_0, a_1 \ldots a_{n-1}, a_n \rangle\!\!\!\!\rangle x, y)^n$ *is of the form* $a_n \phi + a_{n-1}^2 \psi$, *where ψ is the discriminant of the equation of the $(n - 1)^{st}$ degree* $(a_0, a_1 \ldots a_{n-2}, a_{n-1} \rangle\!\!\!\!\rangle x, y)^{n-1}$. For we evidently get the same result whether we put any term $a_n = 0$ in the discriminant; or whether we put $a_n = 0$ in the quantic, and then form the discriminant. But if we make $a_n = 0$ in the quantic, we

get x multiplied by the $(n-1)^{ic}$ written above, and (Art. 106) its discriminant will then be the discriminant of that $(n-1)^{ic}$ multiplied by the square of the result of making in it $x=0$; that is, by the square of a_{n-1}. In like manner we see that the discriminant is of the form $a_0\phi + a_1^2\psi$.*

108. The discriminant being a function of the determinants $x_1 y_2 - x_2 y_1$, &c. must satisfy the two differential equations (Art. 58),

$$na_0\frac{d\Delta}{da_1} + (n-1)a_1\frac{d\Delta}{da_2} + (n-2)a_2\frac{d\Delta}{da_3} + \&c. = 0,$$

$$a_1\frac{d\Delta}{da_0} + 2a_2\frac{d\Delta}{da_1} + 3a_3\frac{d\Delta}{da_2} + \&c. = 0;$$

or, if the original equation had been written with binomial coefficients,

$$na_1\frac{d\Delta}{da_0} + (n-1)a_2\frac{d\Delta}{da_1} + \&c. = 0, \quad a_0\frac{d\Delta}{da_1} + 2a_1\frac{d\Delta}{da_2} + \&c. = 0.$$

Ex. To form the discriminant of $(a_0, a_1, a_2, \ldots \mathbb{X} x, y)^n$, which we suppose arranged according to the powers of a_0. We know (Art. 107) that the absolute term is $a_1^2 D$, where D is the discriminant of $(a_1, a_2, \ldots \mathbb{X} x, y)^{n-1}$. The discriminant then is $a_1^2 D + a_0\phi + a_0^2\psi + \&c.$; operating on this with $a_1\frac{d}{da_0} + 2a_2\frac{d}{da_1} + 3a_3\frac{d}{da_2} + \&c.$, we may equate separately to zero the coefficient of each power of a_0. Thus then the part independent of a_0 is

$$a_1\phi + 4a_1a_2 D + a_1^2\left(2a_2\frac{d}{da_1} + 3a_3\frac{d}{da_2} + \&c.\right)D,$$

or, remembering that $\left(a_2\frac{d}{da_1} + 2a_3\frac{d}{da_2} + \&c.\right)D = 0$, we have

$$\phi = -4a_2 D + a_1\left(a_3\frac{d}{da_2} + 2a_4\frac{d}{da_3} + \&c.\right)D,$$

and the discriminant is

$$(a_1^2 - 4a_0a_2)D + a_1 a_0\left(a_3\frac{d}{da_2} + 2a_4\frac{d}{da_3} + \&c.\right)D + a_3^2\psi + \&c.$$

In like manner, from the coefficient of a_1 we can determine ψ, but the result is not simple enough to seem worth writing down.

---

* This theorem was first published by Joachimsthal; I had, however, previously been led by simple geometrical considerations to the following theorem in which it is included. If a_1 contain a factor z, and if a_0 contain z^2 as a factor, the discriminant will be divisible by z^2. If a_2 contain z as a factor, if a_1 contain z^2, and a_0 contain z^3, the discriminant will in general be divisible by z^6. In like manner, if a_3 contain z; a_2, z^2; a_1, z^3; and a_0, z^4, the discriminant will be divisible by z^{12}, &c.

109. If the discriminant of a binary quantic vanishes, the quantic has equal roots, and the actual values of these roots can be found by a process similar to that employed in Lesson X. Let $U = a_0 x^n + a_1 x^{n-1} + a_2 x^{n-2} + \&c.$ be a quantic whose discriminant vanishes, and having therefore a square factor $(x - a)^2$. Then evidently V, where

$$V = A_0 x^n + A_1 x^{n-1} + A_2 x^{n-2} + \&c.$$

will also be divisible by $x - a$, provided that A_0, A_1, &c. be any quantities satisfying the condition

$$A_0 a^n + A_1 a^{n-1} + A_2 a^{n-2} + \&c. = 0.$$

In this case then we shall have $U + \lambda V$ divisible by $x - a$. Let it $= (x - a)\{(x - a)\,\phi(x) + \lambda \psi(x)\}$. It follows then, from Art. 107, that the discriminant of $U + \lambda V$ is the discriminant of the quantity within the brackets, multiplied by the square of the result of substituting a for x inside the brackets. But this result is $\lambda \psi(a)$. We have proved then that in the case supposed, the discriminant of $U + \lambda V$ is divisible by λ^2.

But since $U + \lambda V$ is derived from U by altering a_0 into $a_0 + \lambda A_0$, &c., the discriminant of $U + \lambda V$ is derived from the discriminant of U by a like substitution, and is therefore

$$\Delta + \lambda \left(A_0 \frac{d\Delta}{da_0} + A_1 \frac{d\Delta}{da_1} + A_2 \frac{d\Delta}{da_2} + \&c. \right) + \lambda^2 (\&c.).$$

By hypothesis $\Delta = 0$. But the discriminant will not be divisible by λ^2 unless the coefficient of λ vanish. Now the relation thus obtained between A_0, A_1, &c. must be identical with the relation $A_0 a^n + A_1 a^{n-1} + \&c. = 0$, which we have already seen is the only relation that need be satisfied by A_0, A_1, &c. in order that the discriminant of $U + \lambda V$ may be divisible by λ^2. We must have therefore the quantities a^n, a^{n-1}, a^{n-2}, &c. respectively proportional to $\dfrac{d\Delta}{da_0}$, $\dfrac{d\Delta}{da_1}$, $\dfrac{d\Delta}{da_2}$, &c. Dividing any one of these terms by that consecutive, we get an expression for a. We may state this result: *When the discriminant vanishes, the several differential coefficients of the discriminant with respect to a_0, a_1, &c. are proportional to the differential coefficients of the quantic with respect to the same quantities.*

110. This result may be confirmed by forming the actual values of $\dfrac{d\Delta}{da_1}$, &c. in terms of the roots; which may be done by solving from the n equations

$$\frac{d\Delta}{da} = \frac{d\Delta}{da_1}\frac{da_1}{da} + \frac{d\Delta}{da_2}\frac{da_2}{da} + \&c.$$

We know the expressions for Δ, a_1, a_2, &c. in terms of the roots (see Art. 61), and therefore from these n equations can find the n sought quantities $\dfrac{d\Delta}{da_1}$, &c. The result will be found to be

$$\frac{d\Delta}{da_n} = \Sigma\,(\beta-\gamma)^2(\gamma-\delta)^2(\delta-\beta)^2$$
$$\times \{(\alpha-\beta)(\alpha-\gamma) + (\alpha-\beta)(\alpha-\delta) + (\alpha-\gamma)(\alpha-\delta)\},$$

where the product of the squares of all the differences, not containing α, is multiplied by the sum of the products ($n-2$ taken together) of the differences which contain α,

$$\frac{d\Delta}{da_{n-1}} = \Sigma\alpha\,(\beta-\gamma)^2(\gamma-\delta)^2(\delta-\beta)^2\{(\alpha-\beta)(\alpha-\gamma)+\&c.\},$$

$$\frac{d\Delta}{da_{n-2}} = \Sigma\alpha^2(\beta-\gamma)^2(\gamma-\delta)^2(\delta-\beta)^2\{(\alpha-\beta)(\alpha-\gamma)+\&c.\},\ \&c.,$$

and the supposition $\alpha=\beta$ reduces these sums to quantities which are in the ratio 1, α, α^2, &c. As in Art. 92, it follows from the theorem of the last article that $\dfrac{d\Delta}{da_p}\dfrac{d\Delta}{da_q} - \dfrac{d\Delta}{da_r}\dfrac{d\Delta}{da_s}$ is divisible by Δ when $p+q=r+s$. If more than two of the roots are equal to each other, all these differentials vanish identically, and we find the equal roots by proceeding to second differentials of the discriminant.

111. We know, from Art. 94, that instead of the functions in the last two articles, which are of an order in the coefficients only one lower than the discriminant, we may substitute functions of an order two lower, and possessing the same property, viz. that they vanish when more than two roots are equal, and that if two roots are equal ($\alpha=\beta$) they are to each other in the ratios 1, α, α^2, &c. If we proceed by Bezout's method of elimination (Art. 80) to eliminate between the two differen-

tials U_1, U_2; the resulting equations of the $(n-2)^{\text{th}}$ degree, when expressed in terms of the roots are $\Sigma(\alpha-\beta)^2(x-\gamma)(x-\delta)=0$, $\Sigma q_1(\alpha-\beta)^2(x-\gamma)(x-\delta)=0$, $\Sigma q_2(\alpha-\beta)^2$, &c.$=0$, where q_1, q_2, &c. are the sum, sum of products in pairs, &c. of all the roots except α and β.* The discriminant is then, by Bezout's method, expressed as a determinant, whose constituents are

$$\Sigma\ (\alpha-\beta)^2,\ \Sigma q_1\ (\alpha-\beta)^2,\ \Sigma q_2\ (\alpha-\beta)^2,\ \&c.,$$
$$\Sigma q_1(\alpha-\beta)^2,\ \Sigma q_1^2\ (\alpha-\beta)^2,\ \Sigma q_1 q_2(\alpha-\beta)^2,\ \&c.,$$
$$\Sigma q_2(\alpha-\beta)^2,\ \Sigma q_1 q_2(\alpha-\beta)^2,\ \Sigma q_2^2\ (\alpha-\beta)^2,\ \&c.,\ \&c.$$

And when the given equation has two roots equal, the first minors of this determinant will, by Art. 94, be in the ratio 1, α, α^2, &c. A somewhat simpler series, possessing the same property, is $\Sigma(\beta-\gamma)^2(\gamma-\delta)^2(\delta-\beta)^2$, $\Sigma\alpha(\beta-\gamma)^2(\gamma-\delta)^2(\delta-\beta)^2$, $\Sigma\alpha^2(\beta-\gamma)^2$, &c.

112. The following proof of the theorem of Art. 109 is applicable to the case of a quantic in any number of variables. For simplicity, we confine ourselves to the case of two independent variables, the method, which is that of Art. 98, being equally applicable in general. Let the coefficients in U be functions of any quantities a, b, &c., and let variations be given to these quantities consistent with the supposition that the discriminant still vanishes, and therefore such that

$$\frac{d\Delta}{da}\delta a + \frac{d\Delta}{db}\delta b + \&c. = 0.$$

And if the effect of this change in a, b, &c. is to alter the singular roots from x, y into $x+\delta x$, $y+\delta y$; since these new values satisfy U_1, U_2, U_3, &c., we must have

$$\frac{dU_1}{da}\delta a + \frac{dU_1}{db}\delta b + \&c. + \frac{dU_1}{dx}\delta x + \frac{dU_1}{dy}\delta y = 0,$$
$$\frac{dU_2}{da}\delta a + \frac{dU_2}{db}\delta b + \&c. + \frac{dU_2}{dx}\delta x + \frac{dU_2}{dy}\delta y = 0,$$
$$\frac{dU_3}{da}\delta a + \frac{dU_3}{db}\delta b + \&c. + \frac{dU_3}{dx}\delta x + \frac{dU_3}{dy}\delta y = 0.$$

---

* The first of these functions of degree $n-2$ is one of the series to which we are led by Sturm's process; but any of the others may be substituted for it, and will give rise to another series possessing the same property.

Multiply these equations by x, y, z respectively and add; then since $nU = xU_1 + yU_2 + zU_3$, the coefficient of δa will be $n\dfrac{dU}{da}$; and since $\dfrac{dU_2}{dx} = \dfrac{dU_1}{dy}$, $\dfrac{dU_3}{dx} = \dfrac{dU_1}{dz}$, the coefficient of δx will be $(n-1)U_1$, which will vanish, since U_1 is satisfied for the singular roots. We get therefore

$$\frac{dU}{da}\delta a + \frac{dU}{db}\delta b + \&c. = 0,$$

and therefore the differentials of Δ with respect to a, b, &c. are proportional to the differentials of U with respect to the same quantities, it being understood that the x, y, z which occur in the latter differentials are the singular roots.

113. The theorem proved for binary quantics (Art. 107) may be extended to quantics in general. Let a be the coefficient of the highest power of any of the variables, b, c, d, &c. those of the terms involving the next highest power; then the discriminant is of the form

$$a\theta + (\phi, \chi, \psi, \&c. \mathbin{\mathchoice{\mathrm{Y}}{\mathrm{Y}}{\mathrm{Y}}{\mathrm{Y}}} b, c, d, \&c.)^2.$$

Thus, for a ternary quantic to which, for greater simplicity, we confine ourselves, if a be the coefficient of z^n; b, c those of $z^{n-1}x$, $z^{n-1}y$, then if in the discriminant we make $a = 0$, the remaining part will be of the form $b^2\phi + bc\psi + c^2\chi$. To prove this: first, let U be any quantic whose discriminant vanishes, V any other satisfied by the singular roots of U, then I say that the discriminant of $U + \lambda V$ will be divisible by λ^2. For let $U = az^n + bz^{n-1}x + \&c.$, $V = Az^n + Bz^{n-1}x + \&c.$, then the coefficient of λ in the discriminant of $U + \lambda V$ will be $A\dfrac{d\Delta}{da} + B\dfrac{d\Delta}{db} + \&c.$, and (Art. 112) $\dfrac{d\Delta}{da}$, &c. are proportional to z^n, $z^{n-1}x$, &c. The coefficient of λ is therefore proportional to the result of substituting the singular roots in V, and therefore vanishes.

Now in the case we are considering, the supposition of $a = 0$, $b = 0$, $c = 0$ must make the discriminant vanish, since then all the differentials vanish for the singular roots $x = 0$, $y = 0$. Any other quantic V will vanish for the same values, provided only $A = 0$. The general form of the discriminant then must be such that if we substitute for b, $b + \lambda B$; for c, $c + \lambda C$, &c., and then make a, b, $c = 0$, the result must be divisible by λ^2; or, in other words, if we put for b, λB; for c, λC, &c., and then make $a = 0$, the result is divisible by λ^2, which was the thing to be proved.

114. Concerning discriminants in general, it only remains to notice that the discriminant of a quadratic function in any number of variables is immediately expressed as a symmetrical determinant. And conversely, from any symmetrical determinant, we may form a quadratic function which shall have that determinant for its discriminant. The simplest notation for the coefficients of a quadratic function is to use a double suffix, writing the coefficients of x^2, y^2, &c., a_{11}, a_{22}, a_{33}, &c., and those of xy, xz, &c., a_{12}, a_{13}; a_{12} and a_{21} being identical in this notation. The discriminant is then obviously the symmetrical determinant

$$\begin{vmatrix} a_{11}, & a_{12}, & a_{13}, & \&c. \\ a_{21}, & a_{22}, & a_{23}, & \&c. \\ a_{31}, & a_{32}, & a_{33}, & \&c. \\ \&c. & & & \end{vmatrix}.$$

LESSON XII.

LINEAR TRANSFORMATIONS.

115. *Invariants.* The discriminant of a binary quantic being a function of the differences of the roots is evidently unaltered when all the roots are increased or diminished by the same quantity. Now the substitution of $x + \lambda$ for x is a particular case of the general *linear transformation*, where, in a homogeneous function, we substitute for each variable a linear function of the variables; as for example, in the case of a binary quantic where we substitute for x, $\lambda x + \mu y$, and for y, $\lambda' x + \mu' y$. It will illustrate the nature of the enquiries in which we shall presently engage if we examine the effect of

LINEAR TRANSFORMATIONS. 93

this substitution on the discriminant of the binary quadratic, $ax^2 + 2bxy + cy^2$. When the variables are transformed, it becomes

$$a(\lambda x + \mu y)^2 + 2b(\lambda x + \mu y)(\lambda' x + \mu' y) + c(\lambda' x + \mu' y)^2;$$

and if we call the transformed equation $a'x^2 + 2b'xy + c'y^2$, we have

$$a' = a\lambda^2 + 2b\lambda\lambda' + c\lambda'^2, \quad c' = a\mu^2 + 2b\mu\mu' + c\mu'^2,$$
$$b' = a\lambda\mu + b(\lambda\mu' + \lambda'\mu) + c\mu\mu'.$$

It can now be verified without difficulty that

$$a'c' - b'^2 = (ac - b^2)(\lambda\mu' - \lambda'\mu)^2;$$

that is to say, the discriminant of the transformed quadratic is equal to the discriminant of the given quadratic multiplied by the square of the determinant $\lambda\mu' - \lambda'\mu$, which is called the *modulus of transformation*.

116. Now, a corresponding theorem is true for the discriminant of any binary quantic. We can see *à priori* that this must be the case, for if a given quantic has a square factor, it will have a square factor still when it is transformed; so that whenever the discriminant of the given quantic vanishes, that of the transformed must necessarily vanish too. The one must therefore contain the other as a factor. The theorem however can be formally proved as follows: Let the original quantic be $(xy_1 - yx_1)(xy_2 - yx_2)$ &c., then (Art. 105) the discriminant is $(x_1y_2 - y_1x_2)^2 (x_1y_3 - y_1x_3)^2$ &c.

Now the linear factor $(xy_1 - yx_1)$ of the given quantic becomes by transformation $y_1(\lambda X + \mu Y) - x_1(\lambda' X + \mu' Y)$, and if we write this in the form $Y_1 X - X_1 Y$, we shall have $Y_1 = \lambda y_1 - \lambda' x_1$, $X_1 = -\mu y_1 + \mu' x_1$. If then the transformed quantic be written as the product of the linear factors $(Y_1 X - X_1 Y)(Y_2 X - X_2 Y)$ &c., we have expressions, as above, for $Y_1, X_1; Y_2, X_2$ &c. in terms of $y_1, x_1; y_2, x_2$ &c. We can then, without difficulty, verify that

$$(Y_1 X_2 - X_1 Y_2) = (\lambda\mu' - \lambda'\mu)(y_1 x_2 - x_1 y_2).$$

It follows immediately that $(Y_1 X_2 - Y_2 X_1)^2 (Y_1 X_3 - Y_2 X_1)^2$ &c. is equal to $(y_1 x_2 - x_1 y_2)^2 (y_1 x_3 - y_3 x_1)^2$ &c. multiplied by a power of $\lambda\mu' - \lambda'\mu$ equal to the number of factors in the expression for the discriminant in terms of the roots. A corresponding theorem is true for the discriminant of a quantic in any number of variables.

What I have called Modern Algebra may be said to have taken its origin from a paper in the *Cambridge Mathematical Journal*, for Nov. 1841, where Dr. Boole established the principles just stated and made some important applications of them. Subsequently Mr. Cayley proposed to himself the problem to determine à *priori* what functions of the coefficients of a given equation possess this property of *invariance;* viz. that when the equation is linearly transformed, the same function of the new coefficients is equal to the given function multiplied by a quantity independent of the coefficients. The result of his investigations was to discover that this property of invariance is not peculiar to discriminants and to bring to light other important functions, (some of them involving the variables as well as the coefficients) whose relations to the given equation are unaffected by linear transformation. In explaining this theory, even where, for brevity, we write only three variables, the reader is to understand that the processes are all applicable in exactly the same way to any number of variables.

117. We suppose then that the variables in any homogeneous quantic in k variables are transformed by the substitution

$$x = \lambda_1 X + \mu_1 Y + \nu_1 Z + \&c.,$$
$$y = \lambda_2 X + \mu_2 Y + \nu_2 Z + \&c.,$$
$$z = \lambda_3 X + \mu_3 Y + \nu_3 Z + \&c., \&c.,$$

and we denote by Δ the *modulus of transformation;* namely, the determinant, whose constituents are the coefficients of transformation, $\lambda_1, \mu_1, \nu_1, \&c., \lambda_2, \mu_2, \nu_2, \&c., \&c.$

Now it is evidently not possible in general so to choose the coefficients $\lambda_1, \mu_1, \&c.$, that a certain given function $ax^n + \&c.$ shall assume, by transformation, another given form $a'X^n + \&c.$ In fact, if we make the substitution in $ax^n + \&c.$, and then equate coefficients, we obtain, as in Art. 115, a series of equations $a' = a\lambda_1^n + \&c.$, the number of which will be equal to the number of terms in the general function of the nth degree in

LINEAR TRANSFORMATIONS. 95

k variables. And to satisfy these equations we have only at our disposal the k^2 constants λ_1, λ_2, &c., a number which will in general be less than the number of equations to be satisfied.* It follows then that when a function $ax^n + $ &c. is capable of being transformed into $a'X^n + $ &c., there will be relations connecting the coefficients a, b, &c., a', b', &c. In fact we have only to eliminate the k^2 constants from any $k^2 + 1$ of the equations $a' = a\lambda_1^n + $ &c., and we obtain a series of relations connecting a, a', &c., which will be equivalent to as many independent relations as the excess over k^2 of the number of equations. Thus in the case of a binary quantic, the number of terms in a homogeneous function of the nth degree is $n + 1$. If then, in any quantic $ax^n + $ &c., we substitute for x, $\lambda_1 X + \mu_1 Y$, and for y, $\lambda_2 X + \mu_2 Y$, and if we then equate coefficients with $a'X^n + $ &c., we have $n + 1$ equations connecting a, a', λ_1, &c., from which, if we eliminate the four quantities λ_1, λ_2, μ_1, μ_2, we get a system equivalent to $n - 3$ independent relations between a, b, a', b', &c. It will appear in the sequel that these relations can be thrown into the form $\phi(a, b, $ &c.$) = \phi(a', b', $ &c.$)$; or, in other words, that there are functions of the coefficients a, b, &c. which are equal to the same functions of the transformed coefficients. The process indicated in this article is not that which we shall actually employ in order to find such functions, but it gives an *à priori* explanation of the existence of such functions, and it shows what number of such functions, independent of each other, we may expect to find.

118. Any function of the coefficients of a quantic is called an *invariant*, if when the quantic is linearly transformed, the same function of the new coefficients is equal to the old function

---

* The number of terms in the general equation of the nth degree homogeneous in k variables is $\dfrac{(n+1)(n+2)...(n+k-1)}{1.2...(k-1)}$; and it is easy to see that the only cases where this number is not greater than k^2 are first; when $n = 2$, when it becomes $\frac{1}{2}k(k+1)$, a number necessarily less than k^2, k being an integer; and secondly, the case $k = 2$, $n = 3$, when both numbers have the same value 4. That is to say, the only cases where a given function can be made by transformation to assume any assigned form, are first, the case of a quadratic function in any number of variables; and secondly, the case of a cubic function homogeneous in two variables.

multiplied by some power of the modulus of transformation; that is to say, when we have

$$\phi(a', b', c', \&c.) = \Delta^p \phi(a, b, c, \&c.).$$

Such a function is said to be an *absolute invariant* when $p = 0$; that is to say, when the function is absolutely unaltered by transformation even though Δ be not $= 1$. If a quantic have two ordinary invariants, it is easy to deduce from them an absolute invariant. For if it have an invariant ϕ, which when transformed becomes multiplied by Δ^p, and another ψ, which when transformed becomes multiplied by Δ^q, then evidently the qth power of ϕ divided by the pth power of ψ will be a function which will be absolutely unchanged by transformation.

It follows, from what has been just said, that a binary quadratic or cubic can have no invariant but the discriminant, which we saw (Art. 116) is an invariant. For if there were a second, we could from the two deduce a relation $\phi(a, b, \&c.) = \phi(a', b', \&c.)$. But we see from Art. 117 that there can be no relation connecting a, b, &c. with a', b', &c., since, with the help of the four constants λ_1, &c. at our disposal, we can transform a given quadratic or cubic, so that the coefficients of the transformed equation may have any values we please. In the same manner we see that a quantic of the second order in any number of variables can have no invariant but the discriminant.

119. In the same manner as we have invariants of a single quantic we may have invariants of a system of quantics. Let there be any number of simultaneous equations $ax^n + \&c. = 0$, $a'x^n + \&c. = 0$, &c., and if when the variables in all are transformed by the same substitution, these become $AX^n + \&c. = 0$, $A'X^n + \&c. = 0$, &c., then any function of the coefficients is an invariant if the same function of the new coefficients is equal to the old function multiplied by a power of the modulus of transformation; that is to say, if

$$\phi(A, B, A', B', A'', \&c.) = \Delta^p \phi(a, b, a', b', a'', \&c.).$$

The simplest example of such invariants is the case of a system of linear equations. The determinant of such a system is an invariant of the system. This is evident at once from the definition of an invariant and from the form in which the fundamental theorem for the multiplication of determinants has been stated at p. 18.

If we are given an invariant of a single quantic, we can derive from it a series of invariants of systems of quantics of the same degree. In order to make the spirit of the method more clear, we illustrate it in the first instance by a simple example. We have seen (Art. 115) that $ac - b^2$ is an invariant of the quadratic $ax^2 + 2bxy + cy^2$, and we shall now thence derive an invariant of a system of two quadratics. Suppose that by a linear transformation $ax^2 + 2bxy + cy^2$ becomes $AX^2 + 2BXY + CY^2$, and $a'x^2 + 2b'xy + c'y^2$ becomes $A'X^2 + 2B'XY + C'Y^2$; then evidently by the same transformation, (k being any constant),

$$(a + ka') x^2 + 2 (b + kb') xy + (c + kc') y^2$$

will become $(A + kA') X^2 + 2 (B + kB') XY + (C + kC') Y^2$.

Forming then the invariant of the last quadratic, we have (Art. 115)

$$(A+kA')(C+kC') - (B+kB')^2 = \Delta^2 \{(a + ka')(c + kc') - (b + kb')^2\}.$$

But since k is arbitrary, the coefficients of the respective powers of k must be equal on both sides of the equation; and therefore we have not only, as we knew before,

$$(AC - B^2) = \Delta^2 (ac - b^2), \quad (A'C' - B'^2) = \Delta^2 (a'c' - b'^2),$$

but also $AC' + A'C - 2BB' = \Delta^2 (ac' + a'c - 2bb')$,

an equation which may also be directly verified by the values of A, B, &c. given Art. 115. We see then that $ac' + a'c - 2bb'$ is an invariant.

By exactly the same method, if we have any invariant of a quantic $ax^n + $ &c., and if we want to form invariants of the system $ax^n + $ &c., $a'x^n + $ &c., we have only to substitute in the given invariant for each coefficient a, $a + ka'$, for b, $b + kb'$, &c., and the coefficient of every power of k in the result will be an invariant. Writing down, by Taylor's theorem, the result of substituting $a + ka'$ for a, &c., the theorem to which we have been led may be stated thus: If we have any invariant of a quantic $ax^n + $ &c., and if we perform on it the operation

$a'\dfrac{d}{da}+b'\dfrac{d}{db}+$ &c., we get an invariant of the system of two quantics ax^n+&c., $a'x^n+$&c. We may repeat the same operation and thus get another invariant of the system, or we may operate with $a''\dfrac{d}{da}+b''\dfrac{d}{db}+$&c., and thus get an invariant of a system of three quantics; and so on. This latter process gives us the invariants which we should find by substituting for a, $a+ka'+la''$, &c., and taking the coefficients of the products of every power of k and l. In the same manner we get invariants of a system of any number of quantics.

120. *Covariants.* A *covariant* is a function involving not only the coefficients of a quantic, but also the variables, and such that when the quantic is linearly transformed, the same function of the new variables and coefficients shall be equal to the old function multiplied by some power of the modulus of transformation; that is to say, if ax^n+&c. when transformed becomes AX^n+&c., a function ϕ will be a covariant* if it is such that

$$\phi(A, B, \&c., X, Y, \&c.) = \Delta^r \phi(a, b, \&c., x, y, \&c.).$$

Every invariant of a covariant is an invariant of the original quantic. This follows at once from the definitions. Let the quantic be ax^n+&c., and the covariant $a'x^m+$&c. which are supposed to become by transformation AX^n+&c., $A'X^m+$&c. Now an invariant of the covariant is a function of its coefficients such that

$$\phi(A', B', \&c.) = \Delta^r \phi(a', b', \&c.).$$

But A', B', &c. by definition can only differ by a power of the modulus from being the same functions of A, B, &c. that a', b', &c. are of a, b, &c. Hence when the functions are both

---

* In the geometry of curves and surfaces, all transformations of coordinates are effected by linear substitutions. An invariant of a ternary or quaternary quantic is a function of the coefficients, whose vanishing expresses some property of the curve or surface independent of the axes to which it is referred, as, for instance, that the curve or surface should have a double point. A covariant will denote another curve or surface, the locus of a point whose relation to the given curve is independent of the choice of axes. Hence the geometrical importance of the theory of invariants and covariants.

expressed in terms of the coefficients of the original quantic and its transformed, we have
$$\psi(A, B, \&c.) = \Delta^q \psi(a, b, \&c.),$$
or the function is an invariant. Similarly, a covariant of a covariant is a covariant of the original quantic.

121. We shall in this and the next article establish principles which lead to an important series of covariants.

If in any quantic u we substitute $x + kx'$ for x, $y + ky'$ for y, &c., where $x'y'z'$ are cogredient to xyz, then the coefficients of the several powers of k, which are all of the form $\left(x' \dfrac{d}{dx} + y' \dfrac{d}{dy} + \&c. \right)^p u$, have been called the first, second, third, &c. *emanants** of the quantic. Now *each of these emanants is a covariant of the quantic*. We evidently get the same result whether in any quantic we write $x + kx'$ for x, &c., and then transform x, x', &c. by linear substitutions, or whether we make the substitutions first and then write $X + kX'$ for X, &c. For plainly
$$\lambda_1 X + \mu_1 Y + \nu_1 Z + k(\lambda_1 X' + \mu_1 Y' + \nu_1 Z')$$
$$= \lambda_1 (X + kX') + \mu_1 (Y + kY') + \nu_1 (Z + kZ').$$

If then u becomes by transformation U, we have proved that the result of writing $x + kx'$ for x, &c. in u, must be the same as the result of writing $X + kX'$ for X, &c. in U, and since k is indeterminate, the coefficients of k must be equal on both sides of the equation; or
$$x' \frac{du}{dx} + y' \frac{du}{dy} + \&c. = X' \frac{dU}{dX} + Y' \frac{dU}{dY} + \&c., \&c. \quad \text{Q.E.D.}$$

122. *If we regard any emanant as a function of x', y', &c. treating x, y, &c. as constants; then any of its invariants will be a covariant of the original quantic when x, y, &c. are considered as variables.*

We have just seen that $x'^p \dfrac{d^p u}{dx^p} + \&c.$ becomes $X'^p \dfrac{d^p U}{dX^p} + \&c.$ when we substitute for x', $\lambda_1 X' + \mu_1 Y' + \&c.$, and for x,

---

* In geometry emanants denote the polar curves or surfaces of a point with regard to a curve or surface.

$\lambda_1 X + \mu_1 Y +$ &c. It is evidently a matter of indifference whether the substitutions for x', &c. and for x, &c. are simultaneous or successive. If then on transforming x', &c. alone, $x'^p \frac{d^p u}{dx'^p} +$ &c. becomes $aX'^p +$ &c., then a, &c. will be such functions of x, &c. as when x, y, &c. are transformed will become $\frac{d^p U}{dX^p}$, &c. Now an invariant of the given emanant considered as a function only of x', y', &c. is by definition such a function of its coefficients as differs only by a power of the modulus from the corresponding function of the transformed coefficients a, b, &c. But since, as we have seen, a, &c. become $\frac{d^p U}{dX^p}$, &c. when x, &c. are transformed, it follows that the given invariant will be a function of $\frac{d^p u}{dx^p}$, &c., which when x, &c. are transformed will differ only by a power of the modulus from the corresponding function of $\frac{d^p U}{dX^p}$, &c. It is therefore by definition a covariant of the quantic.

Thus then, for example, since we have proved (Art. 115) that if the binary quantic $ax^2 + 2bxy + cy^2$ becomes by transformation $AX^2 + 2BXY + CY^2$, then

$$AC - B^2 = \Delta^2 (ac - b^2);$$

it follows now, by considering the second emanant $\left(x' \frac{d}{dx} + y' \frac{d}{dy} \right)^2 u$ of a quantic of any degree, that

$$\frac{d^2 U}{dX^2} \cdot \frac{d^2 U}{dY^2} - \left(\frac{d^2 U}{dX dY} \right)^2 = \Delta^2 \left\{ \frac{d^2 u}{dx^2} \cdot \frac{d^2 u}{dy^2} - \left(\frac{d^2 u}{dx dy} \right)^2 \right\},$$

a theorem of which other demonstrations will be given.

123. In general, if we take the second emanant of a quantic in any number of variables, and form its discriminant, this will be a covariant which is called the *Hessian* of the quantic. It was noticed (Art. 114) that the discriminant of every quadratic function may be written as a determinant. Thus then if, as we have done elsewhere, we use the suffixes 1, 2, &c. to denote differentiation with respect to x, y, &c., so that, for

example, u_{11} shall denote $\dfrac{d^2u}{dx_1^2}$, then the quadratic emanant is $u_{11}x'^2 + 2u_{12}x'y' + \&c.$, and its discriminant, which is the Hessian, is the determinant

$$\begin{vmatrix} u_{11}, & u_{12}, & u_{13}, & \&c. \\ u_{21}, & u_{22}, & u_{23}, & \&c. \\ u_{31}, & u_{32}, & u_{33}, & \&c. \\ \&c. & & & \end{vmatrix}.$$

124. We have seen (Art. 119) that the determinant of a system of linear equations is an invariant of the system. If then, given a system $u, v, w, \&c.$, we take their first emanants $x'u_1 + y'u_2 + z'u_3 + \&c.$, &c., their determinant

$$\begin{vmatrix} u_1, & u_2, & u_3, & \&c. \\ v_1, & v_2, & v_3, & \&c. \\ w_1, & w_2, & w_3, & \&c. \\ \&c. & & & \end{vmatrix}$$

is a covariant of the system. This is the determinant already called the Jacobian (p. 69). The Hessian is the Jacobian of the system of differentials of a single quantic $u_1, u_2, u_3, \&c.$

125. *Contravariants.* When a set of variables $x, y, \&c.$ are linearly transformed, it constantly happens that other variables connected with them are also linearly transformed, but by a substitution different from that which is applied to $x, y, \&c.$ If the equations connecting x, y, z with the new variables be written as before

$$x = \lambda_1 X + \mu_1 Y + \nu_1 Z, \quad y = \lambda_2 X + \mu_2 Y + \nu_2 Z, \quad z = \lambda_3 X + \mu_3 Y + \nu_3 Z,$$

then variables ξ, η, ζ are said to be transformed by the *inverse* substitution, if the new variables, expressed in terms of the old, are

$$\Xi = \lambda_1 \xi + \lambda_2 \eta + \lambda_3 \zeta, \quad H = \mu_1 \xi + \mu_2 \eta + \mu_3 \zeta, \quad Z = \nu_1 \xi + \nu_2 \eta + \nu_3 \zeta,$$

where if in the first substitution the coefficients are the constituents of the determinant $(\lambda_1 \mu_2 \nu_3)$ read horizontally, in the second they are the same constituents read vertically; and where if in the first substitution the old variables are expressed in terms of the new, in the second the new are expressed in terms of the old. Stated thus, it is evident that the relation between the two substitutions is reciprocal. Solving for ξ, η, ζ in terms of Ξ, H, Z, we get (Art. 28)

$$\Delta\xi = L_1\Xi + M_1H + N_1Z, \quad \Delta\eta = L_2\Xi + M_2H + N_2Z,$$
$$\Delta\zeta = L_3\Xi + M_3H + N_3Z,$$

where L_1, M_1, &c. are the minors obtained by striking out from the matrix of the determinant $\lambda_1\mu_2\nu_3$ (the modulus of transformation) the line and column containing λ_1, μ_1, &c.

Sets of variables x, y, z; ξ, η, ζ, supposed to be transformed according to the different rules here explained, are said to be *contragredient* to each other. In what follows, variables supposed to be contragredient to x, y, z are denoted by Greek letters, the letters α, β, γ being usually employed in subsequent lessons. We proceed to explain two of the most important cases in which the inverse substitution is employed.

126. When a function of x, y, z, &c. is transformed by linear substitutions to a function of X, Y, Z, &c., then the differential coefficients, with respect to the new variables, are linear functions of those with respect to the old, but are expressed in terms of them by the *inverse* substitution. We have

$$\frac{d}{dX} = \frac{d}{dx}\frac{dx}{dX} + \frac{d}{dy}\frac{dy}{dX} + \frac{d}{dz}\frac{dz}{dX} + \&c.$$

But from the expressions for x, y, &c. in terms of X, Y, &c., we have

$$\frac{dx}{dX} = \lambda_1, \quad \frac{dy}{dX} = \lambda_2, \quad \frac{dz}{dX} = \lambda_3.$$

Hence then $\quad \dfrac{d}{dX} = \lambda_1 \dfrac{d}{dx} + \lambda_2 \dfrac{d}{dy} + \lambda_3 \dfrac{d}{dz} + \&c.$

Similarly $\quad \dfrac{d}{dY} = \mu_1 \dfrac{d}{dx} + \mu_2 \dfrac{d}{dy} + \mu_3 \dfrac{d}{dz} + \&c., \&c.$

Thus then, according to the definition given in the last article, the operating symbols $\dfrac{d}{dx}$, $\dfrac{d}{dy}$, $\dfrac{d}{dz}$, &c. are *contragredient* to x, y, z, &c.; that is to say, when the latter are linearly transformed, the former will be linearly transformed also, but according to a different rule, viz. the rule explained in the last article.

LINEAR TRANSFORMATIONS. 103

If, as before, u_1, u_2, &c. denote the differential coefficients of u, and U_1, U_2, &c. those of the transformed function U, we have just proved that

$$U_1 = \lambda_1 u_1 + \lambda_2 u_2 + \lambda_3 u_3, \quad U_2 = \mu_1 u_1 + \mu_2 u_2 + \mu_3 u_3, \quad \&c.$$

Consequently, if u_1, u_2, u_3 all vanish, U_1, U_2, U_3 must all vanish likewise. Now we know that u_1, u_2, u_3 all vanish together only when the discriminant of the system vanishes; if then the discriminant of the original system vanishes, we see now that the discriminant of the transformed system must vanish likewise, and therefore that the latter contains the former as a factor, as has been already stated (Art. 116).

127. In plane geometry, if x, y, z be the trilinear co-ordinates of any point, and $x\xi + y\eta + z\zeta = 0$ be the equation of any line, ξ, η, ζ may be called the tangential co-ordinates of that line (see *Conics*, Art. 70). Now, if the equation be transformed to any new system of axes by the substitution $x = \lambda_1 X +$ &c., the new equation of the line becomes

$$\xi(\lambda_1 X + \mu_1 Y + \nu_1 Z) + \eta(\lambda_2 X + \mu_2 Y + \nu_2 Z) + \zeta(\lambda_3 X + \mu_3 Y + \nu_3 Z),$$

so that if the new equation of the right line be written $\Xi X + H Y + Z Z = 0$, we have

$$\Xi = \lambda_1 \xi + \lambda_2 \eta + \lambda_3 \zeta, \quad H = \mu_1 \xi + \mu_2 \eta + \mu_3 \zeta, \quad Z = \nu_1 \xi + \nu_2 \eta + \nu_3 \zeta.$$

In other words, *when the co-ordinates of a point are transformed by a linear substitution, the tangential co-ordinates of a line are transformed by the inverse substitution.* In like manner, in the geometry of three dimensions, the tangential co-ordinates of any plane are contragredient to the co-ordinates of any point. When we transform to new axes, all co-ordinates $xyzw$, $x'y'z'w'$, &c. expressing different points, are cogredient: that is to say, all must be transformed by the same substitution $x = \lambda_1 X +$ &c., $x' = \lambda_1 X' +$ &c., &c. But the tangential co-ordinates of every plane will be transformed by the inverse substitution, which has been just explained.

We shall make frequent use of the principle stated in this article, that

$$x\xi + y\eta + z\zeta = X\Xi + YH + ZZ,$$

where x, y, z being supposed to be changed by the substitution,

$x = \lambda_1 X + \mu_1 Y + $ &c., ξ, η, ζ are supposed to be changed by the inverse substitution $\Xi = \lambda_1 \xi + \lambda_2 \eta + \lambda_3 \zeta$, &c. In other words, in the case supposed, $x\xi + y\eta + z\zeta$ is a function absolutely unaltered by transformation.

128. If a function $ax^n + $ &c. becomes by transformation $AX^n + $ &c., then any function involving the coefficients and those variables which are supposed to be transformed by the inverse substitution, is said to be a *contravariant* if it is such that it differs only by a power of the modulus from the corresponding function of the transformed coefficients and variables: that is to say, if

$$\phi(A, B, \&c., \Xi, H, \&c.) = \Delta^r \phi(a, b, \&c., \xi, \eta, \&c.).$$

Such functions for instance constantly present themselves in geometry. If we have an equation expressing the condition that a line or plane should have to a given curve or surface a relation independent of the axes to which it is referred; as, for example, the condition that the line or plane should touch the curve or surface; then, when we transform to new axes, it is obviously indifferent whether we transform the given relation by substituting for the old coefficients their values in terms of the new, or whether we derive the condition from the transformed equation of the curve by the same rule as that by which it was originally formed. In this way it is seen that such a condition is of such a kind that $\phi(a, b, \xi, \&c.)$ differs only by a factor from $\phi(A, B, \Xi, \&c.)$.

129. Besides covariants and contravariants there are also functions involving both sets of variables, and such as to differ only by a power of the modulus from the corresponding transformed functions: *i.e.* such that

$$\phi(A, B, \&c., X, Y, \&c., \Xi, H, \&c.) = \Delta^p \phi(a, b, \&c., x, y, \&c., \xi, \eta, \&c.).$$

Mr. Sylvester uses the name concomitant as a general word to include all functions whose relations to the quantic are unaltered by linear transformation, and he calls the functions now under consideration *mixed concomitants*. I do not choose to introduce a name on my own responsibility; otherwise I should be inclined to call them divariants. The simplest function of the kind is $x\xi + y\eta + z\zeta$, which we have seen (Art. 127) is transformed to a similar function; and is therefore a concomitant of every quantic whatever.

130. If we are given any invariant I of the quantic
$$a_0 x^n + n a_1 x^{n-1} y + n b_1 x^{n-1} z + \tfrac{1}{2} n(n-1) a_2 x^{n-2} y^2 + \&c.,$$
we can deduce from it a contravariant by the method used in Art. 119. If $a_0 x^n + \&c.$ becomes by transformation $A_0 X^n + \&c.$, then, since $x\xi + \&c.$ becomes $X\Xi + \&c.$, it follows that
$$a_0 x^n + \&c. + k(x\xi + y\eta + z\zeta)^n = A_0 X^n + \&c. + k(X\Xi + Y\mathrm{H} + Z\mathrm{Z})^n.$$
Now an invariant of the original quantic fulfils the condition
$$\phi(A_0, A_1, B_1, \&c.) = \Delta^p \phi(a_0, a_1, b_1, \&c.).$$
Forming then the same invariant of the new quantic, it will be seen that
$$\phi(A_0 + k\Xi^n, A_1 + k\Xi^{n-1}\mathrm{H}, \&c.) = \Delta^p \phi(a_0 + k\xi^n, a_1 + k\xi^{n-1}\eta, \&c.).$$
Since k is arbitrary we may equate the coefficients of like powers of k on both sides of this equation.

But, by Taylor's theorem, these coefficients are all of the form
$$\left(\xi^n \frac{d}{da_0} + \xi^{n-1}\eta \frac{d}{da_1} + \xi^{n-1}\zeta \frac{d}{db_1} + \xi^{n-2}\eta^2 \frac{d}{da_2} + \&c. \right)^q I.$$
We have proved then that they differ only by a power of the modulus from the corresponding function of the transformed equation. They are therefore *contravariants*, since it is assumed all along that ξ, η, ζ are to be transformed by the inverse substitution. Mr. Sylvester has called contravariants formed by this rule, first, second, &c. *evectants* of the given invariant. Thus
$$\xi^n \frac{dI}{da_0} + \xi^{n-1}\eta \frac{dI}{da_1} + \&c.$$
is the first evectant. It is to be observed that in the original quantic the coefficients are supposed to be written *with*, and in the evectant *without*, binomial coefficients. Comparing this article with Art. 119 we see that the function $\xi^n \dfrac{dI}{da_0} + \&c.$ may be considered either as a *contravariant* of the single given quantic, or as an *invariant* of the system obtained by combining with the given quantic the linear function $x\xi + y\eta + z\zeta$. The theory of contravariants therefore may be included under that of invariants.

If we perform the operation $\xi^n \dfrac{d}{da_0} + $ &c. upon any covariant, we obtain a mixed concomitant, for it is proved in the same way that the result, which will evidently be a function involving variables of both kinds, will be transformed into a function of similar form.

Ex. 1. We know that $ac - b^2$ is an invariant of $ax^2 + 2bxy + cy^2$; hence $c\xi^2 - 2b\xi\eta + a\eta^2$ is a contravariant of the same system.

Ex. 2. Similarly $abc + 2fgh - af^2 - bg^2 - ch^2$, being the discriminant, and therefore an invariant of $ax^2 + by^2 + cz^2 + 2fyz + 2gzx + 2hxy$,

$(bc - f^2)\xi^2 + (ca - g^2)\eta^2 + (ab - h^2)\zeta^2 + 2(gh - af)\eta\zeta + 2(hf - bg)\zeta\xi + 2(fg - ch)\xi\eta$

is a contravariant of the same quantic. Geometrically, as is well known, it expresses the tangential equation of the conic represented by the given quantic.

Ex. 3. Given a system of two ternary quadrics $ax^2 + $ &c., $a'x^2 + $ &c., then since $a'(bc - f^2) + $ &c. is an invariant of the system (Art. 119); operating with $\xi^2 \dfrac{d}{da} + $ &c., we find that

$(bc' + b'c - 2ff')\xi^2 + (ca' + c'a - 2gg')\eta^2 + (ab' + a'b - 2hh')\zeta^2$
$+ 2(gh' + g'h - af' - a'f)\eta\zeta + 2(hf' + h'f - bg' - b'g)\zeta\xi + 2(fg' + f'g - ch' - c'h)\xi\eta$

is a contravariant of the system. We might have equally found this contravariant by operating with $a' \dfrac{d}{da} + $ &c. on the contravariant of the last article. Geometrically, it expresses the condition that a line should be cut harmonically by two conics.

131. When the discriminant of a quantic vanishes, it has a set of singular roots $x'y'z'$ [geometrically the co-ordinates of the double point on the curve or surface represented by the quantic]; and in this case the first evectant will be a perfect n^{th} power of $(x'\xi + y'\eta + z'\zeta)$. Since we have seen that the evectant is a function unaltered by transformation, it is sufficient to see what it becomes in any particular case. Now if the discriminant vanishes, the quantic can be so transformed that the new coefficients of x^n, $x^{n-1}y$, $x^{n-1}z$ shall vanish; that is to say, so that the singular root shall be $y = 0$, $z = 0$, [geometrically, so that the point yz shall be the double point]. Now it was proved (Art. 113) that the form of the discriminant is

$$a_0 \rho + a_1^2 \phi + a_1 b_1 \psi + b_1^2 \chi.$$

Evidently then, not only will this vanish when a_0, a_1, b_1 vanish; but also its differentials will vanish with respect to every coefficient except a_0. The evectant then reduces itself to

FORMATION OF INVARIANTS AND COVARIANTS. 107

$\frac{dI}{da_0}$ multiplied by the perfect n^{th} power ξ^n; which is what $(x'\xi + y'\eta + z'\zeta)^n$ becomes when y' and $z' = 0$, and $x' = 1$. Thus then, if the discriminant of a ternary quadric vanish, the quadric represents two lines: the contravariant

$$(bc - f^2) \xi^2 + (ca - g^2) \eta^2 + \&c.$$

becomes a perfect square; and if we identify it with $(x'\xi + y'\eta + z'\zeta)^2$, we get $x'y'z'$ the co-ordinates of the intersection of the pair of lines. If a quantic have two sets of singular roots, all the first differentials of the discriminant vanish, and its second evectant becomes a perfect n^{th} power of

$$(x'\xi + y'\eta + z'\zeta)(x''\xi + y''\eta + z''\zeta),$$

where $x'y'z'$, $x''y''z''$ are the two sets of singular roots. And so on.

LESSON XIII.

FORMATION OF INVARIANTS AND COVARIANTS.

132. HAVING now shown what is meant by invariants, &c. we go on to explain the methods by which such functions can be formed. Three of these methods will be explained in this Lesson; and a fourth in the next Lesson.

Symmetric functions. The following method is only applicable to binary quantics. *Any symmetric function of the differences of the roots is an invariant, provided that each root enters into the expression the same number of times.** It is evident that

---

* If in the equation the highest power of x is written with a coefficient a_0, we have to divide by that coefficient in order to obtain the expression for the sum, &c. of the roots; and all symmetric functions of the roots are fractions containing powers of a_0 in the denominator. When we say that a symmetric function of the roots is an invariant, we understand that it has been made integral by multiplying it by such a power of a_0 as will clear it of fractions; or, what comes to the same thing, if we form the symmetric function on the supposition that the coefficient of x^n is 1, that we make it homogeneous by multiplying each term by whatever power of a^0 may be necessary.

an invariant must be a function of the differences of the roots, since it is to be unaltered when for x we substitute $x + \lambda$. Now the most general linear transformation is evidently equivalent to an alteration of each root α into $\dfrac{\lambda\alpha + \mu}{\lambda'\alpha + \mu'}$. By this change the difference between any two roots $\alpha - \beta$, becomes $\dfrac{(\lambda\mu' - \lambda'\mu)(\alpha - \beta)}{(\lambda'\alpha + \mu')(\lambda'\beta + \mu')}$. In order then that any function of the differences may, when transformed, differ only by a factor from its former value, it is necessary that the denominator should be the same for every term; and therefore the function must be a product of differences, in which each root occurs the same number of times. Thus for a biquadratic, $\Sigma(\alpha - \beta)^2(\gamma - \delta)^2$ is an invariant, because, when we transform, all the terms of which the sum is made up have the same denominator. But $\Sigma(\alpha - \beta)^2$ is not an invariant, the denominator for the term $(\alpha - \beta)^2$ being $(\lambda'\alpha + \mu')^2(\lambda'\beta + \mu')^2$, and for the term $(\gamma - \delta)^2$ being $(\lambda'\gamma + \mu')^2(\lambda'\delta + \mu')^2$.

133. Or perhaps the same thing may be more simply stated by writing the equation in the homogeneous form. We saw (Art. 116) that if we change x into $\lambda x + \mu y$, y into $\lambda' x + \mu' y$, the quantity $x_1 y_2 - x_2 y_1$ becomes $(\lambda\mu' - \lambda'\mu)(x_1 y_2 - x_2 y_1)$, and consequently, that any function of the determinants $x_1 y_2 - x_2 y_1$ &c. is an invariant. Now (Art. 57) any function of the roots expressed in the ordinary way, is changed to the homogeneous form by writing for α, β, &c. $\dfrac{x_1}{y_1}$, $\dfrac{x_2}{y_2}$, &c., and then multiplying by such a power of the product of all the y's as will clear it of fractions. If any function of the differences in which all the roots do not equally occur, be treated in this way, powers of the y's will remain after the multiplication, and the function will not be an invariant. Thus, for a biquadratic $\Sigma(\alpha - \beta)^2$ becomes $\Sigma y_3^2 y_4^2 (x_1 y_2 - x_2 y_1)^2$; but the function $\Sigma(\alpha - \beta)^2(\gamma - \delta)^2$, in the expression for which all the roots occur, becomes $\Sigma(x_1 y_2 - x_2 y_1)^2(x_3 y_4 - x_4 y_3)^2$ which being a function of the determinants only *is* an invariant.

It is proved in like manner, that any symmetric function formed of differences of roots and differences between x and

FORMATION OF INVARIANTS AND COVARIANTS. 109

one or more of the roots is a covariant, provided that each root enters the same number of times into the expression. Thus for a cubic $\Sigma (\alpha - \beta)^2 (x - \gamma)^2$ is a covariant.

134. We can, by the method just explained, form invariants or covariants which shall vanish on the hypothesis of any system of equalities between the roots. Thus, let it be required to form an invariant which shall vanish when any three roots are all equal it is evident that every term must contain some one of the three differences $\alpha - \beta$, $\beta - \gamma$, $\gamma - \alpha$; and in like manner for every other set of three that can be formed out of the roots. Thus, in a biquadratic, there are four sets of three roots: the difference $\alpha - \beta$ belongs to two of these sets, and $\gamma - \delta$ to the other two; therefore $\Sigma (\alpha - \beta)^2 (\gamma - \delta)^2$* is an invariant which will vanish if any set of three roots are all equal. In like manner for a quintic, there are ten sets of three: $\alpha - \beta$ belongs to three sets, $\gamma - \delta$ to three other sets; the remaining sets are $\alpha\gamma\epsilon$, $\alpha\delta\epsilon$, $\beta\gamma\epsilon$, $\beta\delta\epsilon$, two of which contain $\gamma - \epsilon$, and the other two $\delta - \epsilon$. The function then $\Sigma (\alpha - \beta)^4 (\gamma - \delta)^2 (\delta - \epsilon)^2 (\gamma - \epsilon)^2$ is an invariant which will vanish if any set of three roots are all equal. This invariant (Arts. 53, 54) is of the fourth order and its weight is 10.

So, again, if we wish to form a covariant of a biquadratic which shall vanish when two distinct pairs of roots are equal; the expression must contain a difference from each of the pairs $\alpha - \beta$, $\gamma - \delta$; $\alpha - \gamma$, $\beta - \delta$; $\alpha - \delta$, $\beta - \gamma$. Such an expression would be

$$\Sigma (\alpha - \beta)^2 (\beta - \gamma)^2 (\gamma - \alpha)^2 (x - \delta)^4,$$

or $\quad \Sigma (\alpha - \beta)(\alpha - \gamma)(\alpha - \delta)(x - \beta)^2 (x - \gamma)^2 (x - \delta)^2,$

which are covariants of the fourth and sixth degrees respectively in the variables; and of the fourth and third in the coefficients, and every term of which vanishes when two distinct pairs of roots are equal.

135. *Mutual differentiation of covariants and contravariants.* When we say that $\phi(a, b, \xi, \eta, \&c.)$ is a contravariant, ξ, η, &c. may be any quantities which are supposed to be transformed by

---

* $\Sigma (\alpha - \beta)(\gamma - \delta)$ would vanish identically.

the reciprocal substitution. Now we have shown (Art. 126) that the differential symbols $\frac{d}{dx}$, $\frac{d}{dy}$, &c. are so transformed. We may, therefore, in any contravariant substitute these differential symbols for ξ, η, &c., and we shall obtain an operating symbol unaltered by transformation, and which, therefore, if applied either to the quantic itself or to any of its covariants, will give a covariant if any of the variables remain after differentiation; and if not, an invariant. Similarly, if applied to a mixed concomitant, it will give either a contravariant or a new mixed concomitant, according as the variables are or are not removed by differentiation. Or again, in any contravariant instead of substituting for ξ, η, &c., $\frac{d}{dx}$, $\frac{d}{dy}$, &c. and so obtaining an operating symbol, we may substitute $\frac{dU}{dx}$, $\frac{dU}{dy}$, where U is either the quantic itself or any of its covariants, and so obtain a new covariant. The relation between the sets of variables x, y, z, &c., ξ, η, ζ, &c. being reciprocal, we may, in like manner, substitute in any covariant, for x, y, z, &c., $\frac{d}{d\xi}$, $\frac{d}{d\eta}$, $\frac{d}{d\zeta}$, &c., when we get an operative symbol which when applied to any contravariant will give either a new contravariant or an invariant.

Thus then, if we are given any covariant and contravariant, by substituting in one of them, differential symbols and operating on the other we obtain a new contravariant or covariant; which again may be combined with one of the two given at first, so as to generate another; and so on.

136. In the case of a binary quantic this method may be stated more simply. The formulæ for direct transformation being

$$x = \lambda_1 X + \mu_1 Y, \quad y = \lambda_2 X + \mu_2 Y,$$

those for the reciprocal transformation are (Art. 125)

$$\Xi = \lambda_1 \xi + \lambda_2 \eta, \quad H = \mu_1 \xi + \mu_2 \eta,$$

whence $\quad \Delta \xi = \mu_2 \Xi - \lambda_2 H, \quad \Delta \eta = - \mu_1 \Xi + \lambda_1 H,$

which may be written

$$\Delta \eta = \lambda_1 H + \mu_1 (-\Xi); \quad \Delta (-\xi) = \lambda_2 H + \mu_2 (-\Xi).$$

FORMATION OF INVARIANTS AND COVARIANTS. 111

Thus we see that, with the exception of the constant factor Δ, η and $-\xi$ are transformed by exactly the same rules as x and y; and it may be said that y and $-x$ are contragredient to x and y. Thus then in binary quantics, covariants and contravariants are not essentially distinct, and we have only in any covariant to write η and $-\xi$ for x and y when we have a contravariant; or *vice versâ*. In fact, suppose that by transformation any homogeneous function whatever $\phi(x, y)$ becomes $\Phi(X, Y)$; the formulæ just given show that $\phi(\eta, -\xi)$ will become $\dfrac{1}{\Delta^p} \Phi(\mathrm{H}, -\Xi)$, where p is the degree of the function in x and y. If then $\phi(x, y)$ is a covariant; that is to say, a function which becomes by transformation one differing only by a power of Δ from a function of like form in X and Y; evidently $\phi(\eta, -\xi)$ will by transformation become one differing only by a power of Δ from one of like form in Ξ and H; that is to say, it will be a contravariant. For example, the contravariant, noticed (Art. 130, Ex. 1), $c\xi^2 - 2b\xi\eta + a\eta^2$, by the substitution just mentioned, becomes the original quantic.

Instead then of saying that the differential symbols are contragredient to x and y, we may say that they are cogredient to y and $-x$; and if either in the quantic itself or any of its covariants we write $\dfrac{d}{dy}$, $-\dfrac{d}{dx}$ for x and y, we get a differential symbol which may be used to generate new covariants in the manner explained in the last article. Or we may substitute $\dfrac{du}{dy}$, $-\dfrac{du}{dx}$ for x and y, and so get a new covariant. The following examples will sufficiently illustrate this method:

Ex. 1. To find an invariant of a quadratic, or of a system of two quadratics. Suppose that by transformation $ax^2 + 2bxy + cy^2$ becomes $AX^2 + 2BXY + CY^2$, then since we have seen that $\Delta \dfrac{d}{dy}$, $-\Delta \dfrac{d}{dx}$ are transformed by the same rules as x and y, it follows that the operative symbol

$$\Delta^2 \left(a \frac{d^2}{dy^2} - 2b \frac{d^2}{dx\,dy} + c \frac{d^2}{dx^2} \right) \text{ becomes by transformation } \left(A \frac{d^2}{dY^2} - 2B \frac{d^2}{dX\,dY} + C \frac{d^2}{dX^2} \right).$$

If then we operate on the given quadratic itself, we get

$$4\Delta^2 (ac - b^2) = 4(AC - B^2),$$

which shows that $ac - b^2$ is an invariant; or if we operate on $a'x^2 + 2b'xy + c'y^2$ and its transformed, we get

$$2\Delta^2 (ac' + ca' - 2bb') = 2(AC' + CA' - 2BB'),$$

which shows that $ac' + ca' - 2bb'$ is an invariant. We might also infer that
$$a(bx + cy)^2 - 2b(bx + cy)(ax + by) + c(ax + by)^2$$
is a covariant; but this is only the quantic itself multiplied by $ac - b^2$.

Ex. 2. Every binary quantic of even degree has an invariant of the second order in the coefficients. We have only to substitute, as just explained, $\frac{d}{dy}$, $-\frac{d}{dx}$ for x and y, and operate on the quantic itself. Thus for the quartic $(a, b, c, d, e\backslash x, y)^4$, we find that $ae - 4bd + 3c^2$ is an invariant; or for the general quantic $(a_0, a_1...a_{n-1}, a_n\backslash x, y)^n$, we find that $a_0 a_n - n a_1 a_{n-1} + \frac{1}{2}n(n-1) a_2 a_{n-2} - \&c.$ is an invariant; where the coefficients are those of the binomial, but the middle term is divided by two.

If we apply this method to a quantic of odd degree; as, for example, if we operate on the cubic $ax^3 + 3bx^2y + 3cxy^2 + dy^3$, with $d\frac{d^3}{dx^3} - 3c\frac{d^3}{dx^2 dy} + 3b\frac{d^3}{dx dy^2} - a\frac{d^3}{dy^3}$, it will be found that the result vanishes identically. We thus find however that a system of two cubics has the invariant $(ad' - a'd) - 3(bc' - b'c)$. Or, in general, that a system of two quantics of odd degree, $a_0 x^n + \&c.$, $b_0 x^n + \&c.$, has the invariant
$$(a_0 b_n - a_n b_0) - n(a_1 b_{n-1} - a_{n-1} b_1) + \tfrac{1}{2} n(n-1)(a_2 b_{n-2} - a_{n-2} b_2) - \&c.,$$
which vanishes when the two quantics are identical.

137. When, by the method just explained, we have found an invariant of a quantic of any degree, we have immediately by the method of Art. 122, a covariant of any quantic of higher degree. Thus knowing that $ac - b^2$ is an invariant of a quadratic, by forming that invariant of the quadratic emanant of any quantic, we learn that $\frac{d^2u}{dx^2}\frac{d^2u}{dy^2} - \left(\frac{d^2u}{dx\,dy}\right)^2$ is a covariant of any quantic above the second degree. In like manner, from the invariant of a quartic $ae - 4bd + 3c^2$, we infer that for every quantic above the fourth degree,
$$\frac{d^4u}{dx^4}\frac{d^4u}{dy^4} - 4\frac{d^4u}{dx^3 dy}\frac{d^4u}{dx\,dy^3} + 3\left(\frac{d^4u}{dx^2 dy^2}\right)^2$$
is a covariant, &c. In this way we see that a quantic in general has a series of covariants, of the second order in the coefficients, and of the orders $2(n-2)$, $2(n-4)$, $2(n-6)$, &c. in the variables. These covariants may be combined with the original quantic and with each other, so as to lead to new covariants or invariants.

Ex. 1. A quartic has an invariant of the third order in the coefficients. We know that its Hessian
$$(ax^2 + 2bxy + cy^2)(cx^2 + 2dxy + ey^2) - (bx^2 + 2cxy + dy^2)^2,$$
or $(ac - b^2) x^4 + 2(ad - bc) x^3 y + (ae + 2bd - 3c^2) x^2 y^2 + 2(be - cd) xy^3 + (ce - d^2) y^4$,

FORMATION OF INVARIANTS AND COVARIANTS. 113

is a covariant. Operate on this with $(a, b, c, d, e \left(\frac{d}{dy}, -\frac{d}{dx} \right)^4$; and we get seventy-two times

$$ace + 2bcd - ad^2 - eb^2 - c^3,$$

which is therefore an invariant.

Ex. 2. Every quantic of odd degree has an invariant of the fourth order in the coefficients. The quantic has a quadratic covariant $\frac{d^{n-1}u}{dx^{n-1}} \frac{d^{n-1}u}{dy^{n-1}} - \&c.$ of the second order in the coefficients; and the discriminant of this quadratic will be an invariant of the original quantic (Art. 120) and will be of the fourth order in its coefficients. In fact, it is proved in this way that *every* quantic has an invariant of the fourth order; for if we take any of the covariants of this article which are all of even degree, its invariant of the second order will be of the fourth in the coefficients of the original quantic. But when the quantic is of even degree, it *may* happen that the invariant so found is only the square of its invariant of the second order.

Ex. 3. To form the invariant of the fourth order for a cubic.

Its Hessian is $\quad (ax + by)(cx + dy) - (bx + cy)^2$;

or $\quad (ac - b^2)x^2 + (ad - bc)xy + (bd - c^2)y^2.$

Hence $\quad (ad - bc)^2 - 4(ac - b^2)(bd - c^2)$

is an invariant of the cubic. In fact, it is its discriminant

$$a^2d^2 + 4ac^3 + 4db^3 - 3b^2c^2 - 6abcd.$$

138. From any invariant of a binary quantic we can generate a covariant. For from it we can form (Art. 130) the evectant contravariant $\xi^n \frac{dI}{da_0} + \&c.$; and then in this substituting $y, -x$ for ξ and η, we have a covariant. For example, from the discriminant of a cubic which has been just written we form the evectant

$$\xi^3 (ad^2 - 3bcd + 2c^3) - 3\xi^2 \eta (acd + bc^2 - 2b^2d)$$
$$- 3\xi\eta^2 (abd + b^2c - 2ac^2) + \eta^3 (a^2d - 3abc + 2b^3),$$

whence we infer that the cubic has the cubic covariant

$$(a^2d - 3abc + 2b^3, abd + b^2c - 2ac^2, 2b^2d - acd - bc^2, 3bcd - ad^2 - 2c^3 \mathbin{\mathrm{\mathbf{(}}} x, y)^3.$$

139. *The differential equation.*—If n be the order of any binary quantic, θ the order in the coefficients of any of its invariants; then the weight (see Art. 52) of every term in the invariant is constant and $= \tfrac{1}{2}n\theta$. If we alter x into λx, leaving y unchanged; then since this is a linear transformation, the invariant must, by definition, remain unaltered, except that it may be multiplied by a power of λ which is in this case the modulus of transformation. It is proved then precisely as in

Q

114   FORMATION OF INVARIANTS AND COVARIANTS.

Art. 53 that the *weight*, or sum of the suffixes in every term is constant.

Again, the invariant must remain unaltered, if we change x into y, and y into x, a linear transformation, the modulus of which is -1. The effect of this substitution is the same as if for every coefficient a_a we substitute a_{n-a}. Hence the sum of a number of suffixes

$$\alpha + \beta + \gamma + \&c. = (n - \alpha) + (n - \beta) + (n - \gamma) + \&c.,$$

whence $2(\alpha + \beta + \gamma + \&c.) = n\theta$.   Q.E.D.

COR. n and θ cannot both be odd since their product is an even number; or, *a binary quantic of odd degree cannot have an invariant of odd order.*

140. The principle just established enables us to write down immediately the literal part of any invariant whose order is given. For the order being given, the weight is given also. Thus if it were required to form for a quartic an invariant of the third order in the coefficients, the weight must be 6, and the terms of the invariant must be

$$Aa_4a_2a_0 + Ba_4a_1a_1 + Ca_3a_3a_0 + Da_3a_2a_1 + Ea_2a_2a_2,$$

where the coefficients A, B, &c. remain to be determined. The reader will observe that there are as many terms in this invariant as the ways in which the number 6 can be expressed as the sum of three numbers from 0 to 4 inclusive; and generally that there may be as many terms in any invariant as the ways in which its weight $\frac{1}{2}n\theta$ can be expressed as the sum of θ numbers from 0 to n inclusive.

We determine the coefficients from the consideration that since an invariant is to be unaltered by the substitution either of $x+\lambda$ for x, or $y+\lambda$ for y, evidently, as in Art. 58, every invariant must satisfy the two differential equations

$$a_0\frac{dI}{da_1} + 2a_1\frac{dI}{da_2} + 3a_2\frac{dI}{da_3} + \&c. = 0, \quad na_1\frac{dI}{da_0} + (n-1)a_2\frac{dI}{da_1} + \&c. = 0,$$

it being supposed that the original equation has been written with binomial coefficients. In practice only one of these equations need be used; for the second is derived from the first by changing each coefficient a_a into a_{n-a}. It is sufficient then to use one of the equations, provided we take care that the function we form is symmetrical with regard to x and y; that is to say, which does not change (or at most changes sign)* when we change a_a into a_{n-a}. And this condition will always be fulfilled if we take care that the weight of the invariant is that which has been just assigned. Thus then, in the example chosen for an illustration, if we operate on $Aa_4 a_2 a_0 +$ &c. with $a_0 \dfrac{d}{da_1} +$ &c., we get

$$(2B + 2A)\, a_4 a_1 a_0 + (D + 6C + 4A)\, a_2 a_2 a_0$$
$$+ (2D + 4B)\, a_3 a_1 a_1 + (6E + 3D)\, a_2 a_2 a_1 = 0,$$

whence if we take $A = 1$, the other coefficients are found to be $B = -1,\ D = 2,\ C = -1,\ E = -1$, and the invariant is

$$a_4 a_2 a_0 + 2a_3 a_2 a_1 - a_4 a_1 a_1 - a_3 a_3 a_0 - a_2 a_2 a_2.$$

141. In seeking to determine an invariant of given order by the method just explained, we have a certain number of unknown coefficients $A,\ B,\ C$, &c. to determine, and we do so by the help of a certain number of conditions formed by means of the differential equation. Now evidently if the number of these conditions were greater than the number of unknown coefficients, the formation of the invariant would in general be impossible; if they were equal we could form one invariant; if the number of conditions were less, we could form more than one invariant of the given order. We have just seen that the number of terms in the invariant, which is one more than the number of unknown coefficients, is equal to the number of ways in which its weight $\tfrac{1}{2}n\theta$ can be written, as the sum of θ numbers, none being greater than n. But the effect of the operation $a_0 \dfrac{d}{da_1} +$ &c. is evidently to diminish the weight by one; the number of conditions to be fulfilled is therefore equal to the number of ways in which $\tfrac{1}{2}n\theta - 1$ can be expressed

---

* When we change x into y and y into x this is a transformation whose modulus is $\begin{vmatrix} 0, & 1 \\ 1, & 0 \end{vmatrix}$ or -1. Any invariant therefore which when transformed becomes multiplied by an odd power of the modulus of transformation will change sign when we interchange x and y. Such invariants are called *skew* invariants.

as the sum of θ numbers, none exceeding n. Thus, in the example of Art. 140, the number of conditions used to determine A, B, &c. was equal to the number of ways in which 5 can be expressed as the sum of three numbers from 0 to 4 inclusive. To find then generally whether an invariant of a binary quantic of the order θ can be formed, and whether there can be more than one, we must compare the number of ways in which the numbers $\frac{1}{2}n\theta$, $\frac{1}{2}n\theta - 1$ can be expressed as the sum of θ numbers from 0 to n inclusive. On this principle is founded Mr. Cayley's investigation of the number of invariants of a binary quantic, of which we give an account in an Appendix.

142. Similar reasoning applies to covariants. A covariant, like the original quantic, must remain unaltered, when we change x into ρx, and at the same time every coefficient a_a into $\rho^a a_a$. If then the coefficient of any power of x, x^μ in the covariant be $a_a b_\beta c_\gamma$, &c. it is obvious, as before, that $\mu + a + \beta +$ &c. must be constant for every term; and we may call this number the *weight* of the covariant.

Again, in order that the covariant may not change when we alter x into y and y into x, we must have

$$\mu + a + \beta + \gamma + \&c. = (p - \mu) + (n - a) + (n - \beta) + \&c.,$$

where p is the degree of the covariant in x and y; whence if θ be the order of the covariant in the coefficients, we have immediately its weight $= \frac{1}{2}(n\theta + p)$. Thus if it were required to form a quadratic covariant to a cubic, of the second order in the coefficients, $n = 3$, $\theta = p = 2$, and the weight is 4. We have then for the terms multiplying x^2, $a + \beta = 2$ and these terms must be $a_2 a_0$ and $a_1 a_1$. In like manner the terms multiplying xy must be $a_3 a_0$, $a_2 a_1$, and those multiplying y^2 must be $a_3 a_1$, $a_2 a_2$. In this manner we can determine the literal part of a covariant of any order. The coefficients are determined as follows:

143. From the definition of a covariant it follows that we must get the same result whether in it we change x into $x + \lambda y$, or whether we make the same change in the original quantic

FORMATION OF INVARIANTS AND COVARIANTS. 117

and then form the covariant. But this change in the original quantic is equivalent (Art. 58) to changing a_1 into $a_1 + \lambda a_0$, a_2 into $a_2 + 2a_1\lambda + a_0\lambda^2$, &c. Hence in the covariant also the change of x to $x + \lambda y$ must be equivalent to changing a_1 into $a_1 + \lambda a_0$, &c. Let the covariant then be

$$A_0 x^p + p A_1 x^{p-1} y + \tfrac{1}{2} p(p-1) A_2 x^{p-2} y^2 + \&c.$$

Let us express that these two alterations are equivalent, and let us confine our attention to the terms multiplying λ. Then if, as in Art. 60, we use the abbreviation $\dfrac{d}{d\zeta}$ to denote the operation $a_0 \dfrac{d}{da_1} + 2a_1 \dfrac{d}{da_2} + \&c.$, we get

$$\frac{dA_0}{d\zeta} = 0,\ \frac{dA_1}{d\zeta} = A_0,\ \frac{dA_2}{d\zeta} = 2A_1, \&c.,\ \frac{dA_{p-1}}{d\zeta} = (p-1)A_{p-2},\ \frac{dA_p}{d\zeta} = pA_{p-1}.$$

In like manner, writing $\dfrac{d}{d\eta}$ for $na_1 \dfrac{d}{da_0} + (n-1)a_2 \dfrac{d}{da_1} + \&c.$, we have

$$\frac{dA_0}{d\eta} = p A_1,\ \frac{dA_1}{d\eta} = (p-1) A_2, \&c.$$

Thus we see that when we have determined A_0 so as to satisfy the equation $\dfrac{dA_0}{d\zeta} = 0$; in other words, when A_0 is a function of the differences of the roots of the quantic (Art. 58), all the other terms of the covariant are known. The covariant is in fact

$$A_0 x^p + \frac{dA_0}{d\eta} x^{p-1} y + \frac{d^2 A_0}{d\eta^2} \frac{x^{p-2} y^2}{1.2} + \frac{d^3 A_0}{d\eta^3} \frac{x^{p-3} y^3}{1.2.3} + \&c.$$

It will be observed that the weight of the covariant being $\tfrac{1}{2}(n\theta + p)$ the weight of the term A_0 is $\tfrac{1}{2}(n\theta - p)$; since the weight of A_0 together with p makes up the weight of the covariant. This term A_0, whence all the other terms are derived, is called by Mr. M. Roberts the *source* of the covariant. He observes also that the source of the product of two covariants is the product of their sources. For if we multiply the covariant last written by

$$B_0 x^q + \frac{dB_0}{d\eta} x^{q-1} y + \frac{d^2 B_0}{d\eta^2} \frac{x^{q-2} y^2}{1.2} + \&c.,$$

118   FORMATION OF INVARIANTS AND COVARIANTS.

we get, as may be easily seen,

$$A_0 B_0 x^{p+q} + \frac{d(A_0 B_0)}{d\eta} x^{p+r-1} y + \frac{d^2(A_0 B_0)}{d\eta^2} \frac{x^{p+r-2} y^2}{1.2} + \&c.$$

Hence if we know any relation connecting any functions of the differences A_0, B_0, C_0, &c. the same relation will connect the covariants derived from these functions.

Ex. 1. To find the quadratic covariant of a cubic. We have seen (Art. 142) that A_0 is of the form $a_2 a_0 + B a_1 a_1$. Operate on this with $a_0 \frac{d}{da_1} + 2a_1 \frac{d}{da_2}$, and we get $(2 + 2B) a_0 a_1 = 0$, whence $B = -1$ and $A_0 = a_2 a_0 - a_1 a_1$. Operate then with $3a_1 \frac{d}{da_0} + 2a_2 \frac{d}{da_1} + a_3 \frac{d}{da_2}$, and we have $2A_1 = a_2 a_0 - a_2 a_1$. Operate with the same on A_1, and we have $A_2 = a_1 a_3 - a_2 a_2$. The covariant therefore is

$$(a_2 a_0 - a_1 a_1) x^2 + (a_0 a_3 - a_2 a_1) xy + (a_1 a_3 - a_2 a_2) y^2.$$

Ex. 2. To find a cubic covariant of a cubic of the third order in the coefficients. Here $n = 3$, $\theta = 3$, $\frac{1}{2}(n\theta + p) = 6$. The sum then of the suffixes of the coefficient of x^3 will be 3; and this coefficient must be of the form $A a_3 a_0 a_0 + B a_2 a_1 a_0 + C a_1 a_1 a_1$. Operate with $a_0 \frac{d}{da_1} + 2a_1 \frac{d}{da_2} + 3a_2 \frac{d}{da_3}$, and we get

$$(3A + B) a_2 a_0 a_0 + (2B + 3C) a_1 a_1 a_0,$$

whence if we take $A = 1$, we have $B = -3$, $C = 2$, or $A_0 = a_3 a_0 a_0 - 3 a_2 a_1 a_0 + 2 a_1 a_1 a_1$. Operate on this three times successively with $3a_1 \frac{d}{da_0} + 2a_2 \frac{d}{da_1} + a_3 \frac{d}{da_2}$, and we have the remaining coefficients and the covariant is (see Art. 138)

$$(a_3 a_0 a_0 - 3 a_2 a_1 a_0 + 2 a_1 a_1 a_1) x^3 + 3 (a_3 a_1 a_0 - 2 a_2 a_2 a_0 + a_2 a_1 a_1) x^2 y$$
$$+ 3 (2 a_3 a_1 a_1 - a_3 a_2 a_1 - a_3 a_2 a_0) xy^2 + (3 a_3 a_2 a_1 - 2 a_2 a_2 a_2 - a_3 a_3 a_0) y^3.$$

144. We have seen that a quantic has as many covariants of the degree p in the variables and of the order θ in the coefficients, as functions A_0 whose weight is $\frac{1}{2}(n\theta - p)$ can be found to satisfy the equation $\frac{dA_0}{d\zeta} = 0$. And, as in Art. 141, we see that this number is equal to the difference of the ways in which the numbers $\frac{1}{2}(n\theta - p)$ and $\frac{1}{2}(n\theta - p) - 1$ can be expressed as the sum of θ numbers from 0 to n inclusive. It may be remarked that p cannot be odd unless both n and θ are odd. Hence only quantics of odd degree can have covariants of odd degree in the coefficients, and these must also be of odd degree in the variables.

145. The results arrived at (Art. 143) may be stated a little differently. The operation $y \frac{d}{dx}$ performed on any quantic is

FORMATION OF INVARIANTS AND COVARIANTS.   119

equivalent to a certain operation performed by differentiating with respect to the coefficients. Thus, for the quantic $(a_0, a_1, a_2 \ldots \!\!\not\!\!(x, y)^n$, we get the same result whether we operate on it with $y \dfrac{d}{dx}$ or with $a_0 \dfrac{d}{da_1} + 2a_1 \dfrac{d}{da_2} + \&c.$ This latter operation then may be written $\left[y \dfrac{d}{dx} \right]$; and the property already proved for a covariant may be written that we have for it $y \dfrac{d}{dx} - \left[y \dfrac{d}{dx} \right] = 0.$ In other words, that we get the same result whether we operate on the covariant with $y \dfrac{d}{dx}$ or with $a_0 \dfrac{d}{da_1} + 2a_1 \dfrac{d}{da_2} + \&c.$ In his Memoirs on Quantics, Mr. Cayley has started with this property as his *definition* of a covariant; a definition which includes invariants also, since for them we have $y \dfrac{dI}{dx} = 0$, and therefore also $\left[y \dfrac{dI}{dx} \right] = 0.$

146. It can be proved, in like manner, that quantics in any number of variables satisfy differential equations which may be written $y \dfrac{d}{dx} = \left[y \dfrac{d}{dx} \right]$, $z \dfrac{d}{dx} = \left[z \dfrac{d}{dx} \right]$, &c. Thus, for the quantic $(a, b, c, f, g, h \!\!\not\!\!(x, y, z)^2$, we have

$$y \frac{d}{dx} = a \frac{d}{dh} + g \frac{d}{df} + 2h \frac{d}{db}, \quad z \frac{d}{dx} = a \frac{d}{dg} + h \frac{d}{df} + 2g \frac{d}{dc},$$

and every covariant must satisfy these two equations. While every invariant must satisfy the two equations

$$a \frac{dI}{dh} + g \frac{dI}{df} + 2h \frac{dI}{db} = 0, \quad a \frac{dI}{dg} + h \frac{dI}{df} + 2g \frac{dI}{dc} = 0;$$

as may easily be proved from the consideration that the invariant remains unaltered if we substitute for x, $x + \lambda y$ or $x + \mu z$.

LESSON XIV.

SYMBOLICAL REPRESENTATION OF INVARIANTS AND COVARIANTS.

147. It remains to explain a fourth method of finding invariants and covariants, given by Mr. Cayley in 1846 (*Cambridge and Dublin Mathematical Journal*, vol. I., p. 104, and *Crelle*, vol. XXX.); which not only enables us to arrive at such functions, but also affords the basis of a regular calculus by means of which they may be compared and identified.

Let x_1, y_1; x_2, y_2 be any two cogredient sets of variables; then it has been proved (Arts. 126, 116, 135) that $\dfrac{d}{dx_1}, \dfrac{d}{dy_1}, \dfrac{d}{dx_2}, \dfrac{d}{dy_2}$ are transformed by the reciprocal substitution; that $\dfrac{d}{dx_1}\dfrac{d}{dy_2} - \dfrac{d}{dx_2}\dfrac{d}{dy_1}$ is an invariant symbol of operation; and that if we operate with any power of this symbol on any function of x_1, y_1, x_2, y_2, we shall obtain a covariant of that function. We shall use for $\dfrac{d}{dx_1}\dfrac{d}{dy_2} - \dfrac{d}{dx_2}\dfrac{d}{dy_1}$ the abbreviation $\overline{12}$.

Suppose now that we are given any two binary quantics U, V, we can at once form covariants of this system of two quantics. For we have only to write the variables in U with the suffix (1), those in V with the suffix (2), and then operate on the product UV with any power of the symbol $\overline{12}$; when the result must be an invariant or covariant. Thus if

$$U = ax_1^2 + 2bx_1y_1 + cy_1^2; \quad V = a'x_2^2 + 2b'x_2y_2 + c'y_2^2,$$

and if we operate on UV with $\overline{12}^2$, which, written at full length, is

$$\dfrac{d^2}{dx_1^2}\dfrac{d^2}{dy_2^2} + \dfrac{d^2}{dx_2^2}\dfrac{d^2}{dy_1^2} - 2\dfrac{d^2}{dx_1 dy_1}\dfrac{d^2}{dx_2 dy_2},$$

the result is $ac' + ca' - 2bb'$, which is thus proved to be an invariant of the system of equations. In general, it is obvious that the differentials marked with the suffix (1) only apply to U, and those with the suffix (2) only to V; and it is unnecessary to

retain the suffixes after differentiation;* so that $\overline{12}^2$ applied to two quantics of any degree gives the covariant
$$\frac{d^2U}{dx^2}\frac{d^2V}{dy^2} + \frac{d^2U}{dy^2}\frac{d^2V}{dx^2} - 2\frac{d^2U}{dxdy}\frac{d^2V}{dxdy}.$$
If we had operated simply with $\overline{12}$ we should have obtained the Jacobian $\dfrac{dU}{dx}\dfrac{dV}{dy} - \dfrac{dU}{dy}\dfrac{dV}{dx}$, which we saw, Art. 124, was a covariant of the system of quantics.

Similarly the symbol $\overline{12}^3$ applied to two cubics gives the invariant
$$(ad' - a'd) - 3(bc' - b'c),$$
or to any two quantics gives the covariant
$$\frac{d^3U}{dx^3}\frac{d^3V}{dy^3} - 3\frac{d^3U}{dx^2dy}\frac{d^3V}{dxdy^2} + 3\frac{d^3U}{dxdy^2}\frac{d^3V}{dx^2dy} - \frac{d^3U}{dy^3}\frac{d^3V}{dx^3};$$
and so in like manner for the other powers of $\overline{12}$.

148. We can by this method obtain also invariants or covariants of a single function U. It is, in fact, only necessary to suppose in the last article the quantics U and V to be identical. Thus, for instance, in the example of the two quadratics given in the last Article, if we make $a = a'$, $b = b'$, $c = c'$, the invariant $\overline{12}^2$ becomes $2(ac - b^2)$. And, in like manner, the expression there given for the covariant $\overline{12}^2$ of a system U, V, by making $U = V$, gives the covariant of a single quantic
$$\frac{d^2U}{dx^2}\frac{d^2U}{dy^2} - \left(\frac{d^2U}{dxdy}\right)^2.$$

In general, whenever we want by this method to form the covariants of a single function, we resort to this artifice:—We first form a covariant of a system of distinct quantics, and then suppose the quantics to be made identical. And in what

---

* If W be any function containing x_1, y_1; x_2, y_2; we shall get the same result whether we linearly transform these variables, and afterwards omit all the suffixes in the transformed equation; or whether we omit the suffixes first, and afterwards transform x and y. This results immediately from the fact that x_1, y_1; x_2, y_2; x, y are cogredient. It follows then at once that if W written as a function of x_1, y_1; x_2, y_2; be a covariant of U, V; that is to say, if the expression of the coefficients of W in terms of the coefficients of U and V be unaffected by transformation, then W is also a covariant when the suffixes are all omitted.

R

follows, when we use such symbols as $\overline{12}^n$ &c. without adding any subject of operation, we mean to express derivatives of a single function U. We take for the subject operated on, the product of two or more quantics U_1, U_2, &c., where the variables x_1, y_1; x_2, y_2; &c. are written in each respectively, instead of x and y; and we suppose that after differentiation all the suffixes are omitted, and that the variables, if any remain, are all made equal to x and y.

149. From the omission of the suffixes after differentiation, it follows at once that it cannot make any difference what figures had been originally used, and that $\overline{12}^n$ and $\overline{34}^n$ are expressions for the same thing. In the use of this method we have constantly to employ transformations depending on this obvious principle. Thus, we can show that when n is odd, $\overline{12}^n$ applied to a single function vanishes identically. For, from what has been said $\overline{12}^n = \overline{21}^n$; but $\overline{12}$ and $\overline{21}$ have opposite signs, as appears immediately on writing at full length the symbol for which $\overline{12}$ is an abbreviation. It follows then that $\overline{12}^n$ must vanish when n is odd. Thus, in the expansion of $\overline{12}^3$, given at the end of Art. 147, if we make $U = V$, it will obviously vanish identically. The series $\overline{12}^2$, $\overline{12}^4$, $\overline{12}^6$ &c. gives the series of invariants and covariants which we have already found (p. 112). It is easy to see that, when n is even, $\overline{12}^n$ applied to $(a_0, a_1, a_2 \ldots \mathbin{\!)\!} x, y)^n$ gives

$$a_0 a_n - n a_1 a_{n-1} + \tfrac{1}{2} n(n-1) a_2 a_{n-2} - \&c.,$$

where the last coefficient must be divided by two, as is evident from the manner of formation.

150. The results of the preceding Articles are extended without difficulty to any number of functions. We may take any number of quantics U, V, W, &c., and, writing the variables in the first with the suffix (1), those in the second with the suffix (2), in the third with the suffix (3), and so on, we may operate on their product with the product of any number of symbols $\overline{12}^a$, $\overline{23}^\beta$, $\overline{31}^\gamma$, $\overline{14}^\delta$, &c.; where, as before, $\overline{23}$ is an abbreviation for $\dfrac{d}{dx_2}\dfrac{d}{dy_3} - \dfrac{d}{dx_3}\dfrac{d}{dy_2}$, &c. After the differentiation we suppress

the suffixes, and we thus get a covariant of the given system of quantics, which will be an invariant if it happens that no power of x and y appear after differentiation. Any number of the quantics U, V, W, &c., may be identical; and in the case with which we shall be most frequently concerned, viz., where we wish to form derivatives of a single quantic, the subject operated on is $U_1 U_2 U_3$ &c., where U_1 and U_2 only differ by having the variables written with different suffixes.

It is evident that in this method the degree of the derivative in the coefficients will be always equal to the number of different figures in the symbol for the derivative. For if all the functions were distinct, the derivative would evidently contain a coefficient from every one of the quantics U, V, W, &c.; and it will be still true, when U, V, W are supposed identical, that the degree in the coefficients is equal to the number of factors in the product $U_1 U_2 U_3$ &c. which we operate on. Thus, the derivatives considered in the last Article being all of the form $\overline{12}^p$ are all of the second degree in the coefficients.

Again, if it were required to find the degree of the derivative in x and y. Suppose, in the first place, that the quantics were distinct, U being of the degree n, V of the degree n', W of the degree n'', and so on; and suppose that in the operating symbol the figure 1 occurs α times; 2, β times; and so on; then, since U is differentiated α times; V, β times, &c., the result is of the degree $(n-\alpha)+(n'-\beta)+(n''-\gamma)+$ &c. When the quantics are identical, if there are p factors in the product $U_1 U_2 ... U_p$ which we operate on, the degree of the result in x and y will be $np-(\alpha+\beta+\gamma+$ &c.$)$. While again, if there be r factors such as $\overline{12}$ in the operating symbol, it is obvious that $\alpha+\beta+\gamma+$ &c.$)=2r$. It is clear that if we wish to obtain an *in*variant, we must have $\alpha = \beta = \gamma = n$.

151. To illustrate the above principles, we make an examination of all possible invariants of the third degree in the coefficients. Since the symbol for these can only contain three figures, its most general form is $\overline{12}^\alpha.\overline{23}^\beta.\overline{31}^\gamma$; while, in order that it should yield an *in*variant, we must have

$$\alpha + \gamma = \alpha + \beta = \beta + \gamma = n,$$

whence $\alpha = \beta = \gamma$. The general form, then, that we have to examine is $(\overline{12}.\overline{23}.\overline{31})^\alpha$. Again, if α be odd, this derivative vanishes identically; for, as in Art. 149, by interchanging the figures 1 and 2, we have $(\overline{12}.\overline{23}.\overline{31})^\alpha = (\overline{21}.\overline{13}.\overline{32})^\alpha$; but these have opposite signs. It follows, then, that all invariants of the third order are included in the formula $(\overline{12}.\overline{23}.\overline{31})^\alpha$, where α is even. Thus, $\overline{12}^2.\overline{23}^2.\overline{31}^2$ is an invariant of a quartic, since the differentials rise to the fourth degree; $\overline{12}^4.\overline{23}^4.\overline{31}^4$ is an invariant of an octavic; $\overline{12}^6.\overline{23}^6.\overline{31}^6$ of a quantic of the twelfth degree, and so on; only quantics whose degree is of the form $4m$ having invariants of the third order in the coefficients. If we wish actually to calculate one of these, suppose $\overline{12}^2.\overline{23}^2.\overline{31}^2$, I write, for brevity, ξ_1, η_1, &c., instead of $\dfrac{d}{dx_1}$, $\dfrac{d}{dy_1}$, &c. Then we have actually to multiply out

$$(\xi_1\eta_2 - \xi_2\eta_1)^2 (\xi_2\eta_3 - \xi_3\eta_2)^2 (\xi_3\eta_1 - \xi_1\eta_3)^2.$$

In the result we omit all the suffixes, and replace ξ^4 by $\dfrac{d^4U}{dx^4}$ &c.; or, when we operate on a quartic, by a_0 the coefficient of x^4. There are many ways which a little practice suggests for abridging the work of this expansion, but we do not think it worth while to give up the space necessary to explain them; and we merely give the results of the expansion of the three invariants just referred to. $\overline{12}^2.\overline{23}^2.\overline{31}^2$ yields the invariant of a quartic already obtained (Art. 137, Ex. 1, and Art. 140), viz.:—

$$a_4a_2a_0 + 2a_3a_2a_1 - a_4a_1^2 - a_0a_3^2 - a_2^3.$$

$\overline{12}^4.\overline{23}^4.\overline{31}^4$ gives

$a_8(a_4a_0 - 4a_3a_1 + 3a_2a_2) + a_7(-4a_5a_0 + 12a_4a_1 - 8a_3a_2)$

$+ a_6(3a_6a_0 - 8a_5a_1 - 22a_4a_2 + 24a_3a_3) + a_5(24a_5a_2 - 36a_4a_3) + 15a_4a_4a_4.$

And $\overline{12}^6.\overline{23}^6.\overline{31}^6$ gives

$a_{12}(a_6a_0 - 6a_5a_1 + 15a_4a_2 - 10a_3a_3) + a_{11}(-6a_7a_0 + 30a_6a_1 - 54a_5a_2 + 30a_4a_3)$

$\quad + a_{10}(15a_8a_0 - 54a_7a_1 + 24a_6a_2 + 150a_5a_3 - 135a_4a_4)$

$\quad + a_9(-10a_9a_0 + 30a_8a_1 + 150a_7a_2 - 430a_6a_3 + 270a_5a_4)$

$\quad + a_8(-135a_8a_2 + 270a_7a_3 + 495a_6a_4 - 540a_5a_5)$

$\quad + a_7(-540a_7a_4 + 720a_6a_5) - 280a_6a_6a_6.$

152. Though the above-mentioned is the only type of invariants of the third order, there is an unlimited number of covariants, the simplest being $\overline{12}^a.\overline{13}$, which, when expanded, is

$$\frac{d^3U}{dx^3}\frac{d^2U}{dy^2}\frac{dU}{dy} - \frac{d^3U}{dx^2dy}\left(2\frac{d^2U}{dxdy}\frac{dU}{dy} + \frac{d^2U}{dy^2}\frac{dU}{dx}\right)$$
$$+ \frac{d^3U}{dxdy^2}\left(\frac{d^2U}{dx^2}\frac{dU}{dy} + 2\frac{d^2U}{dxdy}\frac{dU}{dx}\right) - \frac{d^3U}{dy^3}\frac{d^2U}{dx^2}\frac{dU}{dx}.$$

When this is applied to a cubic, it gives the evectant obtained already (Art. 138).

The general type of invariants of the fourth order in the coefficients is $(\overline{12.34})^a (\overline{13.24})^\beta (\overline{14.23})^\gamma$. Thus the discriminant of a cubic is expressed in this notation $(\overline{12.34})^2 (\overline{13.24})$; but we cannot afford space to enter into greater details on this subject.

153. The principles just laid down afford an easy proof of a remarkable theorem first demonstrated by M. Hermite, and to which we shall refer as "Hermite's Law of Reciprocity." *The number of invariants of the nth order in the coefficients possessed by a binary quantic of the pth degree is equal to the number of invariants of the order p in the coefficients possessed by a quantic of the nth degree.* We have already proved that if any symbol $12^a.23^b.34^c$ &c. denotes an invariant of the order p of a quantic of the degree n; then the number of different figures 1, 2, 3, &c., is p, and each figure occurs n times. But we might calculate by the method of Art. 132 an invariant $\Sigma(\alpha-\beta)^a(\beta-\gamma)^b(\gamma-\delta)^c$ &c., where we replace each symbol $\overline{34}$ by the difference of two roots $(\gamma-\delta)$. This latter is an invariant of a quantic of the pth degree, since there are by hypothesis p roots; and it is of the degree n in the coefficients of the equation (Art. 54).

Thus, for example, a quadratic has but the single independent invariant $(\alpha-\beta)^2$, though of course every power of this is also an invariant; and the general type of such invariants is $(\alpha-\beta)^m$. Hence, only quantics of even degree have invariants of the second order in the coefficients, and the general symbol for such invariants is $\overline{12}^{2m}$.

So again, cubics have no invariant except the discriminant

$(\alpha-\beta)^2 (\beta-\gamma)^2 (\gamma-\alpha)^2$ and its powers; and the discriminant is of the fourth order in the coefficients. Hence only quantics of the degree $4m$ have cubic invariants whose general type is $\overline{12}^{2m}.\overline{23}^{2m}.\overline{31}^{2m}$. It will be proved that quartics have two independent invariants, one of the second, and one of the third degree, in the coefficients; and, of course, any power of one multiplied by any power of the other is an invariant. Hence, quartics have as many invariants of the pth order as the equation $2x + 3y = p$ admits of integer solutions: this is, therefore, the number of invariants of the fourth order which a quantic of the pth degree can possess.

154. Hermite has proved that his theorem applies also to covariants of any given degree in x and y; that is to say, that an n^{ic} possesses as many such covariants of the pth order in the coefficients as a p^{ic} has of the nth order in the coefficients. For, consider any symbol, $\overline{12}^\lambda.\overline{23}^\mu.\overline{34}^\nu$ &c., where there are p figures, and the figure 1 occurs a times, 2 occurs b times, and so on; then we have proved that the degree of this covariant in x and y is $(n-a)+(n-b)+$&c. But we may form the symmetric function

$$\Sigma (\alpha-\beta)^\lambda (\beta-\gamma)^\mu (\gamma-\delta)^\nu (x-\alpha)^{n-a} (x-\beta)^{n-b} \text{ &c.,}$$

which has been proved (Art. 133) to be a covariant of the quantic of the pth degree, whose roots are α, β, &c. Every root enters into its expression in the degree n, which is therefore the order of the covariant in the coefficients, and it obviously contains x and y in the same degree as before, viz. $(n-a)+(n-b)+$&c. Thus, for example, the only covariants which a quadratic has are some power of the quantic itself multiplied by some power of its discriminant, the general type of which is

$$(\alpha-\beta)^{2p} (x-\alpha)^q (x-\beta)^q,$$

the order of which in the coefficients is $2p+q$, and in x and y is $2q$. Hence we infer that every quantic of the degree $2p+q$ has a covariant of the second degree in the coefficients, and of the degree $2q$ in x and y, the general symbol for such covariants being $\overline{12}^{2p}$. When $q=1$, we obtain the theorem given (p. 113), that every quantic of odd degree has a quadratic covariant.

INVARIANTS AND COVARIANTS. 127

155. This notation affords a complete calculus, by means of which invariants or covariants can be transformed, and the identity of different expressions ascertained. Postponing to a subsequent Lesson the explanation of this, we go on to show how the same notation is to be applied to express derivatives of quantics in three or more variables. If $x_1 y_1 z_1$, $x_2 y_2 z_2$, $x_3 y_3 z_3$, be cogredient sets of variables, then, by the rule for multiplication of determinants, the determinant

$$x_1(y_2 z_3 - y_3 z_2) + x_2(y_3 z_1 - y_1 z_3) + x_3(y_1 z_2 - y_2 z_1)$$

is an invariant, which by transformation, becomes a similar function multiplied by the modulus of transformation. And if in the above we write for x_1, $\dfrac{d}{dx_1}$; for y_2, $\dfrac{d}{dy_2}$; and so on, we obtain an invariantive symbol of operation, which we shall write $\overline{123}$. When, then, we wish to obtain invariants or covariants of any function U, we have only to operate on the product $U_1 U_2 U_3 \ldots U_p$ with the product of any number of symbols $\overline{123}^a$ $\overline{124}^\beta$ $\overline{235}^\gamma$ &c., and after differentiation suppress all the suffixes. Thus, for example, let U_1, U_2, U_3 be ternary quadrics, and let the coefficients in U_1 be a, b, c, $2f$, $2g$, $2h$, as at p. 83, then $\overline{123}^2$ expanded is

$$a(b'c'' + b''c' - 2f'f'') + b(c'a'' + c''a' - 2g'g'') + c(a'b'' + a''b' - 2hh')$$
$$+ 2f(g'h'' + g''h' - a'f'' - a''f') + 2g(h'f'' + h''f' - b'g'' - b''g')$$
$$+ 2h(f'g'' + f''g' - c'h'' - c''h');$$

which, when we suppose the three quantics U_1, U_2, U_3, to be identical, or $a = a' = a''$ &c. reduces to six times

$$abc + 2fgh - af^2 - bg^2 - ch^2.$$

If in the above we replace a, the coefficient of x^2, by $\dfrac{d^2 U}{dx^2}$ &c. we get the expansion of $\overline{123}^2$ as applied to any ternary quantic. This covariant is called the Hessian of the quantic.

It is seen, as at Art. 149, that odd powers of the symbol $\overline{123}$ vanish when it is applied to a single quantic. We give as a further example the expansion of $\overline{123}^4$ applied to the quartic,

$$ax^4 + by^4 + cz^4 + 4(a_2 x^3 y + a_3 x^3 z + b_3 y^3 z + b_1 y^3 x + c_1 z^3 x + c_2 z^3 y)$$
$$+ 6(dy^2 z^2 + ez^2 x^2 + fx^2 y^2) + 12xyz(lx + my + nz).$$

Then $\overline{123}^4$ is

$$abc - 4(ab_2c_2 + bc_1a_3 + ca_2b_1) + 3(ad^2 + be^2 + cf^2) + 4(a_2b_3c_1 + a_3b_1c_2)$$
$$- 12(a_2nd + a_3md + b_1ne + b_3le + c_1mf + c_2lf)$$
$$+ 12(lb_1c_1 + mc_2a_2 + na_3b_3) + 12(dl^2 + em^2 + fn^2) + 6def - 12lmn.$$

156. We can express in the same manner functions containing contragredient variables; for if α, β, γ be any variables contragredient to x, y, z, and therefore cogredient with $\dfrac{d}{dx}$, $\dfrac{d}{dy}$, $\dfrac{d}{dz}$, it follows, as before, that the determinant

$$\alpha\left(\frac{d}{dy_1}\frac{d}{dz_2} - \frac{d}{dy_2}\frac{d}{dz_1}\right) + \beta\left(\frac{d}{dz_1}\frac{d}{dx_2} - \frac{d}{dz_2}\frac{d}{dx_1}\right) + \gamma\left(\frac{d}{dx_1}\frac{d}{dy_2} - \frac{d}{dx_2}\frac{d}{dy_1}\right)$$

(which we shall denote by the abbreviation $\overline{\alpha 12}$) is an invariantive symbol of operation. Thus, if U_1, U_2 be two different quadrics, $\overline{\alpha 12}^2$ is the contravariant called ϕ (*Conics*, p. 330), which expanded is

$$\alpha^2(b'c'' + b''c' - 2f'f'') + \beta^2(c'a'' + c''a' - 2g'g'') + \gamma^2(a'b'' + a''b' - 2h'h'')$$
$$+ 2\beta\gamma(g'h'' + g''h' - a'f'' - a''f') + 2\gamma\alpha(h'f'' + h''f' - b'g'' - b''g')$$
$$+ 2\alpha\beta(f'g'' + f''g' - c'h'' - c''h'),$$

and which, when the two quadrics are identical, becomes the equation of the polar reciprocal of the quadric.

In like manner, the quantic contravariant to a quartic, which I have called S (*Higher Plane Curves*, p. 101), may be written symbolically $\overline{\alpha 12}^4$, and the quantic T in the same place may be written $\overline{\alpha 12}^2 \overline{\alpha 23}^2 \overline{\alpha 31}^2$. In any of these we have only to replace the coefficient of any power of x, x^n by $\dfrac{d^n \cdot}{dx^n}$ to obtain a symbol which will yield a mixed concomitant when applied to a quantic of higher dimensions. Thus $\overline{\alpha 12}^2$ is

$$\alpha^2\left\{\frac{d^2U}{dy^2}\frac{d^2U}{dz^2} - \left(\frac{d^2U}{dy\,dz}\right)^2\right\} + \&c.,$$

which, when applied to a quadric, is a contravariant, but, when applied to a quantic of higher order, contains both x, y, z, as well as the contragredient α, β, γ, and therefore, is a mixed concomitant.

INVARIANTS AND COVARIANTS. 129

In general, if we have the symbolical expression for any invariant of a binary quantic, we have only to prefix a contravariant symbol α to every term, when we shall have a contravariant of a ternary quantic of the same order. And in particular it can be proved that if we take the symbolical expression for the discriminant of a binary quantic, and prefix in this manner a contravariant symbol to each term, we shall have the expression for the polar reciprocal of a ternary quantic.

Thus, the symbol for the discriminant of a binary cubic is $\overline{12}^2.\overline{34}^2.\overline{13}.\overline{24}$, and the polar reciprocal of a ternary cubic is $\overline{\alpha 12}^2.\overline{\alpha 34}^2.\overline{\alpha 13}.\overline{\alpha 24}$, which is obviously of the sixth order in the variables α, β, γ, and of the fourth in the coefficients.

157. If in any contravariant we substitute $\dfrac{d}{dx}$, $\dfrac{d}{dy}$, $\dfrac{d}{dz}$ for α, β, γ, and operate on U, we get a covariant (Art. 135); and the symbol for this covariant is got from that for the contravariant by writing a new figure instead of α. Thus from $\overline{\alpha 23}^2$ is got $\overline{123}^2$, from $\overline{\alpha 23}.\overline{\alpha 24}$, is got $\overline{123}.\overline{124}$ &c. Conversely, if in the symbol for any invariant we replace any figure by a contravariant symbol α, we get the evectant of that invariant. Thus,

$$\overline{123}.\overline{124}.\overline{234}.\overline{314}$$

is an invariant of a cubic, and the evectant of that invariant is

$$\overline{123}.\overline{\alpha 12}.\overline{\alpha 23}.\overline{\alpha 31}.$$

In the case of a binary quantic, this rule assumes a simpler form; for if we substitute a contravariant symbol for 1 in $\overline{12}$, it becomes, when written at full length, $\xi \dfrac{d}{dy} - \eta \dfrac{d}{dx}$, but since ξ and η are cogredient with $-y$ and x, this may be written $x \dfrac{d}{dx} + y \dfrac{d}{dy}$, and may be suppressed altogether, since it only affects the result with a numerical multiplier. Hence, given the symbol for any invariant of a binary quantic, its evectant is got by omitting all the factors which contain any one figure. Thus,

$$\overline{12}^2.\overline{34}^2.\overline{13}.\overline{24}$$

s

being the discriminant of a cubic, its evectant, got by omitting the factors which contain 4, is $\overline{12^2}.\overline{13}$.

If in a contravariant of any quantic we substitute $\dfrac{dU}{dx}, \dfrac{dU}{dy}, \dfrac{dU}{dz}$ for α, β, γ, we also get a covariant, and the symbol for it is obtained from that for the contravariant by writing a *different* new figure in place of every α. Thus, from $\overline{a34}^2$ we get $\overline{134}.\overline{234}$; and so on.

158. In the explanation of symbolical methods which has been hitherto given, I have followed the notation and course of proceeding originally made use of by Mr. Cayley. I wish now to explain some modifications of notation introduced by MM. Aronhold and Clebsch, who have employed these symbolical methods with great success, but who perhaps have scarcely sufficiently recognized the substantial identity of their methods with those previously given by Mr. Cayley. The variables are denoted x_1, x_2, x_3, &c., while the coefficients are denoted by suffixes corresponding to the variables which they multiply. Thus the ternary cubic, the ternary quartic, &c. may be briefly denoted $\Sigma a_{ikl} x_i x_k x_l$, $\Sigma a_{iklm} x_i x_k x_l x_m$, &c., where the numbers i, k, l, m are to receive in succession all the values 1, 2, 3. It will be observed that in this notation $a_{iik} = a_{iki} = a_{kii}$, so that when we form the sums indicated we obtain a quantic written with the numerical coefficients of the binomial theorem. Thus when we form the sum $\Sigma a_{ikl} x_i x_k x_l$, the three terms $a_{112} x_1 x_1 x_2$, $a_{121} x_1 x_2 x_1$, $a_{211} x_2 x_1 x_1$ are identical, as in like manner are the six terms

$$a_{123} x_1 x_2 x_3, \ a_{132} x_1 x_3 x_2, \ a_{213} x_2 x_1 x_3, \ a_{231} x_2 x_3 x_1, \ a_{312} x_3 x_1 x_2, \ a_{321} x_3 x_2 x_1,$$

so that the sum written at length would be

$$a_{111} x_1 x_1 x_1 + a_{222} x_2 x_2 x_2 + a_{333} x_3 x_3 x_3 + 3 a_{112} x_1 x_1 x_2 + \ldots + 6 a_{123} x_1 x_2 x_3.$$

And so in like manner in general. Now M. Aronhold uses, as an abbreviated expression for the quantic in general, $(a_1 x_1 + a_2 x_2 + a_3 x_3 + \ldots)^n$, where after expansion we are to substitute for the products $a_i a_k a_l$, &c. the coefficients a_{ikl}. Thus the ternary cubic given above may be written in the abbreviated form $(a_1 x_1 + a_2 x_2 + a_3 x_3)^3$; the terms

$$a_1 a_1 a_1 x_1 x_1 x_1 + 3 a_1 a_1 a_2 x_1 x_1 x_2 + \&c.$$

INVARIANTS AND COVARIANTS. 131

in the expansion of the cube being replaced by $a_{111}x_1x_1x_1$, $3a_{112}x_1x_1x_2$, &c. The quantic might equally have been written $(b_1x_1+b_2x_2+b_3x_3)^3$, $(c_1x_1+c_2x_2+c_3x_3)^3$, &c., it being understood that we are in like manner to substitute for $b_1b_1b_1$, $c_1c_1c_2$, &c. the coefficients a_{111}, a_{112}, &c. Now the rule given by M. Aronhold for the formation of invariants is to to take a number of determinants, whose constituents are the symbols a_1, a_2, a_3; b_1, b_2, &c., to multiply all together, and after multiplication to substitute for the symbols $a_i a_k a_l$, $b_m b_n b_p$, the coefficients a_{ikl}, a_{mnp}, &c. Thus M. Aronhold first discovered a fundamental invariant of a ternary cubic by forming the four determinants $\Sigma \pm a_1 b_2 c_3$, $\Sigma \pm b_1 c_2 d_3$, $\Sigma \pm c_1 d_2 a_3$, $\Sigma \pm d_1 a_2 b_3$; multiplying all together and then performing the substitutions already indicated. This is the same invariant which, in Mr. Cayley's notation, would be designated as $\overline{123}.\overline{234}.\overline{341}.\overline{412}$.

159. In order to see the substantial identity of the two methods, it is sufficient to observe that by the theorem of homogeneous functions any quantic u of the nth order differs only by a numerical multiplier from $\left(x_1 \dfrac{d}{dx_1} + x_2 \dfrac{d}{dx_2} + x_3 \dfrac{d}{dx_3}\right)^n u$, so that if we write it $(a_1x_1 + a_2x_2 + a_3x_3)^n$, the symbols a_1, a_2, a_3 differ only by a numerical constant from the differential symbols $\dfrac{d}{dx_1}$, &c. And we evidently get the same results whether with Mr. Cayley we form determinants whose constituents are $\dfrac{d}{dx_1}$, $\dfrac{d}{dx_2}$, $\dfrac{d}{dx_3}$, or with Aronhold, whose constituents are a_1, a_2, a_3.

And the artifice made use of by both is the same. If we multiply together a number of differential symbols $\left(\dfrac{d}{dx}+\lambda \dfrac{d}{dy}\right)\left(\dfrac{d}{dx}+\mu \dfrac{d}{dy}\right)$, &c. and operate on U, it is evident the result will be a linear function of differentials of U of an order equal to the number of factors multiplied together; and that in this way we can never get any power higher than the first of any differential coefficient. When then it is required to express symbolically a function involving powers of the differential coefficients, the artifice used by Mr. Cayley was to write the function first with different sets of variables and form such a function as $\left(\dfrac{d}{dx_1}+\lambda\dfrac{d}{dy_1}\right)\left(\dfrac{d}{dx_2}+\mu\dfrac{d}{dy_2}\right)U_1U_2$, and *after* differentiation to make the variables identical. So in like manner Aronhold in his symbolic multiplication uses different symbols which have the same meaning after the multiplication has been performed. By multiplying together symbols a_i, a_k, a_l, &c., we can only get a term such as a_{ikl} of the first degree only in the coefficients. When then we want to express symbolically functions of the coefficients of higher order than the first, the artifice is used of multiplying together different sets of symbols a_i, a_k, a_l; b_i, b_k, b_l, &c., the products $a_ia_ka_l\ b_ib_kb_l,\ c_ic_kc_l$, &c. all equally denoting the coefficient a_{ikl}.

160. Having then so fully explained Mr. Cayley's symbolical method, it is unnecessary to explain at greater length one which only differs from it by notation. Reserving, then, to a subsequent lesson illustrations of the applications of these methods, we only add here Clebsch's proof that every invariant can be expressed in the manner indicated. If we are given any invariant $\phi(a, b, c,$ &c.$)$ of the coefficients of a single quantic, it has been proved, Art. 119, that by performing on it the operation $a'\dfrac{d}{da}+b'\dfrac{d}{db}+$&c., we obtain a simultaneous invariant of a pair of quantics, a', b', c' being the coefficients of the second, answering respectively to a, b, c in the first. In like manner, if we operate on this with $a''\dfrac{d}{da}+b''\dfrac{d}{db}+$&c., we get an invariant of a system of three quantics, and we can repeat this process as long as there remains coefficients a, b, c, &c. in the invariant. Thus, then, when we are given any invariant of a quantic whose order in the coefficients is p, we can derive from it an invariant of a system of p quantics which shall be of the first degree in the coefficients of each quantic. And we can fall back at any time on the original invariant by supposing the p quantics to be identical. For the operation $a\dfrac{d}{da}+b\dfrac{d}{db}+$&c. performed on the invariant could only have effected it with a numerical multiplier. Again, we are at liberty to suppose all the new quantics thus

introduced to be perfect n^{th} powers. And it follows thus, that from any invariant of a single quantic, we can deduce an invariant of the system of p quantics, $(a_1x_1 + a_2x_2 + a_3x_3 + \&c.)^n$, $(b_1x_1 + b_2x_2 + \&c.)^n$, $(c_1x_1 + \&c.)^n$, &c. And, as has been said, we fall back on the original invariant if we substitute for each coefficient in the introduced quantics, the corresponding coefficient in that first given; as, for instance, if we substitute for a_1^n, b_1^n, c_1^n, &c. the coefficient of x_1^n in the original quantic, and so on. Invariants of $(a_1x_1 + \&c.)^n$, $(b_1x_1 + \&c.)^n$, &c. are evidently invariants of the system $a_1x_1 + \&c.$, $b_1x_1 + \&c.$ It is proved then that every invariant of the pth order of a single quantic can be expressed symbolically by means of an invariant of p quantics of the first order. The only thing that remains to be proved is that these last invariants are necessarily functions of the determinants whose constituents are the coefficients of the quantics. For M. Clebsch's proof of this I refer to *Crelle*, vol. LIX., p. 7, since, though not really difficult, it is troublesome to explain and would take up much space. And the same thing can be seen from the differential equation of invariants. Thus, for a system of linear quantics of the first degree, $ax + by$, &c., an invariant must satisfy the equation

$$b\frac{dI}{da} + b'\frac{dI}{da'} + b''\frac{dI}{da''} + \&c. = 0,$$

which integrated as an ordinary partial differential equation gives I a function of $ab' - a'b$, $ab'' - a''b$, &c.; and so in like manner in general.

LESSON XV.

CANONICAL FORMS.

161. Since invariants and covariants retain their relations to each other, no matter how the quantic is linearly transformed, it is plain that when we wish to study these relations it is sufficient to do so by discussing the quantic in the simplest form to which it is possible to reduce it. This is only extending to

quantics in general what the reader is familiar with in the case of ternary and quaternary quantics; since, when we wish to study the properties of a curve or surface, every geometer is familiar with the advantage of choosing such axes as shall reduce the equation of this curve or surface to its simplest form.* The simplest form, then, to which a quantic can, without loss of generality, be reduced is called the *canonical form* of the quantic. We can, by merely counting the constants, ascertain whether any proposed simple form is sufficiently general to be taken as the canonical form of a quantic, for if the proposed form does not, either explicitly or implicitly, contain as many constants as the given quantic in its most general form, it will not be possible always to reduce the general to the proposed form.† Thus, a binary cubic may be reduced to the form $X^3 + Y^3$; for the latter form, being equivalent to $(lx + my)^3 + (l'x + m'y)^3$ contains implicitly four constants, and therefore is as general as $(a, b, c, d\mathfrak{X}x, y)^3$. So, in like manner, a ternary cubic in its most general form contains ten constants; but the form $X^3 + Y^3 + Z^3 + 6mXYZ$, contains also ten constants, since, in addition to the m which appears explicitly, X, Y, Z, implicitly

---

* It must be owned however that as in the progress of analysis greater facility is gained in dealing with quantics in their most general form, the advantage diminishes of reducing them to simpler forms.

† It is not true, however, conversely that a form which contains the proper number of constants is necessarily one to which the general equation may be reduced. For when we endeavour by comparison of coefficients to identify such a form with the general equation, although the number of equations is equal to the number of quantities to be determined, it *may* happen that the constants enter into the equations in such a way that all the equations cannot be satisfied. Thus

$$(x - a)^2 + (y - \beta)^2 = lx + my + n$$

is a form containing five constants, and yet is not one to which the general equation of a ternary quadric can be reduced; since the constants enter the equation in such a way that though we have more than enough to make the coefficients of x and y and the absolute term identical with those in any proposed equation, we have no means of identifying the coefficients of x^2, xy and y^2. A more important example is

$$x^4 + y^4 + z^4 + u^4 + v^4,$$

where z, u, v are linear functions. In the case of a ternary quantic this form contains implicitly fourteen independent constants, and therefore seems to be one to which the quartic in general can be reduced. But Clebsch has shewn that a condition must be fulfilled in order that a quartic should be reducible to this form, namely, the vanishing of a certain invariant.

involve three constants each. This latter, then, may be taken as the canonical form of a ternary cubic, and, in fact, the most important advances that have been recently made in the theory of curves of the third degree are owing to the use of the equation in this simple and manageable form.

162. The quadratic function $(a, b, c\!\!\!\!\!/\, x, y)^2$ can be reduced in an infinity of ways to the form $x^2 + y^2$, since the latter form implicitly contains four constants, and the former only three. In like manner the ternary quadric which contains six constants can be reduced in an infinity of ways to the form $x^2 + y^2 + z^2$, since this last contains implicitly nine constants; and in general a quadratic form in any number of variables can be reduced in an infinity of ways to a sum of squares. It is worth observing, however, that though a quadratic form can be reduced in an infinity of ways to a sum of squares, yet the number of positive and negative squares in this sum is fixed. Thus, if a binary quadric can be reduced to the form $x^2 + y^2$, it cannot also be reduced to the form $u^2 - v^2$, since we cannot have $x^2 + y^2$ identical with $u^2 - v^2$, the factors on the one side of the identity being imaginary, and those on the other being real. In like manner, for ternary quadrics we cannot have $x^2 + y^2 - z^2 = u^2 + v^2 + w^2$, since we should thus have $x^2 + y^2 = z^2 + u^2 + v^2 + w^2$, or, in other words,

$$x^2 + y^2 = z^2 + (lx + my + nz)^2 + (l'x + m'y + n'z)^2 + (l''x + m''y + n''z)^2,$$

and if we make x and $y = 0$, one side of the identity would vanish, and the other would reduce itself to the sum of four positive squares which could not be $= 0$. And the same argument applies in general.

163. We commence by showing that, as has been just stated, a cubic may always be reduced to the sum of two cubes. To do this is in fact to solve the equation, since when the quantic is brought to the form $X^3 + Y^3$, it can immediately be resolved into its linear factors. Now, if the cubic $(a, b, c, d\!\!\!\!\!/\, x, y)^3$ become by transformation $(A, B, C, D\!\!\!\!\!/\, X, Y)^3$, then, since (Art. 122) the Hessian $(ax + by)(cx + dy) - (bx + cy)^2$ is a covariant, it will by the definition of a covariant, be transformed

into a similar function of A, B, C, D, X, Y. That is to say, we must have

$$(ac - b^2) x^2 + (ad - bc) xy + (bd - c^2) y^2$$
$$= (AC - B^2) X^2 + (AD - BC) XY + (BD - C^2) Y^2.$$

Now, if in the transformed cubic, B and C vanish, the Hessian takes the form $ADXY$; and we see at once that we are to take for X and Y the two factors into which the Hessian may be broken up. When we have found X and Y, we compare the given cubic with $AX^3 + DY^3$, and determine A and D by comparison of coefficients.

Ex. To reduce $4x^3 + 9x^2 + 18x + 17$ to the form $AX^3 + DY^3$. The Hessian is

$$(4x + 3)(6x + 17) - (3x + 6)^2,$$

or
$$15x^2 + 50x + 15,$$

whose linear factors are $x + 3, 3x + 1$. Comparing then the given cubic with

$$A(x + 3)^3 + D(3x + 1)^3,$$

we have $A + 27D = 4$, $27A + D = 17$, whence $728D = 91$, $728A = 455$, or A is to D in the ratio of 5 to 1. The given cubic then only differs by a factor (viz. 8) from

$$5(x + 3)^3 + (3x + 1)^3,$$

and it is obvious that the roots of the cubic are given by the equation

$$3x + 1 + (x + 3) \sqrt[3]{(5)} = 0.$$

164. It is evident that every cubic cannot be brought by *real* transformation to the form $AX^3 + DY^3$, for this last form has one real factor and two imaginary; and therefore cannot be identical with a cubic whose three factors are real. The discriminant of the Hessian

$$4(ac - b^2)(bd - c^2) - (ad - bc)^2$$

is, with sign changed, the same as that of the cubic. When the discriminant of the cubic is positive, the Hessian has two real factors, and the cubic one real factor and two imaginary. When it is negative, the Hessian has two imaginary factors, and the cubic three real. When it vanishes, both Hessian and cubic have two equal factors, and it can be directly verified that the Hessian of X^2Y is X^2.*

It is to be observed, that a quantic cannot *always* be reduced to its canonical form. The impossibility of the reduction indi-

---

* In general, when a binary quantic has a square factor, this will also be a square factor in its Hessian, as may be verified at once by forming the Hessian of $x^2\phi$.

CANONICAL FORMS. 137

cates some singularity in the form of the quantic. Thus a cubic having a square factor cannot be brought to the form $Ax^3 + Dy^3$, its simplest form being x^2y.

165. In the same manner as a cubic can be brought to the sum of two cubes, so in general any binary quantic of odd degree $(2n-1)$ can be reduced to the sum of n powers of the $(2n-1)^{\text{th}}$ degree, a theorem due to Mr. Sylvester. For the number of constants in any binary quantic is always one more than its degree, or, in the present case, $2n$; and we have the same number of constants if we take n terms of the form $(lx + my)^{2n-1}$. The actual transformation is performed by a method which is the generalization of that employed (Art. 163). For simplicity, we only apply it to the fifth degree, but the method is general. The problem then is to determine u, v, w, so that $(a, b, c, d, e, f\mathbb{X}x, y)^5$ may $= u^5 + v^5 + w^5$. Now we say that if we form the determinant

$$\begin{vmatrix} ax + by, & bx + cy, & cx + dy \\ bx + cy, & cx + dy, & dx + ey \\ cx + dy, & dx + ey, & ex + fy \end{vmatrix},$$

the three factors of this cubic will be u, v, w. For let

$$u = lx + my, \quad v = l'x + m'y, \quad w = l''x + m''y;$$

then, differentiating the identity

$$(a, b, c, d, e, f\mathbb{X}x, y)^5 = u^5 + v^5 + w^5$$

four times successively with regard to x, and dividing by 120, we get

$$ax + by = l^4 u + l'^4 v + l''^4 w.$$

Similarly differentiating three times with regard to x, and once with regard to y,

$$bx + cy = l^3 mu + l'^3 m'v + l''^3 m''w;$$

and so on.

The determinant, then, written above, may be put into the form

$$\begin{vmatrix} l^4u + l'^4v + l''^4w, & l^3mu + l'^3m'v + l''^3m''w, & l^2m^2u + l'^2m'^2v + l''^2m''^2w \\ l^3mu + l'^3m'v + l''^3m''w, & l^2m^2u + l'^2m'^2v + l''^2m''^2w, & lm^3u + l'm'^3v + l''m''^3w \\ l^2m^2u + l'^2m'^2v + l''^2m''^2w, & lm^3u + l'm'^3v + l''m''^3w, & m^4u + m'^4v + m''^4w \end{vmatrix}$$

Now it will be observed that the coefficients of u in every column are proportional to l^2, lm, m^2. Consequently, if we resolved this determinant, as in Art. 22, into partial determinants, every such determinant which contained two of the u columns would vanish as having two columns the same. And so, in like manner, would any which contained two v or two w columns. The determinant then will be uvw multiplied by a numerical factor.*

When, then, the determinant written in the beginning of this Article has been found, by solving a cubic equation, to be the product of the factors $(x + \lambda y)(x + \mu y)(x + \nu y)$, we know that u, v, w can only differ from these by numerical coefficients, and we may put

$$(a, b, c, d, e, f\mathbb{X}x, y)^5 = A(x + \lambda y)^5 + B(x + \mu y)^5 + C(x + \nu y)^5;$$

and then A, B, C are got from solving any of the systems of simple equations got by equating three coefficients on both sides of the above identity.

The determinant used in this Article is a covariant, which is called the *canonizant* of the given quantic.

166. The canonizant may be written in another, and perhaps simpler form, namely,

$$\begin{vmatrix} y^3, & -y^2 x, & yx^2, & -x^3 \\ a, & b, & c, & d \\ b, & c, & d, & e \\ c, & d, & e, & f \end{vmatrix}.$$

This last is the form in which we should have been led to it if we had followed the course that naturally presented itself, and sought directly to determine the six quantities, A, B, C, λ, μ, ν, by solving the six equations got on comparison of coefficients of the identity last written in Art. 165. For the development of the solution in this form, to which we cannot afford the necessary space here, we refer to Mr. Sylvester's Paper (*Philosophical Magazine*, November, 1851). Meanwhile, the identity of the determinant in this Article with that in the last has been

---

* Viz. $(lm' - l'm)^2 (l'm'' - l''m')^2 (l''m - lm'')^2$.

$$= \begin{vmatrix} l^2 & l'^2 & l''^2 \\ lm & l'm' & l''m'' \\ m^2 & m'^2 & m''^2 \end{vmatrix}^2$$

CANONICAL FORMS. 139

shown by Mr. Cayley as follows. We have, by multiplication of determinants (Art. 22),

$$\begin{vmatrix} y^3, & -y^2x, & yx^2, & -x^3 \\ a, & b, & c, & d \\ b, & c, & d, & e \\ c, & d, & e, & f \end{vmatrix} \times \begin{vmatrix} 1, & 0, & 0, & 0 \\ x, & y, & 0, & 0 \\ 0, & x, & y, & 0 \\ 0, & 0, & x, & y \end{vmatrix}$$

$$= \begin{vmatrix} y^3, & 0, & 0, & 0 \\ 0, & ax+by, & bx+cy, & cx+dy \\ 0, & bx+cy, & cx+dy, & dx+ey \\ 0, & cx+dy, & dx+ey, & ex+fy \end{vmatrix},$$

which, dividing both sides of the equation by y^3, gives the identity required.

167. We have still to mention another way of forming the canonizant. Let this sought covariant be $(A, B, C, D\!\!\!\!\!\Cap\!x, y)^3$, where we want to determine A, B, C, D; then (Art. 136) $(A, B, C, D\!\!\!\!\!\Cap\dfrac{d}{dy}, -\dfrac{d}{dx})^3$ will also yield a covariant. But if this operation is applied to $(x+\lambda y)^n$ where $x+\lambda y$ is a factor in $(A, B, C, D\!\!\!\!\!\Cap x, y)^3$, the result must vanish, since one of the factors in the operating symbol is $\dfrac{d}{dy} - \lambda \dfrac{d}{dx}$. Since, then, the given quantic is by hypothesis the sum of three terms of the form $(x+\lambda y)^5$, the result of applying to the given quantic the operating symbol just written must vanish. Thus, then, we have

$$A(d, e, f\!\!\!\Cap x, y)^2 - B(c, d, e\!\!\!\Cap x, y)^2 + C(b, c, d\!\!\!\Cap x, y)^2$$
$$- D(a, b, c\!\!\!\Cap x, y)^2 = 0,$$

or, equating separately to 0 the coefficients of x^2, xy, y^2, we have

$$Ad - Bc + Cb - Da = 0,$$
$$Ae - Bd + Cc - Db = 0,$$
$$Af - Be + Cd - Dc = 0,$$

whence (Art. 28) A is proportional to the determinant got by suppressing the column A or $\begin{vmatrix} a, & b, & c \\ b, & c, & d \\ c, & d, & e \end{vmatrix}$ and so for B, C, D, which values give for the canonizant the form stated in the last Article.

168. We proceed now to quantics of even degree $(2n)$. Since this quantic contains $2n+1$ terms, if we equate it to a sum of n powers of the degree $2n$, we have one equation more to satisfy than we have constants at our disposal. On the other hand, if we add another $2n^{th}$ power, we have one constant too many, and the quantic can be reduced to this form in an infinity of ways. It is easy, however, to determine the condition that the given quantic should be reducible to the sum of n, $2n^{th}$ powers. Thus, for example, the conditions that a quartic should be reducible to the sum of two fourth powers, and that a sextic should be reducible to the sum of three sixth powers, are respectively the determinants

$$\begin{vmatrix} a, & b, & c \\ b, & c, & d \\ c, & d, & e \end{vmatrix} = 0, \quad \begin{vmatrix} a, & b, & c, & d \\ b, & c, & d, & e \\ c, & d, & e, & f \\ d, & e, & f, & g \end{vmatrix} = 0,$$

and so on. For in the case of the quartic the constituents of the determinant are the several fourth differentials of the quantic, and, expressing these in terms of u and v precisely as in Art. 165, it is easy to see that the determinant must vanish, when the quartic can be reduced to the form $u^4 + v^4$. Similarly for the rest. This determinant expanded in the case of the quartic is the invariant already noticed (see Art. 137, Ex. 1),

$$ace + 2bcd - ad^2 - eb^2 - c^3.$$

169. When this condition is not fulfilled, the quantic is reduced to the sum of n powers, together with an additional term. Thus, the canonical form for a quartic is $u^4 + v^4 + 6\lambda u^2 v^2$. We shall commence with the reduction of the general quartic to its canonical form; the method which we shall use is not the easiest for this case, but is that which shows most readily how the reduction is to be effected in general. Let the product, then, of u, v, which we seek to determine, be $(A, B, C\mathbf{\jmath}x, y)^2$, and let us operate with $\left(A, B, C\mathbf{\jmath}\dfrac{d}{dy}, -\dfrac{d}{dx}\right)^2$ on both sides of the identity $(a, b, c, d, e\mathbf{\jmath}x, y)^4 = u^4 + v^4 + 6\lambda u^2 v^2$.

Now, as before, this operation performed on u^4 and on v^4 will vanish, and when performed on $6\lambda u^2 v^2$, it will be found to give $12\lambda' uv$, where $\lambda' = 2(4AC - B^2)\lambda$. Equating then the coefficients of x^2, xy, and y^2 on both sides, after performing the operation, we get the three equations

$$Ac - 2Bb + Ca = \lambda' A,$$
$$Ad - 2Bc + Cb = \lambda' B,$$
$$Ae - 2Bd + Cc = \lambda' C,$$

whence eliminating A, B, C, we have to determine λ', the determinant

$$\begin{vmatrix} a, & b, & c - \lambda' \\ b, & c + \tfrac{1}{2}\lambda', & d \\ c - \lambda', & d, & e \end{vmatrix} = 0,$$

which expanded is the cubic

$$\lambda'^3 - \lambda'(ae - 4bd + 3c^2) - 2(ace + 2bcd - ad^2 - eb^2 - c^3) = 0,^*$$

the coefficients of which are invariants. Thus, then, we have a striking difference in the reduction of binary quantics to their canonical form, between the cases where the degree is odd and where it is even. In the former case, the reduction is unique, and the system u, v, w, &c. can be determined in but one way. When u is of even degree, however, more systems than one can be found to solve the problem. Thus, in the present instance, a quartic can be reduced in three ways to the canonical form, and if we take for λ' any of the roots of the above cubic, its value substituted in the preceding system of equations enables us to determine A, B, C.

170. If now we proceed to the investigation of the reduction of the quantic $(a_0, a_1, a_2 \ldots \!\!\searrow\!\! x, y)^{2n}$, the most natural canonical form to assume would be $u^{2n} + v^{2n} + w^{2n} + \&c. + \lambda u^2 v^2 w^2$, &c., there being n quantities u, v, w, &c. But the actual reduction to this form is attended with difficulties which have not been overcome, except for the cases $n = 2$ and $n = 4$. But the method used in the last Article can be applied if we take for the canonical form $u^{2n} + v^{2n} + \&c. + \lambda V uvw$ &c., where, if

$$uvw \,\&c. = (A_0, A_1, A_2 \ldots \!\!\searrow\!\! x, y,)^n,$$

---

* N.B.—The discriminant of this cubic is the same as that of the quartic.

142   CANONICAL FORMS.

V is a covariant of this latter function such that when $Vuvw$ &c. is operated on by $\left(A_0, A_1 \ldots \S\dfrac{d}{dy}, -\dfrac{d}{dx}\right)^n$, the result is proportional to the product uvw &c. Suppose, for the moment, that we had found a function V to fulfil this condition, then, proceeding exactly as in the last Article, and operating with the differential symbol last written on the identity got by equating the quantic to its canonical form, we get the system of equations

$$A_0 a_n \;\;- nA_1 a_{n+1} + \tfrac{1}{2}n(n-1) A_2 a_{n+2} - \&\text{c.} = \lambda A_0,$$
$$A_0 a_{n-1} - nA_1 a_n \;\;+ \tfrac{1}{2}n(n-1) A_2 a_{n+1} - \&\text{c.} = \lambda A_1,$$
$$A_0 a_{n-2} - nA_1 a_{n-1} + \tfrac{1}{2}n(n-1) A_2 a_n \;\;- \&\text{c.} = \lambda A_2, \&\text{c.},$$

whence, eliminating $A_0, A_1, A_2,$ &c., we get the determinant

$$\begin{vmatrix} a_n - \lambda, & a_{n+1}, & a_{n+2} & \ldots\ldots\ldots\ldots & a_{2n} \\ a_{n-1}, & a_n + \dfrac{1}{n}\lambda, & a_{n+1}, & \ldots\ldots\ldots\ldots & a_{2n-1} \\ a_{n-2}, & a_{n-1}, & a_n - \dfrac{1.2}{n(n-1)}\lambda, & \ldots & a_{2n-2} \\ \ldots\ldots\ldots\ldots\ldots\ldots\ldots\ldots\ldots\ldots\ldots\ldots \\ a_0, & a_1, & a_2, & \ldots\ldots\ldots\ldots & a_n \mp \lambda^* \end{vmatrix}$$

and having found λ by equating to 0 this determinant expanded (a remarkable equation, all the coefficients of which will be invariants), the equations last written enable us to determine the values of $A_0, A_1,$ &c., corresponding to any of the $n+1$ values of λ.

171. To apply this to the case of the sextic, the canonical form here is $u^6 + v^6 + w^6 + Vuvw$, where, if uvw be

$$(A_0, A_1, A_2, A_3 \mathbin{\S} x, y)^3,$$

V is the evectant of the discriminant of this last quantic, and whose value is written at full length (Art. 138). Now it will

---

* The determinant above written may be otherwise obtained as follows. Let x', y' be cogredient to x, y, and let us form the function

$$\left(x'\dfrac{d}{dx} + y'\dfrac{d}{dy}\right)^n U + \lambda (xy' - yx')^n,$$

which (Arts. 121, 127), we have proved to be linearly transformed into a function of similar form. Take the $n+1$ coefficients of the several powers x'^n, $x'^{n-1}y'$, &c., and from these eliminate linearly the $n+1$ quantities x'^n, x'^{n-1}, &c., and we obtain the determinant in question.

afford an excellent example of the use of canonical forms if we show that in any cubic the result of the operation

$$(a_0, a_1, a_2, a_3 \!\!\! \int \!\! \frac{d}{dy}, -\frac{d}{dx}),$$

performed on the product of the cubic and the evectant just mentioned, will be proportional to the cubic itself. For it is sufficient to prove this, for the case when the cubic is reduced to the canonical form $x^3 + y^3$, in which case the evectant will be $x^3 - y^3$ as appears at once by putting $b = c = 0$, and $a = d = 1$ in the value given, p. 113. The product, then, of cubic and evectant will be $x^6 - y^6$, which, if operated on by $\frac{d^3}{dy^3} - \frac{d^3}{dx^3}$, gives a result manifestly proportional to $x^3 + y^3$. And the theorem now proved being independent of linear transformation, if true for any form of the cubic, is true in general. The canonical form, then, being assumed as above, we proceed exactly as in the last Article, and we solve for λ from the equation

$$\begin{vmatrix} a_0, & a_1, & a_2, & a_3 - \lambda \\ a_1, & a_2, & a_3 + \tfrac{1}{3}\lambda, & a_4 \\ a_2, & a_3 - \tfrac{1}{3}\lambda, & a_4, & a_5 \\ a_3 + \lambda, & a_4, & a_5, & a_6 \end{vmatrix} = 0,$$

which, when expanded, will be found to contain only even powers of λ. If we suppose uvw reduced as above to its canonical form $x^3 + y^3$, the three factors of which are

$$x + y, \quad x + \omega y, \quad x + \omega^2 y,$$

where ω is a cube root of unity, then it is evident from the above that the corresponding canonical form for the sextic is

$$A(x+y)^6 + B(x+\omega y)^6 + C(x+\omega^2 y)^6 + D(x^6 - y^6).$$

It can be proved (see Lesson XVII.) that if u, v, w be the factors of the cubic, then the factors of the evectant used above are $u - v$, $v - w$, $w - u$, so that the canonical form of the sextic may also be written

$$u^6 + v^6 + w^6 + \lambda uvw (u - v)(v - w)(w - u).$$

172. In the case of the octavic the canonical form is

$$u^8 + v^8 + w^8 + z^8 + \lambda u^2 v^2 w^2 z^2,$$

for if we operate on $u^a v^a w^a z^a$ with a symbol formed according to the same method as in the preceding Articles, the result will be proportional to $uvwz$. This will be proved in a subsequent Lesson.

As for higher canonical forms we content ourselves with again mentioning that for a ternary cubic, viz. $x^3 + y^3 + z^3 + 6mxyz$, and that given by Mr. Sylvester for a quaternary cubic, $x^3 + y^3 + z^3 + u^3 + v^3$.

LESSON XVI.

SYSTEMS OF QUANTICS.

173. It still remains to explain a few properties of systems of quantics, to which we devote this Lesson. An invariant of a system of quantics is called a *combinant* if it is unaltered (except by a constant multiplier) not only when the variables are linearly transformed, but also when for any of the quantics is substituted a linear function of the quantics. Thus the eliminant of a system of quantics u, v, w is a combinant. For evidently the result of substituting the common roots of vw in $u + \lambda v + \mu w$ is the same as that of substituting them in u; and the eliminant of $u + \lambda v + \mu w$, v, w is the same as the eliminant of uvw. In addition to the differential equations satisfied by ordinary invariants, combinants must evidently also satisfy the equation
$$\frac{a'dI}{da} + \frac{b'dI}{db} + \frac{c'dI}{dc} + \&c. = 0.$$

It follows from this that in the case of two quantics a combinant is a function of the determinants (ab'), (ac'), (bd'), &c.; in the case of three, of the determinants $(ab'c'')$, &c.; and will accordingly vanish identically, if any two of the quantics become identical. If we substitute for u, v; $\lambda u + \mu v$, $\lambda' u + \mu' v$, every one of the determinants (ab') will be multiplied by $(\lambda \mu' - \lambda' \mu)$; and therefore the combinant will be multiplied by a power of $(\lambda\mu' - \lambda'\mu)$ equal to the order of the combinant in the coefficients of any of the quantics. Similarly for any number of quantics. There may be in like manner combinantive covariants, which are equally covariants when for any of the quantics is substituted a linear function of them. For instance, the Jacobian (Art. 84)

$$\begin{vmatrix} U_1, & U_2, & U_3 \\ V_1, & V_2, & V_3 \\ W_1, & W_2, & W_3 \end{vmatrix},$$

if we substitute for U, $lU+mV+nW$, for V, $l'U+m'V+n'W$, &c. by the property of determinants, becomes the product of the determinants $(lm'n'')$, $(U_1V_2W_3)$. The coefficients of a combinantive covariant are also functions of the determinants (ab'), (ac'); $(ab'c'')$, &c.

174. If $U = (a, b, c \ldots \mathfrak{X}x, y)^n$, $V = (a', b', c' \ldots \mathfrak{X}x, y)^n$ be any two binary quantics of the same degree, then $U + kV$ or $(a + ka', b + kb' \ldots \mathfrak{X}x, y)^n$, where we give different values to k, denotes a system of quantics which are said to form an *involution* with U, V. Now there will be in general $2(n-1)$ quantics of the system, each of which will have a square factor. For the discriminant of a quantic of the nth degree is of the order $2(n-1)$ in the coefficients (Art. 101). If then we substitute $a + ka'$ for a, $b + kb'$ for b, &c., there will evidently be $2(n-1)$ values of k, for which the discriminant will vanish.

If we make $y = 1$ in any of the quantics, it denotes n points on the axis of x. We have just proved that in $2(n-1)$ cases, two of the n points denoted by $U + kV$ will coincide; or, in other words we may say, that there are $2(n-1)$ *double points* in the involution.

When $U + kV$ has a square factor $x - \alpha$, we know that α satisfies the two equations got by differentiation, viz. $U_1 + kV_1 = 0$, $U_2 + kV_2 = 0$, and therefore will satisfy the equation got by eliminating k between them, viz. $U_1V_2 - U_2V_1 = 0$. Now $U_1V_2 - U_2V_1$ which is of the degree $2(n-1)$ is the Jacobian of U, V; and we see that by equating the Jacobian to 0, we obtain

U

the $2(n-1)$ double points of the involution determined by U, V.*

175. *If u and v have a common factor, this will appear as a square factor in their Jacobian.* First let it be observed, that since $nu = xu_1 + yu_2$, $nv = xv_1 + yv_2$, then if we write J for $u_1 v_2 - u_2 v_1$, we shall have $n(uv_2 - vu_2) = xJ$, $n(uv_1 - vu_1) = -yJ$. Differentiating the first of these equations with regard to y, and the second with regard to x, we get

$$n(uv_{22} - vu_{22}) = xJ_2, \quad n(uv_{11} - vu_{11}) = -yJ_1.$$

It follows from the equations we have written, that any value α of x which makes both u and v vanish, will make not only J vanish but also its differentials J_1, J_2, and therefore $x - \alpha$ must be a square factor in J.

Or more directly thus: let $u = \beta\phi$, $v = \beta\psi$, where $\beta = lx + my$; then $u_1 = l\phi + \beta\phi_1$, $u_2 = m\phi + \beta\phi_2$, $v_1 = l\psi + \beta\psi_1$, $v_2 = m\psi + \beta\psi_2$; and $u_1 v_2 - u_2 v_1 = \beta^2(\phi_1\psi_2 - \phi_2\psi_1) + \beta l(\phi\psi_2 - \phi_2\psi) + \beta m(\phi_1\psi - \phi\psi_1)$, whence $(n-1)(u_1 v_2 - u_2 v_1)$

$$= (n-1)\beta^2(\phi_1\psi_2 - \phi_2\psi_1) + \beta(lx + my)(\phi_1\psi_2 - \phi_2\psi_1)$$
$$= n\beta^2(\phi_1\psi_2 - \phi_2\psi_1).$$

It follows from what has been said, that the discriminant of the Jacobian of u, v must contain R their resultant as a factor; since whenever R vanishes, the Jacobian has two equal roots. Thus in the case of two quadratics,

$$(a, b, c\!\!\;\rangle\!\!\;x, y)^2, \quad (a', b', c'\!\!\;\rangle\!\!\;x, y)^2,$$

the Jacobian is $\quad (ab')\, x^2 + (ac')\, xy + (bc')\, y^2$,

whose discriminant is $(ab')(bc') - 4(ac')^2$, which is the eliminant of the two quadratics. In the case of quantics of higher order, the discriminant of the Jacobian will, in addition to the resultant, contain another factor, the nature of which will appear from the following articles.

---

* In like manner, for a ternary quantic, the Jacobian of U, V, W is the locus of the double points of all curves of the system $U + kV + lW$ which have double points. And so in like manner for quantics with any number of variables.

SYSTEMS OF QUANTICS. 147

176. It has been said that we can always determine k, so that $u + kv$ shall have a square factor. But since two conditions must be fulfilled, in order that $u + kv$ may have a cube factor, k cannot be determined so that this shall be the case unless a certain relation connect the coefficients of u and v. *This condition will be of the order* $3(n-2)$ *in the coefficients both of u and v.*

If $(x - \alpha)^3$ be a factor in $u + kv + lw$, $x - \alpha$ will be a factor in the three second differential coefficients, or $x = \alpha$ will satisfy the equations

$$u_{11} + kv_{11} + lw_{11} = 0, \quad u_{12} + kv_{12} + lw_{12} = 0, \quad u_{22} + kv_{22} + lw_{22} = 0,$$

whence eliminating k and l, $x = \alpha$ will satisfy the equation

$$\begin{vmatrix} u_{11}, & v_{11}, & w_{11} \\ u_{12}, & v_{12}, & w_{12} \\ u_{22}, & v_{22}, & w_{22} \end{vmatrix} = 0.$$

If then we use the word treble-point in a sense analogous to that in which we used the word double-point (Art. 174), we see that the equation which has been just written, gives the treble points of the system $u + kv + lw$; and since the equation is of the degree $3(n-2)$, there may be $3(n-2)$ such treble points. But we could find the number of treble points otherwise. Suppose we have formed the condition that $u + kv$ should admit of a treble point, and that this condition is of the order p in the coefficients of a. If in this condition we substitute for each coefficient (a) of u, $a + la''$, we get an equation of the degree p in l; and therefore p values of l will be found to satisfy it. In other words p quantics of the system $u + kv + lw$ will have a treble point. It follows then from what has just been proved that $p = 3(n-2)$. And the same argument proves that the condition in question is of the order $3(n-2)$ in the coefficients of v.

This condition is evidently a combinant; for if it is possible to give such a value to k, as that $u + kv$ shall have a cubic factor, it must be possible to determine k, so that $(u + av) + kv$ shall have a cube factor.

177. If $u + kv$ have a cube factor $(x - \alpha)^3$, then the Jacobian of u and v will contain the square factor $(x - \alpha)^2$. For the two differentials $u_1 + kv_1$, $u_2 + kv_2$ will evidently contain this square

factor, and therefore it will appear also in the Jacobian which may be written $(u_1 + kv_1) v_2 - (u_2 + kv_2) v_1$. If then $S = 0$ be the condition that $u + kv$ may have a cube factor, S will be a factor in the discriminant of the Jacobian, since if $S = 0$ the Jacobian has two equal roots, and therefore its discriminant vanishes.

If R be the resultant, the discriminant of the Jacobian can only differ by a numerical factor from RS. For since the Jacobian is of the degree $2(n-1)$, its discriminant is of the degree $2\{2(n-1)-1\}$ in its coefficients, which are of the first order in the coefficients of both u and v. Now R is of the order n in each set of coefficients, S of the order $3(n-2)$. Both these are factors in the discriminant; and it can have no other since
$$n + 3(n-2) = 2\{2(n-1) - 1\}.$$

178. If we form the discriminant of $u + kv$, this considered as a function of k will have a square factor whenever u and v have a common factor. In fact (Art. 107) the discriminant of $u + kv$ will be of the form $(a + ka')\phi + (b + kb')^2 \psi$. But if u and v have a common factor, we can linearly transform u and v so that this factor shall be y, that is to say, so that both a and a' shall vanish. The discriminant will therefore have the square factor $(b + kb')^2$; and since the form of the discriminant is not affected by a linear transformation of the variables, it always has a square factor in the case supposed.

It follows that if we form the discriminant of $u + kv$, and then the discriminant of this again considered as a function of k, the latter will contain as a factor R the resultant of u and v. For it has been proved that when $R = 0$, the function of k has two equal roots, and therefore its discriminant vanishes. For example, the discriminant of a quadratic $ac - b^2$, becomes by the substitution of $a + ka'$ for a, &c.
$$(ac - b^2) + k(ac' + ca' - 2bb') + k^2(a'c' - b'^2),$$
whose discriminant is
$$(ac - b^2)(a'c' - b'^2) - 4(ac' + ca' - 2bb')^2.$$
But this is only a form in which, as Dr. Boole has shewn, the resultant of the two quadratics $(a, b, c\mathbf{\Sigma}x, y)^2$, $(a', b', c'\mathbf{\Sigma}x, y)^2$ can be written. This form, all the component parts of which

are invariants, is sometimes more convenient than that previously given. In the case of quantics of higher order, the discriminant of the discriminant will have R as a factor, but will have other factors besides.

179. If u have either a cube factor or two distinct square factors, the discriminant of $u + kv$ will be divisible by k^2. For if the discriminant of u be Δ, that of $u + kv$ is

$$\Delta + k\left(a'\frac{d\Delta}{da} + b'\frac{d\Delta}{db} + \&c.\right) + \&c.$$

Now when u has a square factor Δ vanishes; and it appears from the expressions in Art. 110, that if either three roots of u are equal $\alpha = \beta = \gamma$, or two distinct pairs be equal $\alpha = \beta$, $\gamma = \delta$, then all the differentials of Δ, $\frac{d\Delta}{da}$, &c. vanish; and therefore the coefficient of k in the expression just given vanishes. The discriminant therefore contains k^2 as a factor. It is evident hence that if $u + av$ have a cube or two square factors, the discriminant of $u + kv$ will be divisible by $(k-a)^2$; since $u + kv$ may be written $u + av + (k-a)v$. If then as before, $S = 0$ express the condition that the series $u + kv$ may include one quantic having a cube factor; and if $T = 0$ be the condition that it should include one having two square factors, both S and T will be factors in the discriminant with respect to k of the discriminant of $u + kv$. For we have just seen that the discriminant has a square factor if either $S = 0$, or $T = 0$. We proved in the last article that the discriminant has R as a factor; and in fact the discriminant will be, as Mr. Cayley has observed, RS^3T^2. I do not know whether there is any more rigid proof of this, than that we see that there is no other case in which the discriminant of $u + kv$ has a square factor, that we find in the case of the third and fourth degrees that S and T enter in the form S^3, T^2; and that we can thus account for the order in general. For the discriminant of $u + kv$ is of the order $2(n-1)$ in k, and the coefficients are of the order $0, 1 \ldots 2(n-1)$ in the coefficients of either quantic. The discriminant then with respect to k will be of the order $2(n-1)(2n-3)$ in the coefficients of either quantic. But R is of the order n, S of the order

150   SYSTEMS OF QUANTICS.

$3(n-2)$, and it will be proved in a subsequent lesson that T is of the order $2(n-2)(n-3)$, and
$$2(n-1)(2n-3) = n + 9(n-2) + 4(n-2)(n-3).$$

180. It was stated (Art. 173) that every combinant of u, v becomes multiplied by a power of $(\lambda\mu' - \lambda'\mu)$ when we substitute $\lambda u + \mu v$, $\lambda'u + \mu'v$ for u, v. It will be useful to prove otherwise that the eliminant of u, v has this property. First, let it be observed that if we have any number of quantics, one of which is the product of several others, u, v, $ww'w''$, their resultant is the product of the resultants (uvw), (uvw'), (uvw''). For when we substitute the common roots of u, v in the last and multiply the results, we evidently get the product of the results of making the same substitution in w, w', w''. Again, the resultant of u, v, kw is the resultant of u, v, w multiplied by k^{mn} since the coefficients of w enter into the resultant in the degree mn. If now $R(u, v)$ denote the resultant of u, v, which are supposed to be both of the same degree n, we have
$$\mu'^n R(\lambda u + \mu v, \lambda'u + \mu'v) = R(\lambda\mu'u + \mu\mu'v, \lambda'u + \mu'v)^*$$
$$= R\{(\lambda\mu' - \lambda'\mu)u, \lambda'u + \mu'v\} = (\lambda\mu' - \lambda'\mu)^n R(u, \lambda'u + \mu'v)$$
$$= (\lambda\mu' - \lambda'\mu)^n \mu'^n R(u, v),$$
whence $R(\lambda u + \mu v, \lambda'u + \mu'v) = (\lambda\mu' - \lambda'\mu)^n R(u, v)$.

By the same method it can be proved that the eliminant of $\lambda u + \mu v + \nu w$, $\lambda'u + \mu'v + \nu'w$, $\lambda''u + \mu''v + \nu''w$ is $(\lambda\mu'\nu'')^{n^2}$ times that of u, v, w, and so on.

181. If U, V be functions of the orders m and n respectively in u, v, which are themselves functions of x, y of the order p, and if D be the result of eliminating u, v, between U, V; then the result of eliminating x, y between U, V will be D^p times the mn^{th} power of the resultant of u, v. For U may be resolved into the factors $u - \alpha v$, $u - \beta v$, &c., and V into $u - \alpha'v$, $u - \beta'v$, &c. And, Art. 180, the resultant of U, V will be the product of all the separate resultants $u - \alpha v$, $u - \alpha'v$. But one of these is $(\alpha - \alpha')^p R(u, v)$. There are mn such resultants. When

---

* The resultant of $u + kv$, v, being the same as the resultant of u, v, Art. 173, we next subtract μ times the second quantic from the first.

therefore we multiply all together, we get the mn^{th} power of $R(u, v)$ multiplied by the pth power of $(\alpha - \alpha')(\alpha - \alpha'')$, &c. But this last is the eliminant of U, V with respect to u, v.

182. Similarly, let it be required to find the discriminant, with respect to x, y, of U, where U is a function of u, v. First, let it be remarked (see Art. 106) that the discriminant of the product of two binary quantics u, v is the product of the discriminants of u and v multiplied by the square of their resultant.

If then $U = (u - \alpha v)(u - \beta v)$, &c. the discriminant of U will be the product of the discriminants of $u - \alpha v$, $u - \beta v$, &c. by the square of the product of all the separate resultants $u - \alpha v$, $u - \beta v$. But, as before, any of these will be $(\alpha - \beta)^p R(u, v)$. If then m be the degree of U considered as a function of u, v; there will be $\tfrac{1}{2}m(m-1)$ separate resultants, and the square of the product of all will be $(\alpha - \beta)^{2p}(\alpha - \gamma)^{2p}$, &c. $\times R^{m(m-1)}(u, v)$. But $(\alpha - \beta)^2 (\alpha - \gamma)^2$, &c. is the discriminant of U considered as a function of u, v. If then we call this Δ, we have proved that the product of the squares of the separate resultants is $\Delta^p R^{m(m-1)}$. Let us now consider the product of the discriminants of $u - \alpha v$, $u - \beta v$, &c.; this is the result of eliminating α between the discriminant of $u - \alpha v$, which is a quantic of the order $2(p-1)$ in α, and the quantic of the mth order got by substituting $u = \alpha v$ in U. Or this product has been otherwise represented by Mr. Sylvester. If a_0, b_0 be the coefficients of x^p in u, v, then (Art. 104) the resultant of $u - \alpha v$, $u_1 - \alpha v_1$ will be $a_0 - \alpha b_0$ times the discriminant of $u - \alpha v$. But

$$R(u, v) R(u - \alpha v, u_1 - \alpha v_1) = R(u - \alpha v, u_1 v - \alpha v v_1) = R(u - \alpha v, u_1 v - u v_1).$$

Now (Art. 175) $p(u_1 v - u v_1) = yJ$ where J is $u_1 v_2 - u_2 v_1$, and $R(u - \alpha v, y)$ is $a_0 - \alpha b_0$. It appears thus that the discriminant of $u - \alpha v$ differs only by a numerical factor from the resultant of $(u - \alpha v, J)$ divided by $R(u, v)$. The product then of all the discriminants will be the resultant of J and the product $u - \alpha v$, $u - \beta v$, &c.; in other words, the resultant of U, J divided by the mth power of $R(u, v)$. Thus we have Mr. Sylvester's result (*Comptes Rendus*, LVIII., 1078) that the discriminant of U with respect to xy is $\Delta^p R(u, v)^{m^2 - 2m} R(U, J)$. But it will be observed that the result expressed thus is not in its most reduced form since $R(U, J)$ contains the factor $R(u, v)^m$.

183. It is not my purpose in this volume to enter with any detail into the theory of ternary and quaternary quantics. For as all this theory has a geometrical meaning, it will be naturally treated of in geometrical treatises; for instance, in my *Geometry of Three Dimensions*, and in the New Edition, which I hope to publish of my *Higher Plane Curves*. It will be proper however here to shew what corresponds in the case of ternary and quaternary quantics to the theory just explained for systems of binary quantics. Let then u and v be two ternary quantics, and let us suppose that we have formed the discriminant of $u + kv$. Then this discriminant considered as a function of k will have a square factor; in the first place, if the curves represented by u and v touch each other. For we have seen (Art. 113) that if the equation of a curve be $az^n + nbz^{n-1}x + ncz^{n-1}y + \&c. = 0$, its discriminant is of the form $a\theta + b^2\phi + bc\psi + c^2\chi$. The discriminant then of $u + kv$ will be of the form $(a + ka')\theta + (b + kb')^2\phi + \&c.$ But if we take for the point xy, a point common to u and v, both a and a' will vanish; and if we take the line y for the common tangent, both b and b' vanish; and the discriminant will be of the form $(c + kc')^2 \chi$; and therefore will always have a square factor in the case supposed.

184. Again, the discriminant will have a square factor if u have either a cusp or two double points. The discriminant Δ of a ternary quantic is the condition that u_1, u_2, u_3 should have a common system of values. If, however, we have either two double points, or a cusp, u_1, u_2, u_3 will have two systems of common values, distinct or coincident, and therefore (Art. 99) not only will Δ vanish, but also its differentials with respect to all the coefficients of u. The discriminant then of $u + kv$, being in general $\Delta + k\left(\dfrac{a'd\Delta}{da} + \dfrac{b'd\Delta}{db} + \&c.\right) + \&c.$ will in this case be divisible by k^2. And as in Art. 179, it will be divisible by $(k - a)^2$ if the curve $u + av$ have either a cusp or two double points. Let then $R = 0$ be the condition that u and v should

touch; $S=0$ the condition that among the systems of curves $u+kv$ should be included one having a cusp; and $T=0$ the condition that there should be included one having two double points, it has been proved that R, S, T are all factors in the discriminant of the discriminant of $u+kv$, considered as a function of k. In fact this discriminant will be RS^3T^2.

As in Art. 179, I can only prove this by showing that the order of the discriminant is thus accounted for. The discriminant of the ternary quantic $u+kv$ is of the order $3(n-1)^2$ in k, and the extreme terms are of the same order in the coefficients of u and v respectively. When then we form the discriminant of this again considered as a function of k, a specimen term of it will be the product of these two extreme terms raised to the power $3(n-1)^2-1$. Hence the order of this discriminant in the coefficients of both u and v will be $3(n-1)^2(3n^2-6n+2)$. Now it will be proved that R is of the order $3n(n-1)$, S of the order $12(n-1)(n-2)$, T of the order $\frac{3}{2}(n-1)(n-2)(3n^2-3n-11)$. But

$$3(n-1)^2(3n^2-6n+2)$$
$$=3n(n-1)+36(n-1)(n-2)+3(n-1)(n-2)(3n^2-3n-11).$$

185. In this place we only give the proof of the order of R, which, as expressing the condition that the curves represented by u and v should touch, has been called, by Mr. Cayley, their *tact-invariant*.

I gave, in 1856, the following method of finding this invariant (*Quarterly Journal*, Vol. I., p. 339). Form the determinant

$$\begin{vmatrix} u_1, & u_2, & u_3 \\ v_1, & v_2, & v_3 \\ \alpha, & \beta, & \gamma \end{vmatrix},$$

and we have the locus of points whose polars, with respect to u and v, intersect on an arbitrary line $\alpha x+\beta y+\gamma z$. If among these points be one common to both curves, its polars will be the tangents at that point, which can have a point common with $\alpha x+$ &c., only on the supposition either that this line passes through the point common to the two curves, or else

x

that the tangents are identical. If then we eliminate the variables between u, v, and the determinant above written which is of the order $m + n - 2$ in the variables and of the first order in the coefficients of each curve, the result will be R multiplied by the result of elimination between u, v and $ax + \&c$. Now the complete resultant is of the order mn in α, β, γ; it contains the coefficients of u in the order

$$n(m + n - 2) + mn = n(2m + n - 2),$$

and in like manner those of v in the degree $m(2n + m - 2)$. The resultant of u, v, $ax + \&c$. contains α, β, γ in the degree mn, the coefficients of u in the degree n, and those of v in the degree m. Hence R contains the coefficients of u in the degree $n(2m + n - 3)$ and those of v in the degree $m(2n + m - 3)$. When m and n are equal we get the number already stated $3n(n-1)$.

This result may be otherwise found as follows: The order of the tact-invariant in the coefficients of v is evidently the same as the number of curves of the form $u + \lambda v + \mu w$ which can be drawn to touch u, where w is another curve of the same order as v. But the point of contact with u is easily seen to be a point on the Jacobian of u, v, w. And this being a curve of the order $(m + 2n - 3)$ meets u in $m(m + 2n - 3)$ points.*

186. The theorem given, Art. 106, for the discriminant of the product of two binary quantics cannot be extended to ternary quantics; for the discriminant of the product of two will, in this case, vanish identically. In fact, the discriminant is the condition that a curve should have a double point; and a curve made up of two others must have double points; namely, the intersections of the component curves. Or, without any geometrical considerations, the discriminant of uv is the condition that values of the variables can be found to satisfy simultaneously the differentials $uv_1 + vu_1$, $uv_2 + vu_2$, &c. But these will all be satisfied by any values which satisfy

---

* I reserve for a geometrical treatise an account of the modifications which the theorems of this chapter receive when u and v are not the most general quantics of their degree, but denote curves having double points.

simultaneously u and v; and such values can always be found when there are more than two variables.

But the theorem of Art. 106 may directly be extended to tact-invariants. The condition that u should touch a compound curve vw will evidently be fulfilled if u touch either v or w, or go through an intersection of either. For an intersection counts, as has been said, as a double point on the complex curve; and a line going through a double point of a curve is to be considered doubly as a tangent. Hence if $T(u, v)$ denote the tact-invariant of u, v, we have

$$T(u, vw) = T(u, v)\, T(u, w)\, \{R(u, v, w)\}^2,$$

where $R(u, v, w)$ is the resultant of u, v, w. And the result may be verified by comparing the order in which the coefficients of u, v, or w occur in these invariants. Thus for the coefficient of u_1 we have

$$(n+p)(n+p+2m-3) = n(n+2m-3) + p(p+2m-3) + 2np.$$

187. What was said, Art. 185, as to the tact-invariant of two ternary quantics may be extended almost word for word to that of a system of three quaternary quantics, or of $k-1$, k-ary quantics. Thus the tact-invariant of three quaternary quantics expresses the condition that two of the mnp points of the surfaces represented by them coincide; or, in other words, that at any point of intersection the three tangent planes to the surfaces have a common line. And it is found, as in Art. 185, by dividing the resultant of v, v, w, and

$$\begin{vmatrix} u_1, & u_2, & u_3, & u_4 \\ v_1, & v_2, & v_3, & v_4 \\ w_1, & w_2, & w_3, & w_4 \\ \alpha, & \beta, & \gamma, & \delta \end{vmatrix}$$

by the resultant of u, v, w and $\alpha x + \beta y + \&c$. In this way it is seen that the tact-invariant contains the coefficients of u in the degree $np(2m+n+p-4)$, a result which is confirmed by observing that this is the number of points in which the intersection of vw meets the Jacobian of u, u', v, w. See *Geometry of Three Dimensions*, p. 438. And, in like manner, in general the tact-invariant of $k-1$, k-ary quantics u, v, w, &c. contains the coefficients of n in the degree found by multiplying the product of n, p, &c. by $2m + n + p +$ &c. $- k$.

188. In order to complete the subject we give here the theory of the tact-invariant of two quaternary quantics,* although we shall have to use principles which are to be established in a subsequent Lesson. The same argument as that used before shews that the order, in the coefficients of U, of the tact-invariant of U, W is the same as the number of quantics of the system $U + \lambda V$ which can touch W; U and V being of the same order. But comparing the coefficients of the tangent planes of $U + \lambda V$ and W we see that the point of contact must satisfy the system, equivalent to two conditions,

$$\begin{vmatrix} U_1, & U_2, & U_3, & U_4 \\ V_1, & V_2, & V_3, & V_4 \\ W_1, & W_2, & W_3, & W_4 \end{vmatrix} = 0.$$

It will be shewn in the Lesson on the order of systems of equations, that this system is of the order

$$\lambda^2 + \mu^2 + \nu^2 + \mu\nu + \nu\lambda + \lambda\mu,$$

where λ, μ, ν are the orders of U_1, V_1, W_1, &c., that is, in the present case $m - 1$, $m - 1$, $n - 1$. The order is then

$$n^2 + 2mn + 3m^2 - 4n - 8m + 6.$$

And the required order of the tact-invariant is n times this number; since the point of contact is got by combining with these conditions $W = 0$, which it must also satisfy.

189. There is now not the least difficulty in forming the general theory of the class of invariants we have been considering, to which Mr. Sylvester proposes to give the name of *osculants*. Let there be i quantics, U, V, W, &c. in k variables; then the osculant is the condition that for the same system of values which satisfy U, V, &c. the tangential quantics $xU_1' + yU_2' +$ &c., &c. shall be connected by an identical relation

$$\lambda (xU_1' + \&c.) + \mu (xV_1' + \&c.) + \nu (xW_1' + \&c.) + \&c. = 0.$$

---

* See *Geometry of Three Dimensions*, p. 439.

In other words, the osculant is the condition that the equations $U=0$, $V=0$, &c., and also the system

$$\begin{Vmatrix} U_1, & U_2, & U_3, & \&c. \\ V_1, & V_2, & V_3, & \&c. \\ W_1, & W_2, & W_3, & \&c. \end{Vmatrix} = 0$$

can be simultaneously satisfied. This latter system having k columns and i rows is equivalent to $k-i+1$ equations; therefore this system combined with the given i equations is apparently equivalent to $k+1$ equations in k variables. It is really, however, only equivalent to k equations; for writing $U=0$ in the form $xU_1 + yU_2 + \&c. = 0$, and similarly for V, &c., we see that when the system of determinants is satisfied, and all but one of the quantics U, V, W, &c., the remaining one must be satisfied also. The system then being equivalent to k equations in k variables cannot be simultaneously satisfied unless a certain condition be fulfilled. The order of this condition, in the coefficients of U, is found by the same method as in the last article. We write for U, $U+\lambda u$, and we examine how many values of the variables can simultaneously satisfy the $i-1$ equations V, W, &c., and the system equivalent to $k-1$ equations

$$\begin{Vmatrix} U_1, & U_2, & U_3, & \&c. \\ u_1, & u_2, & u_3, & \&c. \\ V_1, & V_2, & V_3, & \&c. \\ W_1, & W_2, & W_3, & \&c. \end{Vmatrix} = 0.$$

The order of the $i-1$ equations V, W, &c. is the product of their degrees n, p, &c.; and the order of the osculant in the coefficients of U is the product of this number by the order of the system of determinants, which is found by the rule given in the subsequent Lesson on the order of systems of equations.

When we are given but one quantic, the osculant is the discriminant; when we are given k quantics in k variables, the osculant is the resultant. The theorem of Art. 106 may be extended to osculants in general; viz. that if we form the osculant of $k-1$ quantics in k variables, and if the last be the product of two quantics U, V, then the osculant of the entire system will be the product of the osculant of the system of the other

158   APPLICATIONS TO BINARY QUANTICS.

$k-2$ with U, that of the system of $k-2$ with V, and the square of the resultant of all the quantics.

It would, next in order, be proper to discuss the form of the discriminant of the Jacobian of $\lambda U + \mu V + \nu W$, where U, V, W are ternary quantics, and that of the discriminant of the discriminant with respect to λ, μ, ν. This theory, when U, V, W are quadrics, will be given in Lesson XVIII. In other cases the latter discriminant vanishes identically. I do not know the general theory of the former.

LESSON XVII.

APPLICATIONS TO BINARY QUANTICS.

190. HAVING now explained the most essential parts of the general theory, we wish in this lesson to illustrate its application by examining in some of the simplest cases the different invariants and covariants which a quantic may possess. The method of Art. 117 shows that a binary quantic has in general $n-2$ independent invariants. For it was there proved, that there may be $n-3$ *absolute* invariants; and since we see, as in Art. 118, that from any two ordinary invariants an absolute invariant can be deduced, the number of ordinary invariants is at most one more. If we have two invariants S and T of the same degree; then $S + aT$ (where a is any numerical factor) will of course be also an invariant. We do not consider invariants included under this form as *new* invariants, nor as essentially distinct from the invariants S and T. Nor again, if S were of the second degree, and T of the third, would $S^3 + aT^2$ be said to be a new invariant essentially distinct from S and T. But if another invariant were not rationally expressible in terms of S and T, but only connected with them by an equation such as $R^2 = S^3 + aT^2$, then we speak of R as a new invariant distinct from S and T, though not independent of them. Thus then, though the number of *independent* invariants of a quantic is, as has been said, $n-2$, the number of *distinct* invariants is unlimited after we pass the sixth degree.

APPLICATIONS TO BINARY QUANTICS.   159

If we have a system of two quantics, of the degrees m and n respectively; proceeding still by the method of Art. 117, the number of equations to be satisfied being $m+n+2$, and the number of constants at our disposal being still only 4, we find that there are $m+n-2$ absolute, and $m+n-1$ ordinary, independent invariants of the system. That is to say, there will be in general three new independent invariants of the system in addition to the $m-2$ and $n-2$ independent invariants of the quantics considered separately. If one of the quantics be either linear or a quadratic, there will be only two new independent invariants. This is because the number $n-2$ does not express the number of independent invariants of a linear or quadratic quantic, that number being in both cases one more; that is to say, 0 in the one case and 1 in the other. The number of other invariants will therefore be one less than in the general case.

The invariants of any quantic and of the linear system $x\xi + y\eta + $ &c. (see Art. 130), may be regarded as contravariants of the given quantic; and in binary quantics contravariants may be reduced to covariants by changing ξ and η into y and $-x$. It has been shewn then that a binary quantic has in addition to its invariants, only two independent covariants; or, since we may take the quantic itself for one of these covariants, that all covariants may be expressed (though not necessarily expressed rationally) in terms of the quantic, its invariants, and one covariant.

We now proceed to enumerate the fundamental invariants and covariants of all the most simple systems.

191. *The quadratic.* We have already stated the principal points in the theory of the quadratic form $(a, b, c\!\!\;)\!(x, y)^2$. It has but one independent invariant (Art. 118), viz. the discriminant $ac - b^2$. Any other invariant must be a power of this $(ac - b^2)^m$. We have already showed (Art. 153) that it follows by Hermite's law of reciprocity, that only quantics of even degree have invariants of the second order whose symbol is $\overline{12}^{2m}$. If we make $y=1$ in the quantic, it denotes geometrically a system of two points on the axis of x, and the vanishing of the discriminant expresses the condition that these points should coincide (Art. 174).

In like manner the system of two quadratics

$$(a, b, c\!\!\not{\ }\!x, y)^2, \quad (a', b', c'\!\!\not{\ }\!x, y)^2,$$

has the invariant $\overline{12}^2$ or $ac' + a'c - 2bb'$. When each quantic is taken to represent a pair of points in the manner just stated, the vanishing of this invariant expresses the condition (see *Conics*, p. 291) that the four points should form a harmonic system, the two points represented by each quantic being conjugate to each other. We have also proved (Art. 174) that the covariant $\overline{12}$ (or the Jacobian of the system) represents geometrically the foci of the system in involution determined by the four points.

The eliminant of the system may be written in either of the forms

$$(ac' - ca')^2 + 4(ba' - b'a)(bc' - b'c),$$

or $\quad (ac' + ca' - 2bb')^2 - 4(ac - b^2)(a'c' - b'^2).$

Lastly, given a system of three quadratics

$$(a, b, c\!\!\not{\ }\!x, y)^2, \quad (a', b', c'\!\!\not{\ }\!x, y)^2, \quad (a'', b'', c''\!\!\not{\ }\!x, y)^2,$$

the vanishing of the determinant

$$\begin{vmatrix} a, & b, & c \\ a', & b', & c' \\ a'', & b'', & c'' \end{vmatrix}$$

expresses the condition that the three pairs of points represented by the quadratics shall form a system in involution (*Conics*, p. 296).

192. *The cubic.* We come next to the concomitants of the cubic

$$(a, b, c, d\!\!\not{\ }\!x, y)^3.$$

It has but one invariant (Art. 118), viz. the discriminant

$$a^2d^2 + 4ac^3 + 4db^3 - 3b^2c^2 - 6abcd.$$

The Hessian $\overline{12}^2$ is

$$(ac - b^2)x^2 + (ad - bc)xy + (bd - c^2)y^2,$$

which may also be written as a determinant

$$\begin{vmatrix} a, & b, & c \\ b, & c, & d \\ y^2, & -xy, & x^2 \end{vmatrix}.$$

APPLICATIONS TO BINARY QUANTICS. 161

This has the same discriminant as the cubic itself, see Art. 164. If the roots of the cubic be α, β, γ, then the Hessian is $\Sigma (x-\alpha)^2 (\beta-\gamma)^2$ (see Art. 133). The cubic covariant $\overline{12}^2.\overline{13}$, or the evectant of the discriminant, is (see Art. 138)

$(a^2d-3abc+2b^3,\ abd+b^2c-2ac^2,\ 2b^2d-acd-bc^2,\ 3bcd-ad^2-2c^3\!\!\ \!\!\ \!\!\ )\!(x,y)^3.$

This cubic may be geometrically represented as follows:—If we take the three points represented by the cubic itself, and take the fourth harmonic of each with respect to the other two, we get three new points which will be the geometrical representation of the covariant in question. This theorem is suggested by its being evident on inspection that if the given cubic take the form $xy(x+y)$, then $x-y$ will be a factor in the covariant, as appears by making $a=d=0$, $b=c=1$ in its equation. But $x+y$, $x-y$ are harmonic conjugates with respect to x and y. Now, if $\alpha, \beta, \gamma, \delta$ denote the distances from the origin of four points on the axis of x, any harmonic or anharmonic relation between them is expressed by the ratio of the products $(\alpha-\beta)(\gamma-\delta)$ and $(\alpha-\gamma)(\beta-\delta)$: and this ratio (see Art. 132) is unaltered by a linear transformation; that is, when for each distance α we substitute $\dfrac{\lambda\alpha+\mu}{\lambda'\alpha+\mu'}$. Such relations, then, being unaltered by linear transformations, if proved to exist in one case, exist in general. We find that the other factors in the evectant of $xy(x+y)$ are $x+2y$, $2x+y$, so that our result may be written symmetrically that the evectant of xyz (where x, y, z are connected by the linear relation $x+y+z=0$) is $(x-y)(y-z)(z-x)$. These considerations lead us to the expression for the factors of the covariant in terms of the roots of the given cubic: for if δ be the distance from the origin of the point conjugate to α with respect to β and γ; solving for δ from the equation $\dfrac{2}{\alpha-\delta}=\dfrac{1}{\alpha-\beta}+\dfrac{1}{\alpha-\gamma}$ we get $\delta=\dfrac{\alpha\beta+\alpha\gamma-2\beta\gamma}{2\alpha-\beta-\gamma}$, whence the covariant must be

$\{(2\alpha-\beta-\gamma)x+(2\beta\gamma-\alpha\beta-\alpha\gamma)y\}\{(2\beta-\alpha-\gamma)x$
$+(2\gamma\alpha-\beta\gamma-\beta\alpha)y\}\{(2\gamma-\alpha-\beta)x+(2\alpha\beta-\gamma\alpha-\gamma\beta)y\},$

as may be verified by actual multiplication and substitution in terms of the coefficients of the equation.

193. When we wish to establish any relation between the preceding covariants and invariants, we use the canonical forms, which are, for $U = ax^3 + dy^3$, the discriminant $D = a^2d^2$; the Hessian $H = adxy$; and the cubicovariant, $J = ad(ax^3 - dy^3)$. Thus we can prove that the discriminant of J is the cube of the discriminant of U, for the discriminant of J in its canonical form is a^6d^6. Again, we have been led, by Art. 190, to foresee that J is not independent of U and H; and we can easily establish, by the help of the canonical form, the relation connecting them, due to Mr. Cayley, viz.

$$J^2 - DU^2 = -4H^3.$$

Mr. Cayley has used this equation to solve the cubic U, or, in other words, to resolve it into its linear factors. For, since $J^2 - DU^2$ is a perfect cube, we are led to infer that the factors $J \pm U\sqrt{D}$ will also be perfect cubes, and, in fact, the canonical form shows that they are $2a^2dx^3$ and $2ad^2y^3$. Now, since $xa^{\frac{1}{3}} + yd^{\frac{1}{3}}$ is one of the factors of the canonical form, it immediately follows that the factor in general is proportional to

$$(U\sqrt{D} + J)^{\frac{1}{3}} + (U\sqrt{D} - J)^{\frac{1}{3}},$$

a linear function which evidently vanishes on the supposition $U = 0$.

Ex. Let us take the same example as at p. 136, viz., $U = 4x^3 + 9x^2 + 18x + 17$. Here we have $D = 1600$, $J = 110x^3 - 90x^2y - 630xy^2 - 670y^3$, whence

$$U\sqrt{D} + J = 10(3x + y)^3; \quad U\sqrt{D} - J = 50(x + 3y)^3;$$

and the factors are $3x + y + (x + 3y) \sqrt[3]{5}$.

194. *System of cubic and quadratic.* Let these be

$$(a, b, c, d\mathbf{\chi}x, y)^3, \quad (A, B, C\mathbf{\chi}x, y)^2.$$

The simplest invariant of the system is got by combining the quadric with the Hessian of the cubic, and forming the intermediate invariant of this system of two quadratics, when we have

$$I = A(bd - c^2) - B(ad - bc) + C(ac - b^2).$$

The resultant of the system, formed by the method of Arts. 63 or 82, is

$$R = a^2C^3 - 6abBC^2 + 6acC(2B^2 - AC) + ad(6ABC - 8B^3)$$
$$+ 9b^2AC^2 - 18bcABC + 6bdA(2B^2 - AC)$$
$$+ 9c^2A^2C - 6cdBA^2 + d^2A^3.$$

These two may be taken as the fundamental invariants of the system. In comparing other invariants with these it is convenient to take A and $C=0$, which is equivalent to taking for x and y the two factors of the quadratic. I then reduces to $B(bc - ad)$ and R to $-8B^3ad$. The Jacobian of the system is

$$(Ab-Ba)x^3+(2Ac-Bb-Ca)x^2y+(Ad+Bc-2Cb)xy^2+(Bd-Cc)y^3.$$

But it has, besides, linear covariants. Thus, if we substitute differential symbols in the quadratic and operate on the cubic, we get

$$L_1 = (aC - 2Bb + cA)x + (bC - 2cB + dA)y,$$

and if we operate in like manner with this on the quadratic, we get

$$L_2 = \{aBC - b(2B^2 + AC) + 3cAB - dA^2\}x$$
$$+ \{aC^2 - 3bBC + c(AC + 2B^2) - dAB\}y,$$

which expressed in terms of the roots α, β, γ of the cubic and α', β' of the quadric, are

$$\Sigma(\alpha' - \alpha)(\beta' - \beta)(x - \gamma), \quad \Sigma(\alpha' - \alpha)(\beta' - \beta)(\alpha' - \gamma)(x - \beta').$$

The resultant of L_1 and L_2, when we make A and $C=0$, is proportional to B^3bc. If, then, Δ denote the discriminant $AC - B^2$ of the quadratic, we see that the resultant of L_1, L_2, expressed in terms of the fundamental invariants, is $R + 8\Delta I$.

There are other linear covariants got by writing for a, b, &c. the a, b, &c. of the cubi-covariant (as in Art. 138) of the given cubic.

If we eliminate between the linear covariant L_1 and the quadratic, we get the same result as if we eliminate between L_1 and L_2. But, if we eliminate between L_1 and the cubic, we get an invariant distinct from those already given. If we make A and $C=0$, this invariant is $B^3(ac^3 - db^3)$, which we see is not reducible to the previous forms. But its square can without difficulty be reduced to those forms. For, since the discriminant of the cubic is

$$D = a^2d^2 + 4ac^3 + 4db^3 - 3b^2c^2 - 18abcd,$$

sixteen times the square of the invariant now under consideration is

$$B^6(D - a^2d^2 + 3b^2c^2 + 18abcd)^2 - 64B^6adb^3c^3,$$

and we have only to write in this for Bbc, $I+Bad$, for B^2ad, $-\tfrac{1}{8}R$, and for B^2, $-\Delta$, when the whole is expressed in terms of the fundamental invariants.

195. *System of two cubics.* A system of two cubics $(a, b, c, d\!\int\! x, y)^3$, $(a', b', c', d'\!\int\! x, y)^3$ has for its simplest invariant (see Art. 136, Ex. 2) $(ad') - 3(bc')$; which is a combinant, and which we shall refer to as the invariant P. The properties of this system may be studied most conveniently by throwing the equations into the form $Au^3 + Bv^3 + Cw^3$, $A'u^3 + B'v^3 + C'w^3$, a form to which the two cubics can be reduced in an infinity of ways. For the cubics contain four constants each, or eight in all. And the form just written contains six constants explicitly; and u, v, w contains implicitly a constant each, since u stands for $x + \lambda y$, &c. The second form then is equivalent to one with nine constants, that is to say, one constant more than is necessary to enable us to identify it with the general form.

Any three binary quantics of the first degree are obviously connected by an identical relation of the form $\alpha u + \beta v + \gamma w = 0$. We shall suppose the constants α, β, γ to have been incorporated with u, v, w, so as to write the two cubics, $Au^3 + Bv^3 + Cw^3$, $A'u^3 + B'v^3 + C'w^3$, where $u + v + w = 0$.

Putting for w its value, and writing the cubics
$(A, -C, -C, -C, B-C\!\int\! u, v)^3$, $(A'-C', -C', -C', B'-C'\!\int\! u, v)^3$,
and forming the invariant P of the system, we find it to be

$$(AB') + (BC') + (CA').$$

The resultant of the system is found by solving between the equations $Au^3 + Bv^3 + Cw^3 = 0$, $A'u^3 + B'v^3 + C'w^3 = 0$, when we get $u^3 = (BC')$, $v^3 = (CA')$, $w^3 = (AB')$; and, substituting in the identity $u + v + w = 0$, the resultant is

$$(AB')^{\tfrac{1}{3}} + (BC')^{\tfrac{1}{3}} + (CA')^{\tfrac{1}{3}} = 0,$$

or $\quad \{(AB') + (BC') + (CA')\}^3 = 27(AB')(BC')(CA').$

Now, if we denote the two cubics by U and V, it has been proved, Art. 176, that there is an invariant, which we shall call Q, of the third order in the coefficients of each cubic, which expresses the condition of its being possible to determine λ so that $U + \lambda V$ shall be a perfect cube. Now this invariant is

APPLICATIONS TO BINARY QUANTICS. 165

identical with the product $(AB')(BC')(CA')$, which is of the same degree in the coefficients. For if any factor (AB') in this product vanish, $AV - A'U$ evidently reduces to the perfect cube $(AC')w^3$. It follows then that the resultant is of the form $P^3 - 27Q$.

196. If it were required to form directly the invariant Q for the form $(a, b, c, d\chi x, y)^3$, $(a', b', c', d'\chi x, y)^3$; we might proceed as follows. If $U + \lambda V$ can be a perfect cube, its three second differentials can be simultaneously satisfied; or

$$ax + by + \lambda(a'x + b'y) = 0,$$
$$bx + cy + \lambda(b'x + c'y) = 0,$$
$$cx + dy + \lambda(c'x + d'y) = 0.$$

Solving these equations linearly for x, y, λx, λy, and then equating the product of x by λy to the product of y by λx, we get, for the required condition,

$$\begin{vmatrix} a, & b, & a' \\ b, & c, & b' \\ c, & d, & c' \end{vmatrix} \times \begin{vmatrix} a', & b', & b \\ b', & c', & c \\ c', & d', & d \end{vmatrix} = \begin{vmatrix} a, & b, & b' \\ b, & c, & c' \\ c, & d, & d' \end{vmatrix} \times \begin{vmatrix} a', & b', & a \\ b', & c', & b \\ c', & d', & c \end{vmatrix};$$

or $\{b(bc') + c(ca') + d(ab')\} \times \{a'(cd') + b'(db') + c'(bc')\}$
$= \{b'(bc') + c'(ca') + d'(ab')\} \times \{a(cd') + b(db') + c(bc')\}$,

or $(bc')^3 + (ca')^2(cd') + (db')^2(ab') - 3(ab')(bc')(cd')$
$\qquad\qquad -(ad')(bc')^2 - (ad')(ab')(cd')$.

If, as in the last article, we give a, b, &c. the values $A, -C, -C$, &c. this would become $-(AB')(BC')(CA')$. If then to twenty-seven times this quantity we add $\{(ad') - 3(bc')\}^3$, we get the resultant in the form

$(ad')^3 - 9(ad')^2(bc') + 27(ca')^2(cd') + 27(db')^2(ab')$
$\qquad\qquad - 81(ab')(bc')(cd') - 27(ad')(ab')(cd')$,

a result which agrees with that of Art. 76, it being remembered that there the cubics were written without binomial coefficients.

197. We have, in Art. 195, formed the invariant P of the system $Ax^3 + By^3 + Cz^3$, $A'x^3 + B'y^3 + C'z^3$, by first reducing them to functions of two variables, and then calculating the

value of $(ad')-3(bc')$. We shall for the sake of establishing a useful general principle, give another way of making the same calculation. We know that we may substitute in any binary quantic $\frac{d}{dy}$, $-\frac{d}{dx}$ for x and y, and so obtain an invariantive operative symbol. Now when this change is made in a function expressed in terms of x, y, z, where z is $-(x+y)$, z will become $\frac{d}{dx}-\frac{d}{dy}$. And when the operation is performed on a function similarly expressed, since its differential with respect to x will be $\frac{d}{dx}+\frac{d}{dz}\frac{dz}{dx}$, or in virtue of the relation between x, y, z, $\frac{d}{dx}-\frac{d}{dz}$, we see that the rule may be expressed, that we may substitute in any covariant for x, y, z respectively

$$\frac{d}{dy}-\frac{d}{dz},\ \frac{d}{dz}-\frac{d}{dx},\ \frac{d}{dx}-\frac{d}{dy},$$

and so obtain an operative symbol which we may apply to any covariant expressed in terms of x, y, z, without first reducing it to a function of two variables. Thus, in the present case, we find the invariant P by operating on $A'x^3+B'y^3+C'z^3$, with

$$A\left(\frac{d}{dy}-\frac{d}{dz}\right)^3+B\left(\frac{d}{dz}-\frac{d}{dx}\right)^3+C\left(\frac{d}{dx}-\frac{d}{dy}\right)^3,$$

and the result only differs by a numerical factor from $(AB')+(BC')+(CA')$, as we found before.

In like manner we find that, in the symbolical notation, $\overline{12}$ as applied to a function expressed in terms of x, y, z, denotes

$$\left(\frac{d}{dx_1}\frac{d}{dy_2}-\frac{d}{dx_2}\frac{d}{dy_1}\right)+\left(\frac{d}{dy_1}\frac{d}{dz_2}-\frac{d}{dy_2}\frac{d}{dz_1}\right)+\left(\frac{d}{dz_1}\frac{d}{dx_2}-\frac{d}{dz_2}\frac{d}{dx_1}\right).$$

198. The Jacobian of the system

$$Ax^3+By^3+Cz^3,\ A'x^3+B'y^3+C'z^3,$$

may be found by means of the formula last given, or directly as follows. In virtue of the relation connecting z with x and y, the differentials of U with respect to x and y are proportional to Ax^2-Cz^2, By^2-Cz^2; and $U_1V_2-U_2V_1$, is

$$(AB')x^2y^2+(BC')y^2z^2+(CA')z^2x^2.$$

APPLICATIONS TO BINARY QUANTICS. 167

This is a biquadratic, for which we may calculate the two invariants noticed in p. 112, viz.

$S = ae - 4bd + 3c^2$, $T = ace + 2bcd - ad^2 - eb^2 - c^3$.

Putting in for z^2, $(x+y)^2$, and multiplying the Jacobian by six to avoid fractions, we get

$a = 6(CA')$, $b = 3(CA')$, $e = 6(BC')$, $d = 3(BC')$,

$c = (BC') + (CA') + (AB') = P$,

whence $S = 3P^2$, $T = 54Q - P^3$. We shall presently show that the discriminant of a biquadratic is $S^3 - 27T^2$. The discriminant of the Jacobian therefore, is proportional to $Q(P^3 - 27Q)$, which agrees with Art. 177.

199. If in general we form any invariant of $U + \lambda V$, and then form any invariant of this again considered as a function of λ, the result will be a combinant of the system U, V; that is to say, it will not be altered if we substitute $lU + mV$, $l'U + m'V$ for U, V. For, by this substitution, we get the corresponding invariant of $(l + \lambda l')U + (m + \lambda m')V$, which is equivalent to a linear transformation of λ, by which the invariants of the function of λ will not be altered. If then, in the case of two cubics, the discriminant of $U + \lambda V$ be $A + 4B\lambda + 6C\lambda^2 + 4D\lambda^3 + E\lambda^4$, and if it be required to calculate the invariants of this biquadratic, we may without loss of generality, take instead of U and V two quantics of the system $U + \lambda V$ which have square factors, taking x and y for these factors; and so write $U = ax^3 + 3bx^2y$, $V = 3cxy^2 + dy^3$. For this system we have $P = ad - 3bc$, $Q = b^2c^2(ad - bc)$; the resultant $P^3 - 27Q$ being $a^2d^2(ad - 9bc)$. Now, for this form, the biquadratic is $4ac^3\lambda + (a^2d^2 - 6abcd - 3b^2c^2)\lambda^2 + 4db^3\lambda^3$; or multiplying by six to avoid fractions $A = E = 0$, $B = 6ac^3$, $D = 6db^3$, $C = a^2d^2 - 6abcd - 3b^2c^2 = P^2 - 12b^2c^2$. Hence,

$S = 3C^2 - 4BD = 3P(P^3 - 24Q)$;

$T = 2BCD - C^3 = -(P^6 - 36P^3Q + 216Q^2)$,

whence the discriminant of the biquadratic $S^3 - 27T^2$ is proportional to $Q^3(P^3 - 27Q)$ which agrees with Art. 179. We infer from Art. 117, by counting the constants, that a system of two cubics cannot have more than five independent invariants;

and we see here that P and Q can be expressed, though not rationally, in terms of the five A, B, C, D, E.

200. We may also conveniently use the form $ax^3 + 3bx^2y$, $3cxy^2 + dy^3$, in examining the relations of invariants which are not combinants; but it will be necessary to verify by more general forms, that the relations found to exist in this particular case are true in general. The invariants are all of even order in the coefficients of the system; and there is no invariant of the second order but the combinant P. We may form several invariants of the fourth order, as follows. Form the Hessian of $U + \lambda V$, which will be

$$(\alpha, \beta, \gamma \vert x, y)^2 + \lambda (\alpha', \beta', \gamma' \vert x, y)^2 + \lambda^2 (\alpha'', \beta'', \gamma'' \vert x, y)^2,$$

where the absolute term and the coefficient of λ^2 are the Hessians of U and V respectively; and the coefficient of λ is an intermediate covariant. Now, we may form the quadratic invariant of any one, or any pair of these invariants, $4\alpha\gamma - \beta^2$, $2(\alpha\gamma' + \alpha'\gamma) - \beta\beta'$, &c., and these will all be invariants of the system. If we remember that the discriminant of the Hessian is the same as that of the quantic, and therefore identify

$$(\beta + \lambda\beta' + \lambda^2\beta'')^2 - 4(\alpha + \lambda\alpha' + \lambda^2\alpha'')(\gamma + \lambda\gamma' + \lambda^2\gamma''),$$

with $(A, B, C, D, E \vert 1, \lambda)^4$, we identify all these invariants with A, B, D, E except the two $\beta\beta'' - 2(\alpha\gamma'' + \gamma\alpha'')$, $\beta'^2 - 4\alpha'\gamma'$, which we shall call J, K, and which we find to be connected by the relation $2J + K = 6C$.

In the form $ax^3 + 3bx^2y + \lambda(3cxy^2 + dy^3)$, the Hessian is

$$(-b^2 + \lambda ac) x^2 + \lambda (ad - bc) xy + (\lambda bd - \lambda^2 c^2) y^2;$$

and we find $J = -2b^2c^2$, $K = a^2d^2 - 6abcd + b^2c^2$; and, using the value $6C = a^2d^2 - 6abcd - 3b^2c^2$, we have $6C = P^2 + 6J$, $K = P^2 + 4J$, relations which are without difficulty verified in general.

Lastly, if we form the cubicovariant, as in Art. 138, of $U + \lambda V$, the coefficients of the powers of λ give us four covariants of the same kind, viz. those of U and V, and two intermediate covariants. And if we form the combinant P of each pair of these cubics, we get six invariants of the sixth order in the coefficients. Four of these will, however, be only PA, PB, PD, PE. The remaining pair, viz. the combinants formed with the two extreme and the two intermediate covariants are respectively

$$L = b^3c^3 = 2Q + PJ \text{ and } M = -P^2 + 24Q.*$$

201. *The quartic.* We come next to the quartic, which, as we have seen, p. 112, has the two invariants

$$S = ae - 4bd + 3c^2 \text{ and } T = ace + 2bcd - ad^2 - eb^2 - c^3.$$

We have shown (Art. 169) that the quartic may be reduced to the canonical form $x^4 + 6mx^2y^2 + y^4$, and for this form these invariants are $S = 1 + 3m^2$, $T = m - m^3$.

These invariants, expressed as symmetric functions of the roots, are

$$S = \Sigma (\alpha - \beta)^2 (\gamma - \delta)^2, \quad T = \Sigma (\alpha - \beta)^2 (\gamma - \delta)^2 (\alpha - \gamma)(\beta - \delta),$$

or, more conveniently,

$$T = \{(\alpha - \beta)(\gamma - \delta) - (\alpha - \gamma)(\delta - \beta)\} \{(\alpha - \gamma)(\delta - \beta)$$
$$- (\alpha - \delta)(\beta - \gamma)\} \{(\alpha - \delta)(\beta - \gamma) - (\alpha - \beta)(\gamma - \delta)\}.$$

In the latter form it is easy to see that $T = 0$ is the condition that the four points represented by the quartic should form a harmonic system (*Higher Plane Curves*, p. 192). It was stated (Art. 168) that $T = 0$ is the condition that the quartic can be reduced to the form $x^4 + y^4$,† and that T can be expressed as a determinant

$$\begin{vmatrix} a, & b, & c \\ b, & c, & d \\ c, & d, & e \end{vmatrix}.$$

If M be the modulus of transformation, then (Art. 118) S and T become by transformation M^4S, M^6T, respectively; and the ratio $S^3 : T^2$ is *absolutely* unaltered by transformation.

202. *Every invariant of a quartic can be expressed as a rational function of S and T.* Since the quartic can be reduced to the canonical form $x^4 + 6mx^2y^2 + y^4$, by linear transformation,

---

* By substituting in the expression for the resultant $P^2 - 27Q$, the value of Q, $\frac{1}{4}(L - PJ)$, just obtained, we get an expression for the resultant given by Mr. Warren, *Quarterly Journal*, vol. VI., p. 237.

† Mr. Sylvester gives the name *catalecticant* to the invariant which expresses that a quantic of order $2n$ can be reduced to the sum of n powers of the degree $2n$.

which does not affect the invariant, it is sufficient to prove the theorem for that form. Now, in the first place, we say that any invariant which vanishes when $m = 0$, will also vanish when $m = \pm 1$. For the form $x^4 + y^4$ is changed, by the transformation of $x+y$ and $x-y$ for x and y, into the form $x^4 + 6x^2y^2 + y^4$, and by the change of $x+y \sqrt{(-1)}$, $x-y \sqrt{(-1)}$ for x and y, into $x^4 - 6x^2y^2 + y^4$. Thus, then, we see that if m be a factor in any invariant, $m^2 - 1$ will also be a factor; or the invariant will be divisible by $m - m^3$, that is to say by T.

Let us take now any invariant expressed in terms of the general coefficients a, b, c, d, e; and we say that if it does not vanish when we make $b = 0$, $c = 0$, $d = 0$, the part remaining must be a power of ae. For it evidently must be a symmetrical function of a and e; and it cannot be of any form such as $a^k + e^k$, since the weight of every term must be the same. Let this part remaining then be $a^k e^k$, and let us subtract from the given invariant $(ae - 4bd + 3c^2)^k$ or S^k: the remainder, then, will evidently vanish when we make $b = 0$, $c = 0$, $d = 0$; or, in the case of the canonical form for which b and d are always $= 0$, it will vanish when $m = 0$. By what has been proved then it must be divisible by T: that is to say, we have proved that the invariant is of the form $S^k + T\phi$. But, by the same argument, we prove that ϕ is of the form $S^{k'} + T\psi$; and so on: so that, by repeating the process, the invariant can altogether be expressed rationally as a function of S and T.

203. *To express the discriminant in terms of S and T.* It has been already remarked (Art. 107) that the discriminant of a quantic must vanish, if the first two coefficients a and b vanish; for, in that case, the quantic has a square factor being divisible by y^2. On the other hand it is also true that any invariant which vanishes when a and b are made $= 0$, must contain the discriminant as a factor. Such an invariant, in fact, would vanish whenever the quantic had any square factor $(x - \alpha y)^2$: for, by linear transformation, the quantic could be brought to a form in which this factor was taken for y, and in which therefore the coefficients a and $b = 0$. But an invariant which vanishes whenever any two roots of the quantic are equal, must when expressed in terms of the roots contain as a factor the

difference between every two roots: that is to say, must contain the discriminant as a factor.

It is easy now, by means of S and T, to construct an invariant which shall vanish when we make a and $b=0$. For on this supposition S becomes $3c^2$ and T becomes $-c^3$; therefore $S^3 - 27 T^2$ vanishes. Now this invariant of the sixth order in the coefficients is of the same order as that which we know (Art. 101) the discriminant to be. It must therefore be the discriminant itself, and not the product of the discriminant by any other invariant. The discriminant is therefore

$$(ae - 4bd + 3c^2)^3 - 27(ace + 2bcd - ad^2 - eb^2 - c^3)^2.$$

We can in various ways verify this result. For instance, it appears, from Art. 182,* that the discriminant of the canonical form $x^4 + 6mx^2y^2 + y^4$ is the square of the discriminant of the quadratic $x^2 + 6mxy + y^2$; that is to say, is $(1 - 9m^2)^2$. But

$$(1 - 9m^2)^2 = (1 + 3m^2)^3 - 27(m - m^3)^2.$$

204. We may also derive this result from the form $P^3 - 27Q$ of the resultant of a system of two cubics given, Art. 195. The combinant P of the system U_1, U_2,

$$ax^3 + 3bx^2y + 3cxy^2 + dy^3, \quad bx^3 + 3cx^2y + 3dxy^2 + ey^3,$$

is $(ae - bd) - 3(bd - c^2)$; that is to say, is no other than the invariant S of the biquadratic. And Q, if calculated, will be found to be a perfect square, namely, the square of the invariant T. We can give an *à priori* reason why Q should in this case be a perfect square. For $Q=0$ expresses the condition that for some value of μ, $U_1 + \mu U_2$ shall be a perfect cube. If this be so, we can (Art. 126) linearly transform so that it shall be the differential with regard to x which shall be the perfect cube: suppose $(x + \lambda y)^3$. Then the quantic itself must be of the form $(x + \lambda y)^4 + cy^4$; and in this case we have also $U_2 - \lambda U_1$ a perfect cube. Thus $Q=0$ expresses the condition that the quartic

---

* We may also see this directly, thus: The resultant of $ax^k + by^k$, $a'x^k + b'y^k$ is the kth power of $ab' - a'b$, since the substitution of each root of the first equation in the second gives $ab' - a'b$. Now the discriminant of $ax^4 + 6cx^2y^2 + ey^4$ is the resultant of $ax^3 + 3cxy^2$, $3cx^2y + ey^3$. If we substitute $x = 0$ in the second, and $y = 0$ in the first, we get results e, a, respectively, and the resultant of $ax^2 + 3cy^2$, $3cx^2 + ey^2$ is $(ae - 9c^2)^2$. The discriminant is therefore $ae(ae - 9c^2)^2$.

shall be the sum of two fourth powers; and, in this case, $\lambda U_1 + \mu U_2$ can *in two ways* be made a perfect cube.

We can also easily apply the theory of a system of two cubics by writing the quartic under a form more general than the canonical form, viz. $Ax^4 + By^4 + Cz^4$, where $x + y + z = 0$. In this case, then, we have $a = A + C$, $e = B + C$, $b = c = d = C$, and we easily calculate $S = BC + CA + AB$, $T = ABC$. But if we equate to nothing the two differentials, viz. $Ax^3 - Cz^3$, $By^3 - Cz^3$, we get x^3, y^3, z^3 respectively proportional to BC, CA, AB; and, substituting in $x + y + z = 0$, we get the discriminant in the form

$$(BC)^{\frac{1}{3}} + (CA)^{\frac{1}{3}} + (AB)^{\frac{1}{3}} = 0,$$

or $(BC + CA + AB)^3 - 27 A^2 B^2 C^2 = 0$, or $S^3 - 27 T^2 = 0$.

205. From the expression just given for the discriminant of a quartic in terms of S and T, can be derived the relation (Art. 193) which connects the covariants of a cubic.

If we multiply two quantics together, the invariants of the compound quantic will be invariants of the system formed by the two components. If then we multiply a quantic by $x\xi + y\eta$, the invariants of the compound will (Art. 130) be contravariants of the original quantic; and if we then change ξ and η into y and $-x$, will be covariants of it. If we apply this process to a cubic, the coefficients of the quartic so formed will be

ay, $\frac{1}{4}(3by - ax)$, $\frac{1}{2}(cy - bx)$, $\frac{1}{4}(dy - 3cx)$, $-dx$;

and the invariants S and T of this quartic are found to be the covariants $-\frac{3}{4}H$, $\frac{1}{16}J$ of the cubic. But the discriminant of the product of any quantic by $x\xi + y\eta$, by Art. 106, becomes, when treated thus, the discriminant of U multiplied by U^2. Expressing then the discriminant of the compound quartic in terms of its S and T, we get the relation connecting the H, J, and discriminant of the cubic.

206. The Hessian of the quartic is the evectant of T, and is

$(ac - b^2) x^4 + 2(ad - bc) x^3 y + (ae + 2bd - 3c^2) x^2 y^2$
$\qquad\qquad\qquad + 2(be - cd) xy^3 + (ce - d^2) y^4,$

which, for the canonical form, is

$$m(x^4 + y^4) + (1 - 3m^2) x^2 y^2.$$

If a quartic have a square factor x^2, this will be also a factor in the Hessian. For the second differential U_{22} contains x^2, and U_{12} contains x, therefore x^2 will be a factor in $U_{11}U_{22} - U_{12}^2$. If then a quartic have two square factors, both will be factors in the Hessian, which, being of the fourth degree, can therefore differ only by a numerical factor from the quartic itself. In fact, if a quartic have two square factors, by taking these for x^2 and y^2, the quartic may be reduced to the form cx^2y^2: but, by making a, b, d, e all $= 0$, the Hessian, as given above, reduces to $-3c^2x^2y^2$.

Thus then, by expressing that a quartic differs only by a factor from its Hessian, we get the system of conditions that the quartic shall have two square factors, viz.

$$\frac{ac-b^2}{a} = \frac{ad-bc}{2b} = \frac{ae+2bd-3c^2}{6c} = \frac{be-cd}{2d} = \frac{ce-d^2}{e},$$

a system equivalent to two conditions, as may be verified in different ways.

We have, in Art. 134, given other ways of forming these conditions. One is, to form the covariant

$$\Sigma (\alpha-\beta)(\alpha-\gamma)(\alpha-\delta)(x-\beta)^2(x-\gamma)^2(x-\delta)^2,$$

every term of which must vanish if any two pairs of roots become respectively equal. This covariant expanded is

$(a^2d + 2b^3 - 3abc, a^2e + 2abd - 9ac^2 + 6b^2c, 5abe - 15acd + 10b^2d,$
$10b^2e - 10ad^2, -5ade + 15bce - 10bd^2, -ae^2 - 2bde + 9c^2e - 6cd^2,$
$3cde - 2c^3 - be^2)(x, y)^6$.

Now this covariant, which we shall call J, is no other than the Jacobian of the quantic and its Hessian, which must vanish identically when these two only differ by a factor; and the coefficients in J are, only in a different form, the conditions already written. Again, we have said (Art. 134) that in the same case the covariant $\Sigma(\alpha-\beta)^2(\beta-\gamma)^2(\gamma-\alpha)^2(x-\delta)^4$ vanishes identically. But this, it will be found, is the same as $3TU - 2SH$; and we can easily verify that this covariant vanishes when the quartic has two square factors; for, making a, b, d, e all $= 0$, U reduces to $6cx^2y^2$, H to $-3c^2x^2y^2$, T to $-c^3$, and S to $3c^2$. Thus, then, we see that in the system of conditions given above, the common value of the fractions is $\dfrac{3T}{2S}$.

207. We have already shown (Art. 169) how to reduce a quartic to its canonical form, a problem in which is included that of the solution of the equation, since when it has been reduced to the form $ax^4 + 6cx^2y^2 + ey^4$, it can be solved like a quadratic. The reduction may also be effected by means of the values given for S and T. Imagine the variables transformed by a linear transformation whose modulus is unity, and so that the new b and d shall vanish: then we have $S = ae + 3c^2$, $T = ace - c^3$; and the new c is given by the equation $4c^3 - Sc + T = 0$. We get the x and y which occur in the canonical form from the equations

$$U = ax^4 + 6cx^2y^2 + ey^4, \quad H = acx^4 + (ae - 3c^2)x^2y^2 + cey^4,$$

whence $\quad cU - H = (9c^2 - ae)x^2y^2.$

Our process then, is to solve for c from the cubic just given; then with one of these values of c to form $cU - H$ which will be found to be a perfect square. Taking the square root and breaking it up into its factors we find the new x and y, and consequently know the transformation by means of which the given quartic can be brought to the canonical form. Having got it to the form $ax^4 + 6cx^2y^2 + ey^4$, we can of course, if we please, make the coefficients of x^4 and y^4 unity by writing x^2 and y^2 for $x^2 \sqrt{(a)}$, and $y^2 \sqrt{(e)}$.

Ex. Solve the equation
$$x^4 + 8x^3y - 12x^2y^2 + 104xy^3 - 20y^4 = 0.$$

We have here $S = -216$, $T = -756$, and our cubic is $4c^3 + 216c = 756$, of which $c = 3$ is a root. The Hessian is
$$H = -6x^4 + 60x^3y + 72x^2y^2 + 24xy^3 - 636y^4,$$
$$3U - H = 9(x^4 - 4x^3y - 12x^2y^2 + 32xy^3 + 64y^4) = 9(x^2 - 2xy - 8y^2)^2.$$

The variables then of the canonical form are $X = x + 2y$, $Y = x - 4y$, which give $6x = 4X + 2Y$, $6y = X - Y$; whence substituting in the given quartic the canonical form is found to be $3X^4 + 2X^2Y^2 - Y^4$. The roots then are given by the equations
$$(x + 2y)\sqrt{(3)} = x - 4y, \quad (x + 2y)\sqrt{(-1)} = x - 4y.$$

208. Mr. Cayley has given the root of the quartic in a more symmetrical form. Let c_1, c_2, c_3 be the roots of the cubic of the last article, and it has been shown that $H - c_1U$, $H - c_2U$, $H - c_3U$ are respectively perfect squares; the square roots being of course of the second degree in x and y. But further

$$(c_2 - c_3)(H - c_1U)^{\frac{1}{2}} + (c_3 - c_1)(H - c_2U)^{\frac{1}{2}} + (c_1 - c_2)(H - c_3U)^{\frac{1}{2}},$$

APPLICATIONS TO BINARY QUANTICS. 175

is also a perfect square, and its root is one of the factors of the quartic. It is only necessary to prove that this quantity is a perfect square, for it evidently vanishes when $U=0$. We work with the canonical form, taking for simplicity a and $e=1$. Now if we solve the equation $4z^3 - z(1 + 3c^2) + c - c^3 = 0$, we find the three roots to be c, $-\tfrac{1}{2}(c+1)$, $-\tfrac{1}{2}(c-1)$; and the three corresponding values of $H - cU$ are

$$(1 - 9c^2) x^2 y^2, \quad \tfrac{1}{2}(3c+1)(x^2+y^2)^2, \quad \tfrac{1}{2}(3c-1)(x^2-y^2)^2.$$

Now in order that any quantity of the form

$$\alpha xy + \beta (x^2 + y^2) + \gamma (x^2 - y^2)$$

may be a perfect square, we must obviously have $\alpha^2 = 4(\beta^2 - \gamma^2)$, which is verified when

$$\alpha^2 = 1 - 9c^2, \quad \beta^2 = \tfrac{1}{8}(3c-1)^2(3c+1), \quad \gamma^2 = \tfrac{1}{8}(3c+1)^2(3c-1).$$

Ex. If this method be applied to the last example, the other values of c are $\tfrac{1}{4}\{-3 \pm 9\sqrt{(-3)}\}$; and the squares of the linear factors of the quartic are given in the form

$$-2\sqrt{(3)}\{x^2 - 2xy - 8y^2\} \pm \tfrac{1}{4}\{1 - \sqrt{(-3)}\}[\{1 + \sqrt{(-3)}\} x^2 + \{10 - 2\sqrt{(-3)}\} xy - \{2 - 10\sqrt{(-3)}\} y^2\}]$$

$$\pm \tfrac{1}{4}\{1 + \sqrt{(-3)}\} [\{1 - \sqrt{(-3)}\} x^2 + \{10 + 2\sqrt{(-3)}\} xy - \{2 + 10\sqrt{(-3)}\} y^2].$$

209. It remains to distinguish the cases in which the transformation to the canonical form is made by a real or by an imaginary substitution. The discriminant of the canonical form is, as we have seen (p. 171, note), $ae(ae - 9c^2)^2$: and since the sign of the discriminant is unaffected by linear transformation, we see that whenever the discriminant is positive, a and e of the canonical form have like signs; and when the discriminant is negative, unlike signs. Now the form $ax^4 + 6cx^2y^2 + ey^4$ evidently resolves itself into two factors of the form, either $(x^2 + \lambda y^2)(x^2 + \mu y^2)$ or $(x^2 - \lambda y^2)(x^2 - \mu y^2)$; that is to say, the quartic has either four imaginary roots or four real roots. On the contrary, if a and e have opposite signs, the two factors are of the form $(x^2 + \lambda y^2)(x^2 - \mu y^2)$, or the quartic has two real and two imaginary roots. Hence then when the discriminant is negative; that is to say, when S^3 is less than $27 T^2$, the quartic has two real roots and two imaginary; and when the discriminant is positive, it has either four real or four imaginary roots.* Now

---

* The signs of the invariants do not enable us to distinguish the case of four real roots from that of four imaginary; but the application of Sturm's theorem

the discriminant of the equation $4c^3 - Sc + T = 0$ is $27T^2 - S^3$, therefore (Art. 164) when S^3 is less than $27T^2$, the equation in c has one root real and two imaginary; in the other case it has three real roots. The transformation therefore can only be effected in one real way, when the quartic has two real and two imaginary roots. It is easy to see that if a and e have like signs, in which case the equation can be brought to the form $x^4 + 6mx^2y^2 + y^4$, it can by two other linear transformations be brought to the same form; for write $x + y$ and $x - y$ for x and y, and we have $(1 + 3m) x^4 + 6 (1 - m) x^2y^2 + (1 + 3m) y^4$. Write $x + y \sqrt{(-1)}$, and $x - y \sqrt{(-1)}$ for x and y, and we have $(1 + 3m) x^4 + 6 (m - 1) x^2y^2 + (1 + 3m) y^4$. Thus then when the quartic has four real or four imaginary roots, though there are three real values for c, one of these corresponds to imaginary values of x and y; and there are only two real ways of making the transformation.

The same thing may be also seen thus. Imagine the quartic to have been resolved into two real quadratic factors

$$(a, b, c \unicode{x0028}\!\!\unicode{x2510} x, y)^2, (a', b', c' \unicode{x0028}\!\!\unicode{x2510} x, y)^2;$$

then these two factors U, V can by simultaneous transformation be brought to the form $AX^2 + BY^2, A'X^2 + B'Y^2$, where X^2 and Y^2 are the values of $\lambda U + V$ corresponding to the two values of λ given by the equation

$$(ac - b^2) \lambda^2 + (ac' + ca' - 2bb') \lambda + (a'c' - b'^2) = 0.$$

In order that the values of λ should be real, we must have the eliminant of the two quadratics positive, or

$$(\alpha - \alpha')(\alpha - \beta')(\beta - \alpha')(\beta - \beta')$$

positive. Thus then, when the quantic has four real roots, if we take for α and β the two greatest roots, and for α' and β' the two least, or again, if we take for α and β the two extreme roots, and for α' and β' the two mean roots, we get real values for λ. In the remaining case we get imaginary values. If

---

shews that (the discriminant being positive), when the roots are all real, both the quantities $b^2 - ac$ and $3aT + 2(b^2 - ac) S$ are positive, while if either is negative the four roots are imaginary. (Cayley, *Quarterly Journal*, vol. IV., p. 10).

APPLICATIONS TO BINARY QUANTICS.   177

either of the quadratics has imaginary roots, the resultant of the two is positive, and the values of λ real.

210. Quartics have another covariant which is the cubico-variant J of the first emanant, or symbolically is $\overline{12}^2.\overline{13}$. It is of the third degree in the coefficients and the sixth in the variables. For the form $x^4 + 6mx^2y^2 + y^4$, this covariant is $(1-9m^2)xy(x^4-y^4)$. It is the Jacobian of the quartic and its Hessian. In general we saw (Art. 174) that if U and V be two quartics, six values of λ can be found, such that $U + \lambda V$ shall have a square factor, and that these six factors are the factors of the Jacobian. When V is the Hessian of U, this theory is a little modified. We get instead *three* values for λ, such that $\lambda U - H$ shall contain each *two* square factors, these values being given by the equation $4\lambda^3 - \lambda S + T = 0$. But, as before, these six factors will be the factors of the Jacobian: that is to say, the covariant J has for factors the x and y of the three canonical forms. Geometrically the matter may be stated as follows: Four points on a line determine three different systems in involution (because either the point B, C, or D may be taken as conjugate to A), and the foci of these three systems are determined by the covariant J.

Since, by Art. 207, the square of the product of one set of x and y of the canonical form is determined by $c_1 U - H$, we have J^2 proportional to $(c_1 U - H)(c_2 U - H)(c_3 U - H)$; or (from the equation which determines c) to $4H^3 - SHU^2 + TU^3$. By calculating with the canonical form, the actual value of the latter quantity is found to be $-J^2$.

211. Since H is a covariant of U, it follows that if α and β be any constants, $\alpha U + 6\beta H$* will be a covariant of U, whose invariants also will be invariants of U. The following are the values of the S, T, and discriminant R, of this form.

$$S(\alpha U + 6\beta H) = S\alpha^2 + 18 T\alpha\beta + 3 S^2 \beta^2,$$
$$T(\alpha U + 6\beta H) = T\alpha^3 + S^2 \alpha^2 \beta + 9 S T \alpha \beta^2 + (54 T^2 - S^3)\beta^3,$$
$$R(\alpha U + 6\beta H) = R(\alpha^2 - 9 S \alpha \beta^2 - 54 T \beta^3)^2.$$

---

* The numerical coefficient is added in order to avoid fractions in the following formulæ.

A A

The last is a perfect square, because, as we have just mentioned, instead of six cases where $aU+6\beta H$ has a square factor, we have three cases where it has two square factors.

Hermite has noticed that if we call G the function of α, β, $\alpha^3 - 9S\alpha\beta^2 - 54T\beta^3$, then the values just given for the S and T of $aU+6\beta H$ are respectively the Hessian and the cubicovariant of G. The discriminant of G differs only by a numerical factor from the discriminant of U.

The covariants of $aU+6\beta H$ are also covariants of U. Its Hessian is

$$(\alpha\beta S + 9\beta^2 T) U + (\alpha^2 - 3\beta^2 S) H,$$

which is the Jacobian, with respect to α, β, of G and $aU+6\beta H$. Since J is a combinant of the system U, H, the J of $aU+6\beta H$ will be the same, multiplied however by the numerical factor G. The Hessian of J is $S^2U^2 - 36TUH + 12SH^2$; which is the resultant of $aU+6\beta H$ and the Hessian of G. Mr. Cayley has thrown this into the form

$$\left(SU - \frac{18T}{S} H\right)^2 + \frac{12}{S^2}(S^3 - 27T^2) H^2,$$

showing that it is a perfect square when the discriminant of U vanishes.

212. It may be remarked here, that, by the principle established Art. 137, theorems concerning invariants and covariants of a quantic give at once theorems concerning the covariants of quantics of higher degree. Thus, it having been just proved that the Hessian of the Hessian of a quartic is of the form $\alpha TU + \beta SH$, we can infer that the same is true of the Hessian of the Hessian of any quantic. For if we form the Hessian of $u_{11}u_{22} - u_{12}^2$, this involves the second, third, and fourth differentials of u. But, by the equations $(n-3)u_{111} = xu_{1111} + yu_{1112}$, &c., we can express the second and third differentials in terms of the fourth, and so write the Hessian as a function of fourth differentials only, and of the x and y which we have introduced, and which, it will be found, enter in the fourth degree. It will then be a covariant of the quartic emanant. Now every covariant of a quartic is a function of U and H (Art. 190), and when the covariant is of the fourth degree it must be a linear

APPLICATIONS TO BINARY QUANTICS. 179

function of these quantities. This covariant then will be of the form $\alpha TU + \beta SH$, when S and T are invariants of the quartic emanant, but covariants, as in Art. 137, of the higher quantic.

213. *System of two quartics.* We purpose now to enumerate the combinants of a system of two quartics. If we form the S and T of $\lambda U + \mu V$, which we may write

$$S\lambda^2 + \Sigma\lambda\mu + S'\mu^2, \quad T\lambda^3 + t\lambda^2\mu + t'\lambda\mu^2 + T'\mu^3,$$

the invariants of this quadratic and cubic will be combinants of the system U, V (Art. 199).*

Thus the discriminant of the quadratic is without difficulty found to be

$$(ae')^2 + 16(bd')^2 + 12(ac')(ce') - 48(bc')(cd') - 8(ab')(de') - 8(ad')(be'),$$

which we shall refer to as the combinant A.

Combinants of the same order in the coefficients may be found by other processes. Thus the Jacobian of U and V is

$$(ab')\,x^6 + 3\,(ac')\,x^5 y + \{3\,(ad') + 6\,(bc')\}\,x^4 y^2 + \{(ae') + 8\,(bd')\}\,x^3 y^3$$
$$+ \{3\,(be') + 6\,(cd')\}\,x^2 y^4 + 3\,(ce')\,xy^5 + (de')\,y^6,$$

and their combinantive covariant $\overline{12}^3$ is

$$\{(ad') - 3\,(bc')\}\,x^2 + \{(ae') - 2\,(bd')\}\,xy + \{(be') - 3\,(cd')\}\,y^2.$$

Now the discriminant of the latter covariant, and the quadratic invariant, as at p. 112, of the former will be combinants of the same order as A, but will not be identical with it. If we write the new combinant B,

$$(bd')^2 + (ac')(ce') - (bc')(cd') - (ab')(de') - (ad')(cd') - (bc')(be'),$$

then the quadratic invariant of the Jacobian will be found to be $A + 48B$, and the discriminant of the other covariant to be $A - 12B$.

---

* It has been mentioned that the eliminant of two quadrics may be written either as the discriminant of

$$(ac - b^2)\lambda^2 + (ac' + ca' - 2bb')\lambda\mu + (a'c' - b'^2)\mu^2,$$

or of the Jacobian $\quad (ab')\,x^2 + (ac')\,xy + (bc')\,y^2;$

but it may be added that the former expression is linearly transformed into the latter by taking $\lambda b + \mu b'$, $\lambda a + \mu a'$ for x and y. Similar transformations may be used to facilitate the following calculations.

180   APPLICATIONS TO BINARY QUANTICS.

214. In writing these combinants in terms of the determinants (ab'), &c., I find it convenient to use the abbreviations $(ab') = \alpha$, $(de') = \alpha'$, $(ad') = \beta$, $(be') = \beta'$, $(ac') = \lambda$, $(ce') = \lambda'$, $(bc') = \mu$, $(cd') = \mu'$, $(ae') = \gamma$, $(bd') = \delta$. We have then

$$A = 12\lambda\lambda' - 48\mu\mu' + \gamma^2 + 16\delta^2 - 8\alpha\alpha' - 8\beta\beta',$$

$$B = \lambda\lambda' - \mu\mu' - \beta\mu' - \beta'\mu + \delta^2 - \alpha\alpha'.$$

The resultant of U, V, found by expanding the determinant of p. 67, is

$$R = 1296\lambda^2\lambda'^2 - 3456\,(\alpha\mu\lambda'^2 + \alpha'\mu'\lambda^2) - 1152\,(\alpha\beta\lambda'^2 + \alpha'\beta'\lambda^2)$$
$$- 72\gamma^2\lambda\lambda' - 576\gamma\delta\lambda\lambda' + 9216\alpha\alpha'\mu\mu' + 96\gamma\,(\beta^2\lambda' + \beta'^2\lambda)$$
$$+ 288\gamma\,(\alpha\beta'\lambda' + \alpha'\beta\lambda) + 1536\delta\,(\alpha\beta'\mu' + \alpha'\beta\mu) + 3072\alpha\alpha'\,(\beta\mu' + \beta'\mu)$$
$$+ \gamma^4 - 48\alpha\alpha'\gamma^2 - 16\beta\beta'\gamma^2 - 256\alpha\alpha'\gamma\delta + 512\alpha^2\alpha'^2$$
$$- 256\,(\alpha\beta'^3 + \alpha'\beta^3) - 4096\alpha\alpha'\delta^2.$$

In terms of the preceding combinants can be expressed the combinant which we have called T (p. 149), but which, in order to avoid confusion, we shall now call C; and which expresses the condition that a quartic of the system $U + \lambda V$ can have two square factors. Such a combinant must vanish if V reduce to the single term $c'x^2y^2$. In such a case $\alpha, \alpha', \beta, \beta', \gamma, \delta$ all vanish; and we have $A = 12\lambda\lambda' - 48\mu\mu'$, $B = \lambda\lambda' - \mu\mu'$, $R = 1296\lambda^2\lambda'^2$; hence we see that $(A - 48B)^2 - R$ is a combinant which vanishes on this supposition. And since it is of the same order that we have seen, Art. 179, that T must be, it is identical with it. Using the values already given for A, B, R, we find that $(A - 48B)^2 - R = 128\,C$, where

$$C = -27\gamma(\lambda\mu'^2 + \lambda'\mu^2) + 18(\beta^2\mu'^2 + \beta'^2\mu^2) + 18\delta^2\lambda\lambda' + 36\gamma\delta\mu\mu' - 36\alpha\alpha'\mu\mu'$$
$$+ 18\gamma\delta\lambda\lambda' - 9\gamma^2\mu\mu' - 3\gamma\,(\alpha\beta'\lambda' + \alpha'\beta\lambda) - 24\delta^2\,(\beta\mu' + \beta'\mu)$$
$$- 6\delta\,(\beta^2\lambda' + \beta'^2\lambda) - 6\delta\,(\alpha\beta'\lambda' + \alpha'\beta\lambda) + 2\,(\alpha\beta'^3 + \alpha'\beta^3) + \alpha\alpha'\gamma^2$$
$$+ 4\alpha^2\alpha'^2 - 2\alpha\alpha'\gamma\delta + 4\gamma\delta^3 + 16\alpha\alpha'\delta^2 + 8\delta^4.$$

Again, if we form the invariant which we called I (Art. 194) of the quadratic and cubic of Art. 213, it will be found that

$$(A + 48B)(A - 16B) - R = -128I.$$

215. The other combinants of the system which we shall notice are D, the resultant of the cubic and quadratic, and E

APPLICATIONS TO BINARY QUANTICS. 181

the discriminant of the cubic. D is the invariant which we have called S, p. 149. It may be mentioned that besides the methods already indicated for calculating that invariant in general, the following may be used. It is required to find the condition that λ can be determined so that the three expressions $u_{11} + \lambda v_{11}$, $u_{12} + \lambda v_{12}$, $u_{22} + \lambda v_{22}$ can be made to vanish together. Now we may multiply each of these by the $2(n-2)$ terms x^{2n-5}, &c. of a quantic of the degree $2n-5$, and so obtain $6(n-2)$ equations, from which we can eliminate dialytically the $6(n-2)$ quantities x^{3n-7}, &c., λx^{3n-7}, &c., and so obtain S in the form of a determinant. In the case of the system of two quartics

$D = -16\lambda^3\lambda'^3 + 48\lambda^2\lambda'^2\mu\mu' + 6\lambda\lambda'\mu^2\mu'^2 + .16\mu^3\mu'^3 + 27\lambda^2\mu'^4 + 27\lambda'^2\mu^4$
$+ 36\alpha\lambda\mu\lambda'^3 + 36\alpha'\lambda'\mu'\lambda^3 + 12\lambda^2\lambda'^2(\beta\mu' + \beta'\mu) - 96\alpha\mu^2\lambda'^2\mu'$
$- 96\alpha'\mu'^2\lambda^2\mu - 6\gamma\lambda\mu\mu'^3 - 6\gamma\mu^3\lambda'\mu' - 36\delta\lambda^2\lambda'\mu'^2 - 36\delta\lambda\mu^2\lambda'^2$
$- 48\alpha\mu\mu'^4 - 48\alpha'\mu^4\mu' - 24\delta\lambda\mu\mu'^3 - 24\delta\lambda'\mu'\mu^3 + 24\alpha\beta\lambda\lambda'^3$
$+ 24\alpha'\beta'\lambda^3\lambda' - 18\alpha\beta'\lambda'^2\mu^2 - 18\beta\alpha'\lambda^2\mu'^2 - 6\delta^2\lambda^2\lambda'^2 + 87\gamma^2\mu^2\mu'^2$
$- \lambda\lambda'\mu\mu'(162\alpha\alpha' + 90\beta\beta') - 36\alpha\beta\mu'^4 - 36\alpha'\beta'\mu^4 + 96\delta^2\lambda\lambda'\mu\mu'$
$+ (228\alpha\alpha' - 60\beta\beta')\mu^2\mu'^2 - 4\alpha'^2\gamma\lambda'^3 - 4\alpha'^2\gamma\lambda^3 - 16\alpha\beta^2\lambda'^2\mu'$
$- 16\alpha'\beta'^2\lambda^2\mu - 30\alpha\alpha'^2\lambda^2\mu' - 30\alpha^2\alpha'\lambda'^2\mu - 50\alpha\alpha'\lambda\lambda'(\beta\mu' + \beta'\mu)$
$- 20\delta\lambda\lambda'(\lambda\beta\alpha' + \lambda'\alpha\beta') + 2\gamma\mu\mu'(\lambda\beta\alpha' + \lambda'\alpha\beta') + 48\delta^2\alpha\mu\lambda'^2$
$+ 48\delta^2\alpha'\mu'\lambda^2 + 24\alpha\beta'^2\mu\mu'^2 + 24\alpha'\beta^2\mu^2\mu' + 56\alpha\beta\beta'\mu'^3 + 56\alpha'\beta\beta'\mu^3$
$+ 240\alpha\alpha'\beta\mu\mu'^2 + 240\alpha\alpha'\mu^2\mu'\beta' + 32\alpha\delta^2\mu'^3 + 32\alpha'\delta^2\mu^3 + 3\lambda^2\beta^2\alpha'^2$
$+ 3\lambda'^2\alpha^2\beta'^2 + 24\alpha\alpha'(\beta^2\mu'^2 + \beta'^2\mu^2) - 6\alpha\alpha'\gamma^2\mu\mu' - 12\alpha^2\alpha'^2\lambda\lambda'$
$- 30\alpha\alpha'\beta\beta'\lambda\lambda' + 60\alpha\alpha'\lambda\lambda'\delta^2 + 12\beta\beta'\lambda\lambda'\delta^2$
$+ \{84\alpha^2\alpha'^2 + 120\alpha\alpha'\beta\beta' - 12\beta^2\beta'^2\}\mu\mu' - 192\alpha\alpha'\delta^2\mu\mu' + 48\gamma\delta^3\mu\mu'$
$- 48\delta^4\lambda\lambda' + 6\gamma\alpha\alpha'^2\beta\lambda + 6\gamma\alpha^2\alpha'\beta'\lambda' + 48\alpha^2\alpha'^2(\beta\mu' + \beta'\mu)$
$+ 24\alpha\alpha'\delta(\beta^2\mu' + \beta'^2\mu) - 96\alpha\alpha'\delta^2(\beta\mu' + \beta'\mu) - \alpha^2\alpha'^2\gamma^2 - 4\alpha\alpha'^2\beta^3$
$- 4\alpha^2\alpha'\beta'^3 + 8\alpha^3\alpha'^3 - 4\gamma\delta^2\alpha^2\alpha'^2 - 48\alpha^2\alpha'^2\delta^2 - 16\delta^3\alpha\alpha'\beta\beta' + 64\alpha\alpha'\delta^4$.

216. In studying the relations of these combinants, we may, as in Art. 199, suppose one quartic to want the first two terms, and the other the last two; that is, we may write

$$U = ax^4 + 4bx^3y + 6cx^2y^2, \quad V = 6c'x^2y^2 + 4dxy^3 + ey^4.$$

182   APPLICATIONS TO BINARY QUANTICS.

To save room, we write $ae = l$, $bd = m$, $cc' = n$, $cad^2 + c'eb^2 = p^2$. We find then

$A = (l - 4m)^2 + 12n(l - 4m)$,

$B = m^2 - p^2 + n(l - m)$,

$R = l^2(l - 16m) + 96l^2p^2 - 72nl^2(l + 8m) + 1296l^2n^2$.

$I = -l^2m^2 + 4lm^3 + p^2(l^2 - 2lm - 8m^2) + 6p^4$
$\quad - n(l^3 + 6lm^2 - 16m^3) - 9nlp^2 + 12n^2(l^2 + lm - 2m^2)$,

$D = -n^2\{9lm^2(l - 4m) - 6p^2(2l^2 - 7lm - 4m^2) - 27p^4 + 16n(l - m)^3\}$,

$E = -l^2m^4 + 2lm^2p^2(l + 2m) - (l + 2m)^2p^4 - 2nlm^2(l + 2m)^2$
$\quad + 4p^6 + 2np^2(l + 2m)^3 - 18np^2lm^2 - n^2(l + 2m)^4$
$\quad + 36n^2lm^2(l + 2m) - 6np^4(l + 2m) - 6n^2p^2(l + 2m)^2$
$\quad + 27n^2p^4 + 4n^3(l + 2m)^5 - 108n^3lm^2$.

By the help of these values we can verify the equation
$$16B^3 - AB^2 - 2IB + E = D.$$

Now if we take for the fundamental invariants of the system, the coefficients of the cubic and quadratic of Art. 213, the combinants A, D, E, I are explicitly given as functions of these quantities; and the present equation gives B in terms of the same quantities, and shows that all the combinants we have used can be expressed in terms of the same fundamental invariants.

The Jacobian, with this form, wants the extreme terms. There is no difficulty therefore in calculating its discriminant, and thus verifying the theorem of Art. 177.

217. I have also sometimes found it convenient to suppose each quartic to be the sum of two fourth powers, so that for each the invariant T vanishes. Let the quartics be $au^4 + bv^4$, $a'w^4 + b'z^4$, where u is $\alpha_1 x + \beta_1 y$, &c. We use (12) to denote $\alpha_1\beta_2 - \alpha_2\beta_1$, and we use the abbreviations

$$(12)(34) = L, \quad (13)(42) = M, \quad (14)(23) = N;$$

where it will be observed that we have identically $L + M + N = 0$. Now the invariant S is got by substituting $\dfrac{d}{dy}$, $-\dfrac{d}{dx}$, for x, y in the quartic and then operating on itself. If we operate in this way with u upon u the result vanishes; but if we operate on v the result is (12). We find then at once that the S of $\lambda U + \mu V$ is

$$\lambda^2 ab\,(12)^4 + \lambda\mu\{aa'\,(13)^4 + ab'\,(14)^4 + ba'\,(23)^4 + bb'\,(24)^4\} + \mu^2 a'b'\,(34)^4.$$

The combinant then which we have called A is

$$\{aa'\,(13)^4 + ab'\,(14)^4 + ba'\,(23)^4 + bb'\,(24)^4\}^2 - 4aba'b'L^4.$$

In the same case B is found to be $-aba'b'L^2MN$.

The invariant T is found by operating with the Hessian of a quartic on itself. But here the Hessian of U is $ab\,(12)^2\,u^2v^2$. We find then that the T of $\lambda U + \mu V$, is

$$\lambda^2\mu\{aba'\,(12)^2\,(13)^2\,(23)^2 + abb'\,(12)^2\,(14)^2\,(24)^2\}$$
$$+ \lambda\mu^2\{a'b'a\,(13)^2\,(14)^2\,(34)^2 + a'b'b\,(23)^2\,(24)^2\,(34)^2\}.$$

Hence, we have immediately

$$E = -a^2b^2a'^2b'^2L^4\{aa'N^2(13)^4 + ab'M^2(14)^4 + ba'M^2(23)^4 + bb'N^2(24)^4\}^2,$$

$$D = -a^2b^2a'^2b'^2L^6\{a^2a'^2N^2\,(13)^8 + a^2b'^2M^2\,(14)^8 + b^2a'^2M^2\,(23)^8$$
$$+ b^2b'^2N^2\,(24)^8 - 2MNa^2a'b'\,(13)^4\,(14)^4 - 2MNb^2a'b'\,(23)^4\,(24)^4$$
$$- 2MNa'^2ab\,(13)^4\,(23)^4 - 2MNb'^2ab\,(14)^4\,(24)^4$$
$$+ 2M^2N^2aba'b'\,(M^2 + N^2 - 2L^2)\},$$

$$I = -aba'b'L^4[a^2a'^2N^2(13)^6 + a^2b'^2M^2(14)^6 + b^2a'^2M^2(23)^6 + b^2b'^2N^2(24)^6$$
$$+ (M^2+N^2-2L^2)\{a^2a'b'\,(13)^4\,(14)^4 + b^2a'b'\,(23)^4\,(24)^4$$
$$+ a'^2ab\,(13)^4\,(23)^4 + b'^2ab\,(14)^4\,(24)^4\}$$
$$+ 2M^2N^2\,(M^2 + N^2 - 4L^2)\,aba'b'],$$

by the help of which values we can verify the equation already obtained.

218. *The Quintic.* In studying the quintic we constantly use the canonical form $ax^5 + by^5 + cz^5$, (where $x + y + z = 0$), to which it has been shown (Art. 165) that the general equation may be reduced. Differentiating with regard to x and y successively, we have $u_1 = ax^4 - cz^4$, $u_2 = by^4 - cz^4$. It is evident that the resultant of these two will be the discriminant of the quintic, and that the combinants of this system will be invariants of the quintic. These invariants are then immediately found from the expressions in the last article, where we must write for a and b, a and $-c$, for a' and b', b and $-c$. We have (24),

and therefore $M = 0$; $(13) = 1$, $(12) = -1$, $(34) = -1$, $(14) = -1$, $(23) = 1$. We observe then at once that B vanishes. We can see by counting constants, that any two cubics can be brought by linear transformation to be the two differentials of a single quartic; but two quartics cannot be similarly brought to be the differentials of a single quintic, unless the condition $B = 0$ be fulfilled. The combinant A in like manner becomes

$$b^2c^2 + c^2a^2 + a^2b^2 - 2abc(a+b+c).$$

This, which we shall call J, is the simplest invariant of the quintic, and it may be obtained in other ways. The quintic, it will be observed, has two covariants of the second order in the coefficients, viz. the Hessian $\overline{12}^2$, which for the canonical form is $bcy^3z^3 + caz^3x^3 + abx^3y^3$, and a covariant quadratic $\overline{12}^4$, the S of the quartic emanant, which in the same case is $bcyz + cazx + abxy$. If the quintic be written in the general form $(a, b, c, d, e, f\sharp x, y)^5$, these covariants are respectively

$(ac - b^2)x^6 + 3(ad - bc)x^5y + 3(ae + bd - 2c^2)x^4y^2 + (af + 7be - 8cd)x^3y^3$
$\quad + 3(bf + ce - 2d^2)x^2y^4 + 3(cf - de)xy^5 + (df - e^2)y^6;$

$(ae - 4bd + 3c^2)x^2 + (af - 3be + 2cd)xy + (bf - 4ce + 3d^2)y^2.$

Now we get invariants differing only by a numerical factor, whether we form the discriminant of the latter covariant, or the quadratic invariant $\overline{12}^6$ of the Hessian. In either way we obtain the general value of J, viz.

$a^2f^2 - 10abef + 4acdf + 16ace^2 - 12ad^2e + 16b^2df$
$\quad + 9b^2e^2 - 12bc^2f - 76bcde + 48bd^3 + 48c^3e - 32c^2d^2.$

219. The discriminant of the quintic may be obtained either from the theory of two quartics, or by direct elimination between the two differentials $ax^4 - cz^4$, $by^4 - cz^4$. When these vanish together, we may take abc as the common value of ax^4, by^4, cz^4; whence $x = (bc)^{\frac{1}{4}}$, $y = (ca)^{\frac{1}{4}}$, $z = (ab)^{\frac{1}{4}}$. Substituting in $x + y + z = 0$, we get the discriminant in the form

$$(bc)^{\frac{1}{4}} + (ca)^{\frac{1}{4}} + (ab)^{\frac{1}{4}} = 0,$$

or $\{b^2c^2 + c^2a^2 + a^2b^2 - 2abc(a+b+c)\}^2 - 128a^2b^2c^2(bc + ca + ab) = 0$.

Thus then we are led to the form for the discriminant $J^2 - 128K$,

APPLICATIONS TO BINARY QUANTICS.

where K is the invariant, of the eighth order in the coefficients, which for the canonical form is $a^2b^2c^2(bc+ca+ab)$.

This latter invariant may be otherwise defined as follows: We have given (pp. 137, 138) expressions for the canonizant, namely the covariant cubic whose three roots are the x, y, z of the canonical form. This covariant, which is the T of the quartic emanant, expanded for the general form is

$$(ace - ad^2 - eb^2 + 2bcd - c^3)x^3 + (acf - ade - b^2f + bce + bd^2 - c^2d)x^2y$$
$$+ (adf - ae^2 - bcf + bde + c^2e - cd^2)xy^2$$
$$+ (bdf - be^2 - c^2f + 2cde - d^3)y^3.$$

This for the canonical form reduces to $abcxyz$. Now if substituting in the usual way differential symbols for the variables, we operate with the square of this canonizant on the Hessian we get the invariant K; as we can easily verify by the canonical form. Another way in which K can be found is by forming the invariant I, as in Art. 194, of the covariant quadratic $bcyz + cazx + abxy$, and the canonizant. In any of these ways the general value of K is found to be

$a^3cdf^3 - a^3f^2ce^2 - a^2f^3b^2d - 3a^2f^2d^2e - 3a^2f^3bc^2 + 5a^3fde^3 + 5af^3b^3c$

$\quad - 2a^3e^5 - 2b^2f^3 + a^2f^2b^2e^2 + 11a^2f^2bcde - 5a^2fbce^3 - 5af^2b^3de$

$\quad + 12a^2f^2bd^3 + 12a^2f^2c^3e - 30a^2fbd^2e^2 - 30af^2b^2c^2e + 15a^2bde^4$

$\quad + 15b^4cef^2 - 21a^2f^2c^2d^2 - 34a^2fc^2de^2 - 34af^2b^2cd^2 + 22a^2c^2e^4$

$\quad + 22b^4d^2f^2 + 78a^2fcd^3e + 78af^2bc^3d - 48a^2cd^2e^3 - 48b^3c^2df^2$

$\quad - 27a^2fd^5 - 27af^3c^5 + 18a^2d^4e^2 + 18b^2c^4f^2 + 133afb^2e^2cd$

$\quad - 54ab^3ce^4 - 54b^4de^2f - 18afb^2d^3e - 18afbc^2e^3 + 3ab^2d^2e^3$

$\quad + 3b^3c^2e^2f - 220afbec^2d^2 + 106abc^2de^3 + 106b^3cd^2ef + 93afbcd^4$

$\quad + 93afc^4de - 30abe^2cd^3 - 30b^2ec^2df - 9abed^5 - 9bec^5f - 38ac^4e^3$

$\quad - 38b^3d^4f - 42afc^3d^3 + 8ac^3d^2e^2 + 8b^2c^2d^2f + 6ac^2d^4e + 6bc^4d^2f$

$\quad + 27b^4e^4 - 81b^3e^2cd + 38b^2e^2d^3 + 38b^2e^2c^3 + 25b^2e^2c^2d - 57b^2ecd^4$

$\quad - 57be^2c^4d + 18b^2d^5 + 18c^4e^2 + 74bec^3d^3 - 24bc^2d^5 - 24c^2d^2e + 8c^3d^4$.

The value of the discriminant in general can be derived hence, or else, as I originally obtained it, from the formula (Art. 214), for the resultant of two quartics. We thus find

$$R = a^4f^4 - 20a^3f^3be - 120a^3f^2cd + 160\,(a^3f^2ce^2 + a^3f^3b^2d)$$
$$+ 360\,(a^3f^2d^2e + a^3f^3bc^2) - 640\,(a^3fde^3 + af^3b^3e) + 256\,(a^3e^5 + b^5f^3)$$
$$- 10a^2f^2b^2e^2 - 1640a^2f^2becd + 320\,(a^2fbe^3c + af^2b^3ed)$$
$$- 1440a^2f^2\,(bd^3 + c^3e) + 4080\,(a^2fbe^2d^2 + af^2b^2ec^2)$$
$$- 1920\,(a^2be^4d + b^4ecf^2) + 2640a^2f^2c^2d^2 + 4480\,(a^2fc^3de^2 + af^2b^2cd^3)$$
$$- 2560\,(a^2c^2e^4 + b^4d^2f^2) - 10080\,(a^2fcd^3e + af^2bc^3d)$$
$$+ 5760\,(a^2cd^2e^3 + b^2c^3df^2) + 3456\,(a^3d^5f + af^3c^5)$$
$$- 2160\,(a^2d^4e^2 + b^2c^4f^2) - 180afb^3e^3 - 14920afb^2e^2cd$$
$$+ 7200\,(ab^2e^4c + b^4e^2df) + 960af(b^2ed^3 + be^2c^3)$$
$$- 600\,(ab^2e^3d^2 + b^2e^2c^2f) + 28480afbec^2d^2 - 16000\,(abe^3c^2d + b^3ecd^2f)$$
$$- 11520af(bcd^4 + c^4de) + 7200\,(abe^2cd^3 + b^2ec^3df)$$
$$+ 6400\,(ac^4e^3 + b^3d^4f) + 5120afc^3d^3 - 3200\,(ae^2c^3d^2 + b^2c^2d^3f)$$
$$- 3375b^4e^4 + 9000b^3e^3cd - 4000\,(b^3e^2d^3 + b^2e^3c^3) + 2000b^2e^2c^2d^2.$$

The discriminant may also be expressed as follows: Let
$$A = a^2f^2 - 34afbe + 76afcd - 32ace^2 - 32b^2df - 12aed^2 - 12bc^2f$$
$$+ 225b^2e^2 - 820becd + 480\,(bd^3 + c^3e) - 320c^2d^2;$$
$$B = 3a^2f^2 - 22afbe - 12afcd + 64\,(ace^2 + b^2df)$$
$$- 36\,(aed^2 + bc^2f) - 45b^2e^2 + 20becd;$$
$$C = a^2fe + 2afbd - 9abe^2 - 9afc^2 + 32acde - 18ad^3 + 6b^2cf$$
$$- 15bec^2 + 10bcd^2,$$
$$D = 3a^2df - 2a^2e^2 - 9afbc + abed + 18ac^2e - 12acd^2 + 6b^3f$$
$$- 15b^2ec + 10b^2d^2;$$

and let C', D' be the functions complemental to C and D, (where all these functions vanish if three roots be equal), then three times the discriminant is

$$AB + 64\,CC' - 64DD'.$$

220. Quintics have also an invariant of the twelfth degree, which may be most simply defined as the discriminant of the canonizant. For the canonical form for which the canonizant is $abcxyz$, the discriminant is $-a^4b^4c^4$. And, in general, the discriminant is $-L$, where the following is the value of L as calculated by M. Faà de Bruno. To save space in printing we

APPLICATIONS TO BINARY QUANTICS. 187

omit the complementary terms. Thus $(a^4 c^2 de^y f^3)$ stands for $a^4 c^2 de^y f^3 + a^3 b^y cd^2 f^4$.

$L = a^4c^2d^2f^4 - 2(a^4c^2def^3) + (a^4c^2e^4f^2) - 6(a^4cd^3ef^3) + 16(a^4cd^2e^3f^2)$
$\quad - 14(a^4cde^5f) + 4(a^4ce^7) + 4(a^4d^5f^3) - 11(a^4d^4e^2f^2) + 10(a^4d^3e^4f)$
$\quad - 3(a^4d^2e^6) + 4a^3b^2cde^2f^3 - 2(a^3b^2ce^4f^2) + 6(a^3b^2d^3ef^3)$
$\quad - 16(a^3b^2d^2e^3f^2) + 14(a^3b^2de^5f) - 4(a^3b^2e^7) + 50a^3bc^2d^2ef^3$
$\quad - 82(a^3bc^2de^3f^2) + 32(a^3bc^2e^5f) - 36(a^3bcd^4f^2) + 30(a^3bcd^3e^2f^2)$
$\quad + 30(a^3bcd^2e^4f) - 24(a^3bcde^6) + 28(a^3bd^5ef^2) - 50(a^3bd^4e^3f)$
$\quad + 22(a^3bd^3e^5) + 16(a^3c^4e^3f^2) + 22a^3c^3d^3f^3 + 50(a^3c^3d^2e^2f^2)$
$\quad - 16(a^3c^3de^4f) - 16(a^3c^3e^6) - 54(a^3c^2d^4ef^2) - 46(a^3c^2d^3e^3f)$
$\quad + 60(a^3c^2d^2e^5) + 6(a^3cd^6f^2) + 70(a^3cd^5e^2f) - 56(a^3cd^4e^4)$
$\quad - 18(a^3d^7ef) + 14(a^3d^6e^3) + a^2b^4e^4f^2 + 132a^2b^3cde^2f^3 - 50(a^2b^3ce^4f)$
$\quad + 14(a^2b^3d^3e^2f^2) - 60(a^2b^3d^2e^4f) + 30(a^2b^3de^6) - 168a^2b^2c^2d^3e^2f^2$
$\quad + 48(a^2b^2c^2de^4f) + 4(a^2b^2c^2e^6) + 48(a^2b^2cd^4ef^2) + 2(a^2b^2cd^3e^3f)$
$\quad - 6(a^2b^2cd^2e^5) - 62(a^2b^2d^6f^2) + 90(a^2b^2d^5e^2f) - 39(a^2b^2d^4e^4)$
$\quad - 112(a^2bc^4e^4f) - 82a^2bc^3d^2ef^2 + 170(a^2bc^3d^2e^3f) + 104(a^2bc^3de^5)$
$\quad + 108(a^2bc^2d^5f^2) + 42(a^2bc^2d^4e^2f) - 298(a^2bc^2d^3e^4) - 242(a^2bcd^6ef)$
$\quad + 294(a^2bcd^5e^3) + 72(a^2bd^8f) - 78(a^2bd^7e^2) + 164(a^2c^5de^3f)$
$\quad + 24(a^2c^5e^5) - 63a^2c^4d^3f^2 - 394(a^2c^4d^2e^2f) - 194(a^2c^4d^2e^4)$
$\quad + 324(a^2c^3d^4ef) + 440(a^2c^3d^4e^3) - 78(a^2c^3d^7f) - 428(a^2c^2d^6e^2)$
$\quad + 180(a^2cd^8e) - 27(a^2d^{10}) + 18ab^5e^5f - 38ab^4cde^4f + 36(ab^4ce^6)$
$\quad + 204(ab^4d^3e^3f) - 102(ab^4d^2e^5) - 308(ab^3c^2de^5) - 42ab^3c^2d^3e^3f$
$\quad - 674(ab^3cd^4e^2f) + 590(ab^3cd^3e^4) + 128(ab^3d^5ef) - 138(ab^3d^5e^3)$
$\quad + 4(ab^2c^4e^5) + 652(ab^2c^3d^2e^4) + 714ab^2c^2d^5e^2f + 498(ab^2c^2d^5ef)$
$\quad - 1246(ab^2c^2d^4e^3) - 224(ab^2cd^7f) + 516(ab^2cd^6e^2) - 48(ab^2d^8e)$
$\quad - 136(abc^5de^4) - 1078abc^4d^4ef - 206(abc^4d^3e^3) + 342(abc^3d^5f)$
$\quad + 804(abc^3d^5e^2) - 506(abc^2d^7e) + 90(abcd^9) - 16(ac^7e^4)$
$\quad + 220(ac^6d^3e^3) - 106ac^5d^5f - 392(ac^5d^4e^2) + 222(ac^4d^6e) - 40(ac^3d^8)$
$\quad - 27b^6e^6 + 234b^5cde^5 - 32(b^5d^3e^4) - 713b^4c^2d^2e^4 + 246(b^4cd^4e^3)$
$\quad - 4(b^4d^6e^2) + 866b^3c^3d^3e^3 - 550(b^3c^2d^5e^2) + 56(b^3cd^7e) + 4(b^3d^9)$
$\quad - 139b^2c^4d^4e^2 + 354(b^2c^3d^6e) - 83(b^2c^2d^8) - 330bc^5d^5e$
$\quad + 72(bc^4d^7) - 16c^6d^6$.

188   APPLICATIONS TO BINARY QUANTICS.

On inspecting this invariant it will be seen that it vanishes if b, c, d all vanish. Consequently the form $ax^5 + 5exy^4 + fy^5$, to which Mr. Jerrard has shown that the quintic can be brought by a non-linear transformation, is one to which no quintic can be brought by linear substitution unless $L = 0$.

221. We take J, K, L as the fundamental invariants of the quintic, and we proceed to show how all its other invariants can be expressed in terms of these. In the first place, it will be observed that the interchange either of x and y, or of x and z, is a linear transformation whose modulus is -1. Hence, if any invariant is such that when transformed it is multiplied by an even power of the modulus of transformation, it must, for the canonical form, be unaltered by any interchange of a, b, c; that is to say, it must be a symmetric function of these quantities. If the invariant is multiplied by an odd power of the modulus, it must, for the canonical form, be such as to change sign when any two of the quantities a, b, c are interchanged; it must therefore be of the form $(a-b)(b-c)(c-a)$ multiplied by a symmetric function of a, b, c. Now an invariant is in transformation multiplied by a power of the modulus equal to its weight. And (Art. 139) the weight of an invariant of the quintic whose order is n, is $\tfrac{5}{2}n$. A quintic cannot have an invariant of odd order in the coefficients. If the order is a multiple of 4 the weight is an even number, and the sign of the invariant is unaltered by the interchange of x and y. If the order be not divisible by 4, the invariant is what we have called *skew*, that is to say, such as to change sign when x and y are interchanged. Let us first examine the former kind, which, we have seen, must, for the canonical form be symmetric functions of a, b, c. Now, since $J = (bc + ca + ab)^2 - 4abc(a+b+c)$, $K = a^2b^2c^2(bc + ca + ab)$, $L = a^4b^4c^4$, (from which we infer $H = \tfrac{1}{4}(K^2 - JL) = a^5b^5c^5(a+b+c)$,*) it follows that if we are given any quintic, and transform it to the canonical form by a

---

* The reader must be careful to observe that though, in the case of the canonical form, $a^5b^5c^5(a+b+c)$, for example, is divisible by $a^4b^4c^4$, we have no right to infer that in general H is divisible by L, unless in cases where the quotient $abc(a+b+c)$ has been also proved to be an invariant.

substitution whose modulus is unity, the numerical values of the new a, b, c are given by the cubic

$$\alpha^3 - \frac{H}{L^{\frac{1}{2}}}\alpha^2 + \frac{K}{L^{\frac{3}{4}}}\alpha - L^{\frac{1}{4}} = 0.$$

Now the order of any symmetrical function of a, b, c will be equal to its weight in the coefficients of this cubic, and when this weight is a multiple of 4, it is easy to see that the symmetric function is a rational function of J, K, L.

Being given therefore any invariant whose order in the coefficients is a multiple of 4, it has been proved that we can write down a rational function of J, K, L which, for the canonical form, shall have the same value as this invariant, and therefore be always identical with it. And since it would be manifestly absurd to suppose an integral function of the coefficients to be equal to an irreducible fraction, it follows that every non-skew invariant is an integral function of J, K, L.

If we make a, b, c all equal 0, J, K, L all vanish. Hence when three roots of a quintic are all equal, these three invariants vanish.* If we make a, b, e, f all equal 0, J becomes $-32c^2d^2$, and L, $-16c^6d^6$, and therefore $J^3 - 2048L$ vanishes. Quintics therefore which have two pairs of equal roots must not only have the discriminant $= 0$, but also $J^3 = 2048L$.

222. The simplest skew invariant is got by forming the resultant of the quintic $ax^5 + by^5 + cz^5$, and its canonizant $abcxyz$. Substituting successively the three roots of the canonizant in the quintic, and multiplying together, we get for the resultant $a^5b^5c^5(b-c)(c-a)(a-b)$. This invariant therefore is of the eighteenth order. Previous to its discovery by M. Hermite,†

---

* In general all the invariants of a quantic vanish, if more than $\frac{1}{2}n$ of its roots be all equal. For it is easily seen that if half the coefficients, counting from one end, simultaneously vanish, it is impossible to make with the remaining coefficients any term of the proper weight (Art. 139).

† See *Cambridge and Dublin Mathematical Journal*, vol. IX. p. 172. M. Hermite works with a new canonical form, the x and y of which are the two factors of the quadratic covariant. The quintic then is supposed to be such that $ae - 4bd + 3c^2$, $bf - 4ce + 3d^2$ both vanish, and the quadratic covariant reduces to xy. The advantage of this is that the operating symbol thence derived is simply $\frac{d^2}{dx\,dy}$, and some of the covariants obtained by thus differentiating assume a very simple form. Notwithstanding I have preferred to work with Mr. Sylvester's canonical form which I find much more convenient.

the possibility of the existence of skew invariants had not been recognised. I took the trouble to calculate this invariant, and the result is printed (*Philosophical Transactions*, 1858, p. 455), but as it consists of nearly nine hundred terms I cannot afford room for it here. The leading terms are $a^7 d^2 f^6 - a^6 c^5 f^7$; in this, as in every skew invariant, the complementary terms having opposite signs, and the symmetrical terms vanishing. By the argument used in the last article, it is proved that every skew invariant of a quintic must be the product of this invariant I by a rational function of J, K, L.

223. The square of I being of the thirty-sixth degree can be expressed rationally in terms of J, K, L (Art. 221). The actual expression is easily found.

By forming the discriminant of the cubic (Art. 221)

$$\alpha^3 - \frac{H}{L^{\frac{2}{3}}}\alpha^2 + \frac{K}{L^{\frac{1}{3}}}\alpha - L^{\frac{1}{3}},$$

we obtain the product of the squares of the differences of a, b, c in terms of J, K, L, and thus have

$$I^2 L = H^2 K^2 + 18 H K L^2 - 27 L^4 - 4 K^3 L^2 - 4 H^3;$$

or putting for H its value $\frac{1}{4}(K^2 - JL)$, and dividing by L, we have

$$16 I^2 = JK^4 + 8 LK^3 - 2 J^2 LK^2 - 72 JKL^2 - 432 L^3 + J^3 L^2.$$

224. I do not purpose to enter into detail as to the different covariants* of the quintic. The most remarkable, after those already mentioned, are the linear covariants. If we operate twice with the quadratic covariant $bcyz + cazx + abxy$ on the quintic itself, the result will evidently be of the first degree, and for the canonical form will be $abc(bcx + cay + abz)$. If we eliminate between this covariant and the canonizant we get Hermite's invariant I; and if between this linear covariant and the quintic itself we get $I(J^2 - 3K)$. Thus, then, if I vanish, the quintic is immediately soluble, one of the roots being given by the linear covariant, as also if $J^2 = 3K$.

---

* The actual values in general of some of the simplest are given by Mr. Cayley, *Philosophical Transactions*, CXLVI. 125.

APPLICATIONS TO BINARY QUANTICS. 191

By operating with the linear covariant on the quadratic, we get another linear covariant of the seventh order, viz.

$$abc\{x(a^2c^2 - a^2b^2 + ab^2c - abc^2) + y(a^2b^2 - b^2c^2 + abc^2 - a^2bc)$$
$$+ z(b^2c^2 - c^2a^2 + a^2bc - ab^2c)\}.$$

It has been already proved that we can express all the covariants in terms of any two together with the invariants. Accordingly M. Hermite has used the transformation of taking the two linear covariants for x and y, when all the coefficients in the transformed quintic are found to be invariants. The actual values however are not simple, and I have not found any advantage in the use of this form of the equation. The reduction to this form will be impossible in the particular case where these two linear covariants are identical, which will be when their resultant $JK + 9L$ vanishes. By forming the Jacobian of the quadratic covariant and any other covariant, we obtain a covariant of the same degree in the variables as the latter, and of an order two higher in the coefficients. Thus, from the canonizant in this way we get another cubic covariant,

$$abc\{bc(y^2z - yz^2) + ca(z^2x - zx^2) + ab(x^2y - xy^2)\},$$

which might also have been got by operating with the canonizant on the Hessian. Quintics having linear covariants of every odd order above the third, it follows by the principle of reciprocity that all quantics of odd order above the third have linear covariants of the fifth degree in the coefficients.

225. It ought to have been stated earlier that the sign of the discriminant of any quantic enables us at once to determine whether it has an even or odd number of pairs of imaginary roots. Imagine the quantic resolved into its real quadratic factors, then (Art. 106) the discriminant of the quantic is equal to the product of the discriminants of all the quadratics, multiplied by the square of the product of the resultants of every pair of factors. These resultants are all real, and their squares positive, therefore, in considering the sign of the discriminant, we need only attend to the discriminants of the quadratic factors. But the square of the difference of the roots of a quadratic is positive when the roots are real, and negative when they are imaginary. It follows then that the product of the squares of

the differences of the roots of any quantic is positive when it has an even number of pairs of imaginary roots, and negative when it has an odd number. We have been accustomed to write the discriminant giving the positive sign to the product of the two extreme terms. This will have the same sign as the product of the squares of differences of the roots when the order of the quantic is of the form $4m$ or $4m+1$, and the opposite sign when the order is of the form $4m+2$ or $4m+3$. We see then, in the case of the quintic, that if the discriminant be positive, there will be either four imaginary roots or none; and if the discriminant be negative, there will be two imaginary roots. It remains then further to distinguish the cases when all the roots are real, and where only one is so.

226. In order to discriminate between the remaining cases, there are various ways in which we may proceed. The following* are, in their simplest forms, the criteria furnished by Sturm's theorem. Let J be the invariant as before, and

$H = b^2 - ac, \quad S = ae - 4bd + 3c^2, \quad T = ace + 2bcd - ad^2 - eb^2 - c^3,$

$M = a^2e^2 - a^2df + 3abcf - 3abde + 4acd^2 - 4ac^2e - 2b^3f$
$\qquad + 5b^2ce + 2b^2d^2 - 8bc^2d + 3c^4,$

then the leading terms in the Sturmian functions are proportional to $a, a, H, 5HS + 9aT, -HJ + 12SM + 4S^3 - 216T^2$, the last of course being the discriminant; and the conditions furnished by Sturm's theorem to discriminate the cases of four and no imaginary roots, are that when all the roots are real the three quantities $H, 5HS + 9aT, -HJ + \&c.$, must all be positive.

227. We may apply these conditions to the canonical form

$(c-a) x^5 + 5cx^4y + 10cx^3y^2 + 10cx^2y^3 + 5cxy^4 + (c-b) y^5,$

in which case the equality of all but two of the coefficients renders the direct calculation also easy. We easily find then that the constants are $c-a, c-a, ac, -a^2c^2$; and the fourth being essentially negative we need not proceed further, and we

---

* These values are given by Mr. M. Roberts, *Quarterly Journal*, vol. IV. p. 175. The reader who may use Mr. Cayley's tables of Sturmian functions (*Philosophical Transactions*, vol. CXLVII. p. 735) must be cautioned that the fourth and fifth functions are there given with wrong signs.

learn that the equation just written has always imaginary roots. We find then that when the invariant L of a quintic is positive, the roots of the equation cannot be all real. For L being, with sign changed, the discriminant of the canonizant, when L is positive, the roots of the canonizant are all real, and the quintic can be brought to the canonical form by a real transformation.

When L is negative, two factors of the canonizant are imaginary, and the canonical form is

$$a(-2x)^5 + \{c - d\sqrt{(-1)}\}\{x + y\sqrt{(-1)}\}^5 + \{c + d\sqrt{(-1)}\}\{x - y\sqrt{(-1)}\}^5,$$

which expanded is

$$dy^5 + 5cy^4x - 10dy^3x^2 - 10cy^2x^3 + 5dyx^4 + (c - 16a)x^5.$$

Writing for brevity $c^2 + d^2 = r^2$, I find for this form, the Sturmian constants to be, d, d, r^2, r^4, $r^2(-4a^2d^2 + 20acr^2 + 5r^4)$, and it would seem that the discriminant being positive, the roots are all real if d and $-4a^2d^2 + 20acr^2 + 5r^4$ are both positive.*

228. In practice the criteria† furnished by Sturm's theorem are more convenient than any other, because the functions to be calculated are of lower order in the coefficients. It is however theoretically desirable to express these criteria in terms of the invariants, and this is what has been effected by different methods by M. Hermite and by Mr. Sylvester. We proceed briefly to explain the principles of Mr. Sylvester's method, which is highly ingenious. We have seen already that when the invariants J, K, L are given, the a, b, c of the canonical form may be determined by a cubic equation; and we can infer that to every given system of values of J, K, L will correspond some quintic. But to every system of values of J, K, L will not correspond a *real* quintic. In fact we have seen, Art. 223,

---

* I give this result, though suspecting its accuracy, because it seems to me to disagree with the theory derived from the other methods.

† It may be noticed that there is no difficulty in writing down a multitude of criteria which *might* indicate the existence of imaginary roots; for any symmetric function of squares of differences of roots $\Sigma(\alpha - \beta)^2$, &c. must be positive if all the roots are real. We can without difficulty write down such functions which are also invariants; and which, if negative, show that the equation has imaginary roots. But then these may also be positive when the roots are imaginary, and the problem is to find some criterion or system of criteria, some one of which *must* fail to be satisfied when the roots are not all real.

that the J, K, L of every quintic with real coefficients are such that the quantity G is essentially positive; where G is

$$JK^4 + 8LK^3 - 2J^2LK^2 - 72JL^2K - 432L^3 + J^3L^2.$$

For G has been shown to be the perfect square of a real function of the coefficients of the general quintic, viz. $a^7d^5f^6 + \&c.$, this being the eliminant of the quintic and its canonizant, and therefore necessarily real. We may in the above substitute for K its value in the discriminant from the equation $J^2 - 128K = D$, and so write G

$$JD^4 - 4(J^3 + 2^6L)D^3 + (6J^3 - 29.2^{10}L)J^2D^2$$
$$- (4J^6 - 61.2^6J^3L - 9.2^{22}L^2)JD + (J^3 - 2^{11}L)^2(J^3 - 27.2^{10}L).$$

If now, to assist our conceptions, we take J, D, L for the co-ordinates* of a point in space; then $G = 0$ represents a surface; and points on one side of it making G positive answer to real quintics, while points on the other side making G negative† answer to quintics with imaginary coefficients.

229. Now, in the next place, we say that if the coefficients in an equation be made to vary continuously, the passage from real to imaginary roots must take place through equal roots. For, let any quantic $\phi(x)$ become by a small change of coefficients $\phi(x) + \varepsilon\psi(x)$, (where ε is infinitesimal), and let α be a real root of the first, $\alpha + h$ a root of the second; then we have $\phi(\alpha + h) + \varepsilon\psi(\alpha) = 0$; whence, since $\phi(\alpha) = 0$, we have $h\phi'(\alpha) + \varepsilon\psi(\alpha) = 0$, which gives a real value for h. The consecutive root $\alpha + h$ is therefore also real. But if $\phi'(\alpha)$ vanishes as well as $\phi(\alpha)$, the lowest term in the expansion of $\phi(\alpha + h)$ will be h^2, and the value of h may possibly be imaginary. When therefore the original quantic has equal roots, the corresponding roots of the consecutive quantic may be imaginary.

It follows then that if we represent systems of values of J, D, L, by points in space, in the manner indicated in the last article, two points will correspond to quintics having the same

---

* Mr. Sylvester takes L in the usual direction of x, J of y, and D of z.

† Points for which $G = 0$ answer to real quintics, and it is easy to see that in this case the equation is of the recurring form. For we have proved that when $G = 0$ two of the coefficients of the canonical form are equal. The equation is therefore of the form $ax^5 + ay^5 + b(x+y)^5 = 0$.

number of real roots, provided that we can pass from one to the other without crossing either the plane D or the surface G. If points lie on opposite sides of the plane D, we evidently cannot pass from one to the other without having at an intervening point $D=0$, at which point a change in the character of the roots might take place. If two points, both fulfilling the condition G positive, be separated by sheets of the surface G, we can not pass continuously from one of the corresponding quintics to the other; because when on crossing the surface we have G negative, the corresponding quintic has imaginary coefficients. But when two points are not separated in one of these ways, we can pass continuously from one to the other, without the occurrence of any change in the character of the corresponding quintics.

Now Mr. Sylvester's method consists in shewing, by a discussion of the surface G, that all points fulfilling the condition G positive,* may be distributed into three blocks separated from each other either by the plane D or the surface G. And since there may evidently be quintics of three kinds, viz. having four, two, or no imaginary roots, the points in the three blocks must correspond respectively to these three classes. I have not space for the elaborate investigation of the surface G, by which Mr. Sylvester establishes this; but the following is sufficient to enable the reader to convince himself of the truth of his conclusions.

230. One of the three blocks we may dispose of at once, viz. points on the negative side of the plane D, which we have seen (Art. 225) correspond to quintics having two imaginary roots. Next with regard to points for which D is positive. We have seen, in the last article, that a change in the character of the roots only takes place when $D=0$; our attention is therefore directed to the section of G by the plane D. We see at once, by making $D=0$ in the value of G (Art. 228), that the remainder has a square factor, and consequently that the surface G touches D along the curve $J^8 - 2^{11}L$, and cuts it along $J^3 - 27.2^{10}L$. Now, if a surface merely cut a plane, the

---

* Mr. Sylvester calls these facultative points.

line of section is no line of separation between points on the same side of the surface. If, for example, we put a cup on a table, there is free communication between all the points inside the cup and between all those outside it. But if a plane touch a surface, as, for instance, if we place a cylinder on a table, then while there is still free communication between the points inside the cylinder, the line of contact acts as a boundary line, cutting off communication as far as it extends, between points outside the cylinder on each side of the boundary.

Now Mr. Sylvester's assertion is, that if we take the negative quadrant, viz. that for which both J and L are negative, and if we draw in the plane of xy, the curve $J^3 - 2^{11}L$; then all facultative points in that quadrant, lying above the space included between the curve and the axis $L = 0$, form a block completely separated from the rest, and correspond to the case of five real roots.

231. In order to see the character of the surface, I form the discriminant of G considered as a function of K, which I find to be $-L^7(J+27L)^3$. Consequently, when both J and L are negative, the discriminant is negative, and the equation in K has only two real roots. To every system of values, therefore, of J and L correspond two values of K, and consequently two values of D, and the surface is one of two sheets. Now I say that it is the space between these sheets for which G is positive. In fact, since G is $JD^4 + \&c.$, it may be resolved into its factors $J(D-\alpha)(D-\beta)\{(D-\gamma)^2+\delta^2\}$; and since J is supposed to be negative in the space under consideration, D must evidently be intermediate between α and β in order that G should be positive. Now the last term of the equation being $(J^3-2^{11}L)^2(J^3-27.2^{10}L)$, if J^3 be nearly equal to $2^{11}L$, will be of opposite sign to L, or in the present case will be positive. And the coefficient of D^4 being negative, we see that on both sides of the line $J^3 = 2^{11}L$; the values of D are, one positive and the other negative, that is to say, the two sheets of the surface are one above and the other below the plane D. But I say it is the upper sheet which touches D along $J^3 - 2^{11}L$. This may be seen immediately by looking at the sign of the penultimate term in the equation for D, by which we see that when the last term vanishes, the

two roots are 0 and negative. The theory then already explained shows that the curve $J^3 = 2^{11}L$ acts as a boundary line cutting off communication in that direction between facultative points on the upper side of D. But again communication in the other direction is cut off by the plane $L=0$. For when we make L positive, the discriminant becomes positive, and the equation in D has either four real or four imaginary roots. But the first Sturmian constant is proportional to $L(J^3 + 12L)$, which, when J is negative, and L positive and small, is negative. Immediately beyond the plane L, therefore, the equation to determine D has four imaginary roots, or the surface does not exist. The facultative points therefore, lying as they do *within* the surface or between its sheets, are cut off by the plane L, on which the sheets unite, from communication with points beyond it. Thus the isolation of the block under consideration has been proved.

I need enter into equal detail to prove that all other facultative points have free communication *inter se*. The line of contact $2^{11}L - J^3$ is no line of separation in the quadrant where J and L are both positive. For then it is seen, as before, that it is the points *outside* the two sheets which are facultative, and not the points between the surface and touching plane.

The result of this investigation is, that in order to have all the roots real, we must have the quantity $2^{11}L - J^3$ positive;[*] and L negative, which also infers J negative. If either condition fails, our roots are imaginary. It is supposed that in both cases D is positive.

232. We have seen that the cylinder parallel to the axis of z and standing on the curve $2^{11}L - J^3$ does not meet G above the plane D; the two values of z being one 0, the other negative. Any other surface then standing on the same curve and not meeting G would serve equally well as a wall of separation between the two classes of facultative points. For all the

---

[*] Mr. Sylvester has inadvertently stated his condition to be that $2^{11}L - J^3$ is negative. It is easy to see, however, that what he has proved is that this quantity must be positive. For the block which he has described lies on the side of the curve $2^{11}L - J^3$ next to the axis $L=0$. But when L is 0 and J negative, $2^{11}L - J^3$ is positive.

points between the cylinder and this surface would be non-facultative, and therefore irrelevant to the question. Mr. Sylvester has thus seen that we may substitute for the criterion $2^{11}L - J^3$, $2^{11}L - J^3 + \mu JD$, provided that the second represent a surface not meeting G above the plane D. And on investigating within what limits μ must be taken, in order to fulfil this condition, he finds that μ may be any number between 1 and -2.

He avails himself of this to give criteria expressed as symmetrical functions of the roots. In the first place

$$\Sigma (\alpha - \beta)^2 (\beta - \gamma)^2 (\gamma - \alpha)^2 (\delta - \varepsilon)^4$$

is an invariant (Art. 132), and being of the same order and weight as J can only differ from it by a numerical factor: which factor must be negative, since this function is essentially positive; and J we have seen is essentially negative when the roots are all real. And secondly, the symmetric function

$$\Sigma (\alpha-\beta)^2 (\beta-\gamma)^2 (\gamma-\alpha)^2 (\varepsilon-\alpha)^4 (\varepsilon-\beta)^4 (\varepsilon-\gamma)^4 (\delta-\alpha)^4 (\delta-\beta)^4 (\delta-\gamma)^4,$$

(the relation of which to the other may be seen by writing it in the form $D^2 \Sigma (\alpha - \beta)^{-2} (\beta - \gamma)^{-2} (\gamma - \alpha)^{-2} (\delta - \varepsilon)^{-4}$, where D is the discriminant), is also an invariant, and of the twelfth order. It must therefore be of the form $\alpha J^3 + \beta JD + \gamma L$. Now, by using the quintic* $x (x^2 - a^2) (x^2 - b^2)$, the symmetric function may easily be calculated and identified with the invariants; and the result is that its value is proportional to $2^{11}L - J^3 + \frac{4}{5}JD$. Since then the numerical multiplier of JD is within the prescribed limits, it may be used as a criterion; and Mr. Sylvester's result is that the two symmetrical functions mentioned are such that not only are both positive, as is evident, if the roots are all real, but also if both are positive, and D positive, the roots *must* be all real. It ought to be possible to verify this directly by examining the form of these functions in the case of an equation with four imaginary roots.

---

* It is to be observed that though this form may be safely used in this case, it cannot always be safely used. For when a linear factor of a quintic is also a factor in the Jacobian of the remaining quartic, a relation must exist between the invariants, which, however, I have not taken the trouble to calculate, it being obvious that it is of too high a degree in J, D, L to affect the present question.

APPLICATIONS TO BINARY QUANTICS. 199

233. I have also tried to verify these results by examining the invariants of the product of a linear factor and a quartic, $(ax + \beta y)(x^4 + 6mx^2y^2 + y^4)$; these being necessarily covariants of the quartic (Art. 205). The coefficients of the quintic are then $5\alpha, \beta, 3m\alpha, 3m\beta, \alpha, 5\beta$; and I find for the J of the quintic, $48(8SH - 3TU)$, or 48 times

$$(5m + 27m^3)(\alpha^4 + \beta^4) + (8 - 18m^2 - 54m^4)\alpha^2\beta^2.$$

Now the roots of the quartic are all real when m is negative, and when $9m^2$ is greater than 1. On inspection of the value given for J, we see that when m is negative every term but one is negative. Giving then m its smallest negative value $-\frac{1}{3}$, J is negative, viz. $-144(\alpha^2 - \beta^2)^2$; and J is à fortiori negative for every greater negative value of m. Or we may see the same thing by supposing $\beta = 0$, when we have only to look at the coefficient of the highest power of α in $8SH - 3TU$, which is $-8(b^2 - ac)S - 3Ta$. But now if we call the three Sturmian constants A, B, C: viz.

$$A = b^2 - ac, \quad B = 2SA + 3Ta, \quad C = S^3 - 27T^2,$$

the value given for J becomes $-6AS - B$, which is essentially negative when the roots are all real.

The invariant L, according to my calculation, is

$$54(8SH - 3TU)^2 - 6400(S^3 - 27T^2)(4H^3 - SHU + TU^2)$$
$$+ 150(S^3 - 27T^2)U^2(8SH + 15TU) - 4050U^2S^2(2SH - 3TU),$$

whence $2^{11}L - J^2$ differs only by a positive constant multiplier from

$$-128(S^3 - 27T^2)(4H^3 - 3HU + TU^2)$$
$$+ 3(S^3 - 27T^2)U^2(8SH + 15TU) - 81U^2S^2(2SH - 3TU).$$

Writing $\theta = a^2d - 3abc + 2b^3$, the coefficient of the highest power of a in this is

$$128C\theta^2 + 81a^2S^3 + 45a^2CB - 54a^2CSA.$$

All the terms of this but one are positive when the roots are all real, but as there is one negative term, it is not obvious on the face of the formula, that the whole will be positive when the roots are all real. Still less that if this formula be positive and J negative, the roots are necessarily all real. Therefore, although no doubt Mr. Sylvester's rule may be tested by the process here indicated, to do so requires a closer examination of this formula than I am able to give.*

234. It does not enter into the plan of these Lessons to give an account of the researches to which the problem of resolving the quintic has given rise.† The following however finds a place here on account of its connection with the theory of invariants. Lagrange, as is well known, made the solution of a quintic to depend on the solution of a sextic; and it can easily be proved that functions of five letters can be formed capable of six values by transformation of letters. Let 12345 denote any cyclic function of the roots of a quintic; such, for example, as the product

$$(\alpha - \beta)^2 (\beta - \gamma)^2 (\gamma - \delta)^2 (\delta - \epsilon)^2 (\epsilon - \alpha)^2,$$

where evidently 23451 and 15432 would denote the same as 12345; then it can easily be seen that there can be written down in all twelve such cyclic functions. But further these distribute themselves into pairs; and by so grouping them we can form a function capable of only six values; for instance, 12345 + 13524, 12435 + 14523, 13245 + 12534, 13425 + 14532, 14235 + 12543, 14325 + 13542. The actual formation of the

---

* The verification, however, is easy in the particular case $x(x^4 + 6mx^2y^2 + y^4)$. We have then $J = 48m(5 + 27m^2)$, $L = 12m(5 - 9m^2)^4$; $2^{11}L - J^3$ proportional to $m(1 - 9m^2)(50 + 45m^2 + 648m^4 + 729m^6)$. Thus, when m is negative and $9m^2 > 1$, we have J and L negative and $2^{11}L - J^3$ positive. The latter is positive for imaginary roots only when m is positive, but in this case J is positive. The imaginary roots must therefore be detected by one criterion or other.

The discussion of the invariantive characteristics of the reality of the roots of a quintic was originally commenced by M. Hermite in his classical paper in the *Cambridge and Dublin Mathematical Journal* for 1854, and has been resumed by him in his valuable memoir presented to the French Academy this year. His result, translated into the notation we have used, is that the roots are all real when, the discriminant being positive, we have also positive K, $2^{11}L - J^3 + D$, and $K(JL + K^2) - 18L^2$. It seems to me that this result is superseded by the greater simplicity of Mr. Sylvester's criteria. I should have wished however to give an account of M. Hermite's method, as well as of the many other important additions he has made to the theory of quintics. But having neglected to read his papers as they came out in the *Comptes Rendus*, I find, to my great regret, on taking up his memoir now, that I have not time to make myself master of it without delaying indefinitely the publication of this volume. I hope to find a place further on for some account of his application of the theory of invariants to the Tschirnhausen transformation of equations.

† Among the most remarkable of recent discoveries in this subject is the application to it of the theory of elliptic functions by M. Hermite and M. Kronecker.

sextic having these values for its roots is in most cases a work of extreme labour. M. Hermite, however, pointed out that when the function 12345 is the product of the squares of differences written above,* all the coefficients of the corresponding sextic are invariants, and that the calculation therefore is practicable. I have thought it desirable actually to form the equation, because, when the theory of sextics comes to be studied, it will be necessary to ascertain the invariantive characteristics of sextics whose solution depends on that of a quintic; and it may be useful to be in possession of more than one of the sextics which spring out of the discussion of a quintic.† I take the simple example $x^5 + 2mx^3y^2 + xy^4$, of which, since two pairs of roots are equal with opposite signs, the functions of the differences can easily be formed. I find then that the sextic is the product of

$$t^2 + 2^6(m + m^3)t + 2^{10}(m^6 - 2m^4 + 5m^2),$$

by the square of

$$t^2 + 2^4(m^3 + 3m)t + 2^6(m^6 + 5m^4 + 19m^2 - 25).$$

But if we first multiply the quintic by five, its invariants are

$$J = 2^4 m(5 + 3m^2), \quad D = 2^6 . 5^3 (1 - m^2)^2, \quad L = 4m(5 - m^2)^4.$$

To avoid fractions I write $J = 2A$, $D = 250B$, $J^3 - 2^{11}L = 50C$; and then forming the sextic, and expressing its coefficients in terms of the invariants, I obtain

$t^6 + 4At^5 + (6A^2 - 25B)t^4 + (4A^3 + 2C - 30AB)t^3$
$+ t^2(A^4 + 4AC - 17A^2B + \tfrac{625}{4}B^2)$
$+ t(2A^2C - 4A^3B - 7BC + 110AB^2) + C^2 - 4ABC + 20A^2B^2;$

which is a perfect square, as it ought to be, when $D = 0$.‡

---

* In the method of Messrs. Harley and Cockle, the function 12345 is

$$a\beta + \beta\gamma + \gamma\delta + \delta\varepsilon + \varepsilon a,$$

and the sextic chosen is that whose roots are $12345 - 13524$, &c. This has been calculated by Mr. Cayley (*Philosophical Transactions*, 1861, p. 263), and the result is very simple, two terms of the sextic being wanting; but the coefficients are not invariants.

† The form arrived at by M. Kronecker and M. Brioschi is

$$(x - a)^5 (x - 5a) + 10b(x - a)^3 - c(x - a) + 5b^2 - ac = 0.$$

By the help of the formulæ given further on, the invariants of this equation can be calculated, and a, b, c eliminated.

‡ Though the form with which I have worked is a special one, I believe that the result is general; because it seemed to me that the coefficients only admitted of being expressed in terms of the invariants in one way.

235. M. Hermite has studied in detail the expression of the invariants in terms of the roots. He uses the equation transformed so as to want the first and last terms; that is to say, so that one root is 0 and another infinite; and the calculation is thus reduced to forming symmetric functions of the roots of a cubic. I had been led independently to try the same transformation on the problem discussed in the last article, but found that, even when thus simplified, the problem remained a difficult one. It would be necessary to form for a cubic the sextic whose roots are the six values of

$$\alpha^2(\beta-\gamma)^2(\alpha-\gamma)^2 + \beta^2\gamma^2(\alpha-\beta)^2,$$

and then to identify the result with combinations of the forms assumed by the invariants of the quintic when a and f vanish. Of M. Hermite's results I only give his expression for his own invariant I. Let

$$F = (\alpha-\beta)(\alpha-\varepsilon)(\delta-\gamma) + (\alpha-\gamma)(\alpha-\delta)(\beta-\varepsilon),$$
$$G = (\alpha-\beta)(\alpha-\gamma)(\varepsilon-\delta) + (\alpha-\delta)(\alpha-\varepsilon)(\beta-\gamma),$$
$$H = (\alpha-\beta)(\alpha-\delta)(\varepsilon-\gamma) + (\alpha-\gamma)(\alpha-\varepsilon)(\delta-\beta);$$

the continued product of these is symmetrical with respect to all the roots except α; and if we multiply this product by the similar products obtained for the other four roots we get I.

236. *The Sextic.* The theory of the sextic has as yet been but little studied. It has four independent invariants, which we shall call A, B, C, D, of the orders 2, 4, 6, 10 respectively; and a fifth skew invariant, which we shall call E, of the fifteenth order, whose square is a rational and integral function of the other four. The first, A, is the invariant $\overline{12}^6$, formed by the method of p. 112, which for the general sextic is

$$ag - 6bf + 15ce - 10d^2.$$

I have given (Arts. 170, 171) the canonical form of the sextic; but I believe it will be found in practice not less convenient to use (as in Art. 217) the more general form

$$au^6 + bv^6 + cw^6 + dz^6.$$

To this we should be led by the theory of two quintics, which cannot be more simply expressed than as each the sum of four fifth powers. For the form just given, the invariant A is, by proceeding as in Art. 217, found to be

$$ab\,(12)^6 + ac\,(13)^6 + ad\,(14)^6 + bc\,(23)^6 + bd\,(24)^6 + cd\,(34)^6,$$

which we may write $\Sigma ab\,(12)^6$.

The Hessian of the sextic $\overline{12}^2$ is of the eighth degree, the general coefficients being $ac - b^2$, $4\,(ad - bc)$, $6ae + 4bd - 10c^2$, $4af + 16be - 20cd$, $ag + 14bf + 5ce - 20d^2$,* &c.; and for my canonical form is $\Sigma abu^4v^4\,(12)^2$. The sextic has another covariant of the second order in the coefficients, viz. the S of the emanant quartic, which is of the fourth order in the variables, the general coefficients being

$$ae - 4bd + 3c^2,\ \ 2af - 6be + 4cd,\ \ ag - 9ce + 8d^2,\ \ \&\text{c.},$$

this for the canonical form being $\Sigma abu^2v^2\,(12)^4$. To these covariants we may add the covariant sextic, of the third order, which is the T of the quartic emanant, and whose general coefficients are

$$ace + 2bcd - ad^2 - eb^2 - c^3,\ \ 2acf - 2ade - 2b^2f + 2bce + 2bd^2 - 2c^2d,$$

$$acg + 2adf - 3ae^2 - b^2g - 2bcf + 4bde + 2c^2e - 3cd^2,$$

$$2adg - 2aef - 2bcg + 4bdf - 2be^2 - 2c^2f + 6cde - 4d^3,\ \&\text{c.};$$

and which for the canonical form is $\Sigma abc\,(12)^2(23)^2(31)^2 u^2v^2w^2$.

237. We take for the invariant B, that which has been called by Mr. Sylvester the *catalecticant*, which expresses the condition that the sextic should be reducible to the sum of three sixth powers, and is (Art. 168) the determinant

$$\begin{vmatrix} a, & b, & c, & d \\ b, & c, & d, & e \\ c, & d, & e, & f \\ d, & e, & f, & g \end{vmatrix}.$$

This expanded is

$$aceg - acf^2 - ad^2g + 2adef - ae^3 - b^2eg + b^2f^2 + 2bcdg - 2bcef$$
$$- 2bd^2f + 2bde^2 - c^3g + 2c^2df + c^2e^2 - 3cd^2e + d^4.$$

If now we form the quadrinvariant of the Hessian, we find it proportional to $A^2 + 300B$; if that of the covariant S, we find

---

* I have thought it unnecessary to add the terms which may be written down from symmetry.

$A^2 - 36B$; and if we operate on the sextic with the covariant T, we get B. Applying this last process then to the canonical form, we get, for the value of B,

$$abcd (12)^2 (23)^2 (34)^2 (41)^2 (13)^2 (24)^2,$$

which vanishes, as it ought, if any of the quantities a, b, c, d vanishes, or if any two of the four functions u, v, w, z become identical.

238. We take for the form of the fundamental sextinvariant C, that which involves no power higher than the second of the leading coefficient a, and which for the general form is

$a^2d^2g^2 - 6a^2defg + 4a^2df^3 + 4a^2e^3g - 3a^2e^2f^2 - 6abcdg^2 + 18abcefg$
$- 12abcf^3 + 12abd^2fg - 18abde^2g + 6abe^3f + 4ac^3g^2 - 24ac^2e^2g$
$- 18ac^2dfg + 30ac^2ef^2 + 54acd^2eg - 12acd^2f^2 - 42acde^2f$
$+ 12ace^4 - 20ad^4g + 24ad^3ef - 8ad^2e^3 + 4b^3dg^2 - 12b^3efg$
$+ 8b^3f^3 - 3b^2c^2g^2 + 30b^2ce^2g - 24b^2cef^2 - 12b^2d^2eg - 24b^2d^2f^2$
$+ 60b^2de^2f - 27b^2e^4 + 6bc^3fg - 42bc^2deg + 60bc^2df^2 - 30bc^2e^2f$
$+ 24bcd^3g - 84bcd^2ef + 66bcde^3 + 24bd^4f - 24bd^3e^2 + 12c^4eg$
$- 27c^4f^2 - 8c^3d^2g + 66c^3def - 8c^3e^3 - 24c^2d^3f - 39c^2d^2e^2 + 36cd^4e - 8d^6$.

In terms of these the other invariants of the sixth order can be expressed. Thus, the cubinvariant of the covariant quartic is $A^3 - 108AB - 54C$; the cubinvariant of the Hessian is $3A^3 - 100AB + 2750C$; and the quadrinvariant of the sextic covariant is $2AB - C$. The last-named invariant can be easily calculated in the case of the canonical form. We have to operate with $\Sigma abc (12)^2 (23)^2 (31)^2 u^2v^2w^2$ on itself. Now if we operate with $u^2v^2z^2$ on $u^2v^2w^2$ the result is proportional to $(12)^2 MN$, where M and N have the same meaning as in Art. 217; and if with $u^2v^2w^2$ on itself the result is $- (12)^2 (23)^2 (31)^2$. Hence we get for the invariant in question

$$\Sigma a^2b^2c^2 (12)^6 (23)^6 (31)^6 - 2abcd \Sigma ab (12)^6 M^2N^2.$$

239. If a, b, c all vanish, the invariants A, B, C become respectively $- 10d^2, d^4, - 8d^6$. Hence, when the sextic has as factor a perfect cube, the conditions must be fulfilled $A^2 = 100B$,

APPLICATIONS TO BINARY QUANTICS.   205

$4AB = 5C$, $AC = 80B^2$. If we make a, b, f, g all $= 0$, the invariants become

$$15ce - 10d^2, \quad c^2e^2 - 3cde + d^4, \quad -8c^3e^3 - 39c^2d^2e^2 + 36cd^4e - 8d^6;$$

consequently when the sextic has two square factors; in addition to the discriminant, the condition must be satisfied,

$$(A^3 - 300AB + 250C)^2 = 5(A^2 - 100B)^3.$$

240. If we make $b, d, f = 0$ in the equation, the discriminant will be ag multiplied by the square of the discriminant of $(a, 5c, 5e, g\chi x, y)^3$; and if all the terms vanish but a, d, g, the discriminant will be a^2g^2 multiplied by the cube of the discriminant of $(a, 10d, g\chi x, y)^2$. Knowing these terms in the discriminant, the rest can be calculated by means of the differential equation. The resulting value of Δ is

	a^5g^5	+	$37500a^3bef^5$	+	$250000a^3e^6g$
−	$30a^4bfg^4$	+	$1000a^3c^5g^4$	−	$150000a^3e^5f^2$
−	$300a^4ceg^4$	−	$27000a^3c^4dfg^3$	−	$2500a^2b^3dg^4$
+	$375a^4cf^2g^3$	+	$18750a^3c^2e^2g^3$	+	$750a^2b^3efg^3$
−	$300a^4d^2g^4$	+	$16875a^3c^2ef^2g^2$	−	$410a^2b^3f^3g^2$
+	$3000a^4defg^3$	−	$9375a^3c^2f^4g$	−	$7500a^2b^3c^2g^4$
−	$2500a^4df^3g^2$	−	$7500a^3cd^2eg^3$	+	$43500a^2b^3cdfg^3$
+	$1000a^4e^3g^3$	+	$127500a^3cd^2f^2g^2$	+	$16875a^2b^3ce^2g^3$
−	$7500a^4e^2f^2g^2$	+	$30000a^3cde^2fg^2$	−	$54675a^2b^3cef^2g^2$
+	$9375a^4ef^4g$	−	$412500a^3cdef^3g$	+	$25500a^2b^3cf^4g$
−	$3125a^4f^6$	+	$187500a^3cdf^5$	+	$127500a^2b^3d^2eg^3$
+	$375a^3b^2eg^4$	−	$150000a^3ce^4g^2$	−	$171300a^2b^3d^2f^2g^2$
−	$15a^3b^2f^2g^3$	+	$412500a^3ce^3f^2g$	−	$346500a^2b^3de^2fg^2$
+	$3000a^3bcdg^4$	−	$187500a^3ce^2f^4$	+	$616500a^2b^3def^3g$
−	$5550a^3bcefg^3$	+	$30000a^3d^4g^3$	−	$240000a^2b^3df^5$
+	$750a^3bcf^3g^2$	−	$330000a^3d^3efg^2$	+	$7500a^2b^3e^4g^2$
−	$4800a^3bd^2fg^3$	+	$50000a^3d^3f^3g$	−	$23250a^2b^3e^3f^2g$
−	$27000a^3bde^2g^3$	+	$250000a^3d^2e^3g^2$	+	$11250a^2b^3e^2f^4$
+	$43500a^3bdef^2g^2$	+	$675000a^3d^2e^2f^2g$	+	$57000a^2bc^4fg^3$
−	$7500a^3bdf^4g$	−	$375000a^3d^2ef^4$	+	$30000a^2bc^3deg^3$
+	$57000a^3be^3fg^2$	−	$900000a^3de^3fg$	−	$346500a^2bc^3df^2g^2$
−	$97500a^3be^2f^3g$	+	$500000a^3de^5$	−	$596250a^2bc^3e^2fg^2$

206   APPLICATIONS TO BINARY QUANTICS.

$+ 1222500a^2bc^2ef^3g$ $\quad$ $- 9750000a^2cd^3e^2fg$ $\quad$ $- 2190000ab^4cd^3fg^2$
$- 506250a^2bc^2f^5$ $\quad$ $+ 5250000a^2cd^3ef^3$ $\quad$ $+ 4725000ab^3cd^2e^2g^2$
$- 330000a^2bcd^3g^3$ $\quad$ $+ 3750000a^2cd^2e^4g$ $\quad$ $+ 6030000ab^3cd^2ef^2g$
$+ 1590000a^2bcd^2efg^2$ $\quad$ $- 2250000a^2cd^2e^3f^2$ $\quad$ $- 3360000ab^3cd^2f^4$
$- 330000a^2bcd^2f^3g$ $\quad$ $- 1000000a^2d^6g^2$ $\quad$ $- 15337500ab^3cde^2fg$
$+ 750000a^2bcde^3g^2$ $\quad$ $+ 3000000a^2d^5efg$ $\quad$ $+ 8392500ab^3cde^2f^3$
$- 3172500a^2bcde^2f^2g$ $\quad$ $- 1600000a^2d^5f^3$ $\quad$ $+ 5062500ab^3ce^5g$
$+ 1537500a^2bcdef^4$ $\quad$ $- 1250000a^2d^4e^3g$ $\quad$ $- 3037500ab^3ce^4f^2$
$+ 375000a^2bce^4fg$ $\quad$ $+ 750000a^2d^4e^2f^2$ $\quad$ $- 300000ab^3d^4eg^2$
$- 225000a^2bce^3f^3$ $\quad$ $+ 9375ab^4cg^4$ $\quad$ $+ 900000ab^3d^3e^2fg$
$+ 780000a^2bd^4fg^2$ $\quad$ $- 7500ab^4dfg^3$ $\quad$ $- 480000ab^3d^3ef^3$
$- 1350000a^2bd^3e^2g^2$ $\quad$ $- 9375ab^4e^2g^3$ $\quad$ $- 375000ab^3d^2e^4g$
$- 2190000a^2bd^3ef^2g$ $\quad$ $+ 25500ab^4ef^2g^2$ $\quad$ $+ 225000ab^3d^2e^3f^2$
$+ 1200000a^2bd^3f^4$ $\quad$ $- 11520ab^4f^4g$ $\quad$ $- 900000abc^4dg^3$
$+ 4650000a^2bd^2e^3fg$ $\quad$ $- 97500ab^3c^2fg^3$ $\quad$ $+ 375000abc^4efg^2$
$- 2550000a^2bd^2e^2f^3$ $\quad$ $- 412500ab^3cdeg^3$ $\quad$ $- 202500abc^4f^3g$
$- 1500000a^2bde^5g$ $\quad$ $+ 616500ab^3cdf^2g^2$ $\quad$ $+ 4650000abc^3d^2fg^2$
$+ 900000a^2bde^4f^2$ $\quad$ $+ 1222500ab^3ce^2fg^2$ $\quad$ $+ 4875000abc^3de^2g^2$
$- 150000a^2c^4eg^3$ $\quad$ $- 2197800ab^3cef^3g$ $\quad$ $- 15337500abc^3def^2g$
$+ 7500a^2c^4f^2g^2$ $\quad$ $+ 864000ab^3cf^5$ $\quad$ $+ 7087500abc^3df^4$
$+ 250000a^2c^3d^3g^3$ $\quad$ $+ 50000ab^3d^4g^3$ $\quad$ $+ 843750abc^3e^3fg$
$+ 750000a^2c^3defg^2$ $\quad$ $- 330000ab^3d^3efg^2$ $\quad$ $- 506250abc^3e^2f^3$
$+ 37500a^2c^3df^3g$ $\quad$ $+ 83200ab^3d^3f^3g$ $\quad$ $- 9750000abc^2d^3eg^2$
$+ 1062500a^2c^3e^3g^2$ $\quad$ $+ 37500ab^3de^5g^2$ $\quad$ $+ 900000abc^2d^3f^2g$
$- 2821875a^2c^3e^2f^2g$ $\quad$ $+ 511500ab^3de^2f^2g$ $\quad$ $+ 24750000abc^2d^2e^2fg$
$+ 1265625a^2c^3ef^4$ $\quad$ $- 288000ab^3def^4$ $\quad$ $- 13350000abc^2d^2ef^3$
$- 1350000a^2c^2d^3fg^2$ $\quad$ $- 202500ab^3e^4fg$ $\quad$ $- 9375000abc^2de^4g$
$- 3562500a^2c^2d^2e^2g^2$ $\quad$ $+ 121500ab^3e^3f^3$ $\quad$ $+ 5625000abc^2de^3f^2$
$+ 4725000a^2c^2d^2ef^2g$ $\quad$ $+ 412500ab^2c^3eg^3$ $\quad$ $+ 3000000abcd^5g^2$
$- 2062500a^2c^2d^2f^4$ $\quad$ $- 23250ab^2c^3f^2g^2$ $\quad$ $- 9000000abcd^4efg$
$+ 4875000a^2c^2de^3fg$ $\quad$ $+ 675000ab^2c^2d^2g^3$ $\quad$ $+ 4800000abcd^4f^3$
$- 2625000a^2c^2de^2f^3$ $\quad$ $- 3172500ab^2c^2defg^2$ $\quad$ $+ 3750000abcd^3e^3g$
$- 1875000a^2c^2e^5g$ $\quad$ $+ 511500ab^2c^2df^3g$ $\quad$ $- 2250000abcd^3e^2f^2$
$+ 1125000a^2c^2e^4f^2$ $\quad$ $- 2821875ab^2c^2e^3g^2$ $\quad$ $+ 250000ac^5g^3$
$+ 3750000a^2cd^5eg^2$ $\quad$ $+ 7633125ab^2c^2e^2f^2g$ $\quad$ $- 1500000ac^5dfg^2$
$- 300000a^2cd^4f^2g$ $\quad$ $- 3442500ab^2c^2ef^4$ $\quad$ $- 1875000ac^5e^2g^2$

APPLICATIONS TO BINARY QUANTICS. 207

$$
\begin{array}{lll}
+\ 5062500ac^5ef^2g & -\ 288000b^4cdf^3g & +\ 72000000b^3cd^2ef^3 \\
-\ 2278125ac^5f^4 & +\ 1265625b^4ce^3g^2 & +\ 5062500b^3cde^4g \\
+\ 3750000ac^4d^2eg^2 & -\ 3442500b^4ce^2f^2g & -\ 3037500b^3cde^3f^2 \\
-\ 375000ac^4d^3f^2g & +\ 1555200b^4cef^4 & -\ 1600000b^3d^5g^2 \\
-\ 9375000ac^4de^3fg & +\ 1200000b^4d^3fg^2 & +\ 4800000b^3d^4efg \\
+\ 5062500ac^4def^3 & -\ 2062500b^4d^2e^2g^2 & -\ 2560000b^3d^4f^3 \\
+\ 3515625ac^4e^3g & -\ 3360000b^4d^2ef^2g & -\ 2000000b^3d^3e^3g \\
-\ 2109375ac^4e^2f^2 & +\ 1843200b^4d^2f^4 & +\ 1200000b^3d^3e^2f^2 \\
-\ 1250000ac^3d^4g^2 & +\ 7087500b^4de^3fg & -\ 150000b^3c^2g^3 \\
+\ 3750000ac^3d^3efg & -\ 3888000b^4de^2f^3 & +\ 900000b^3c^4dfg^2 \\
-\ 2000000ac^3d^3f^3 & -\ 2278125b^4e^5g & +\ 1125000b^2c^4e^2g^2 \\
-\ 1562500ac^3d^2e^3g & +\ 1366875b^4e^4f^2 & -\ 3037500b^2c^3ef^2g \\
+\ 937500ac^3d^2e^2f^2 & +\ 500000b^3c^5dg^3 & +\ 1366875b^2c^4f^4 \\
-\ 3125b^6g^4 & -\ 225000b^3c^5efg^2 & -\ 2250000b^2c^3d^2eg^2 \\
+\ 37500b^5cfg^3 & +\ 121500b^3c^5f^3g & +\ 225000b^2c^3d^2f^2g \\
+\ 187500b^5deg^3 & -\ 2550000b^3c^4d^2fg^2 & +\ 5625000b^2c^3de^2fg \\
-\ 2400000b^5df^2g^2 & -\ 2625000b^3c^4de^2g^2 & -\ 3037500b^2c^3def^3 \\
-\ 5062500b^5e^2fg^2 & +\ 8392500b^3c^4def^2g & -\ 2109375b^2c^3e^4g \\
+\ 864000b^5ef^3g & -\ 3888000b^3c^4df^4 & +\ 1265625b^2c^3e^3f^2 \\
-\ 331776b^5f^5 & -\ 506250b^3c^4e^3fg & +\ 750000b^2c^2d^4g^2 \\
-\ 187500b^4c^2eg^3 & +\ 303750b^3c^4e^2f^3 & -\ 2250000b^2c^2d^3efg \\
+\ 11250b^4c^2f^2g^2 & +\ 5250000b^3cd^3eg^2 & +\ 1200000b^2c^2d^3f^3 \\
-\ 375000b^4cd^2g^3 & -\ 480000b^3cd^3f^2g & +\ 937500b^2c^2d^2e^2g \\
+\ 1537500b^4cdefg^2 & -\ 13350000b^3cd^2e^2fg & -\ 562500b^2c^2d^2e^2f^2
\end{array}
$$

241. Instead of the discriminant Δ we may use another invariant D, in which no higher power than the fourth of the extreme coefficients a, g appears, and which does not contain the product a^4g^4. The quantity multiplying a^4 in D is $(eg - f^2)^3$; and the relation connecting Δ and D is

$$\Delta = A^5 - 375A^3B - 625A^2C + 3125D.$$

The value of D is

$$
\begin{array}{llll}
a^4e^3g^3 & -\ 12a^3bde^2g^3 & -\ 24a^3be^2f^3g & +\ 6a^3c^2e^3g^3 \\
-\ 3a^4e^2f^2g^2 & +\ 24a^3bdef^2g^2 & +\ 12a^3bef^5 & +\ 6a^3c^2ef^2g^2 \\
+\ 3a^4ef^4g & -\ 12a^3bdf^4g & +\ a^3c^3g^4 & -\ 3a^3c^2f^4g \\
-\ a^4f^6 & +\ 12a^3be^3fg^2 & -\ 12a^3c^2dfg^3 & +\ 6a^3cd^2eg^3
\end{array}
$$

208   APPLICATIONS TO BINARY QUANTICS.

$+ 42a^5cd^2f^2g^2$	$+ 420a^3bcd^2efg^2$	$- 540a^4cd^2e^3f^2$	$+ 2772ab^3cde^3f^3$
$- 24a^5cde^2fg^2$	$- 72a^3bcd^2f^3g$	$- 258a^4d^5g^2$	$+ 1458ab^3ce^5g$
$- 108a^5cdef^3g$	$+ 156a^3bcde^3g^2$	$+ 816a^4d^5efg$	$- 1296ab^3ce^2f^2$
$+ 60a^5cdf^5$	$- 828a^3bcde^2f^2g$	$- 432a^4d^5f^3$	$- 84ab^3d^4eg^2$
$- 27a^5ce^4g^2$	$+ 348a^3bcdef^4$	$- 324a^4d^4e^3g$	$+ 240ab^3d^4f^2g$
$+ 114a^5ce^3f^2g$	$+ 48a^3bce^4fg$	$+ 180a^4d^4e^2f^2$	$- 240ab^3d^3e^2fg$
$- 60a^5ce^2f^4$	$+ 36a^3bce^3f^3$	$+ 3ab^4cg^4$	$- 240ab^3d^3ef^3$
$+ 5a^5d^4g^3$	$+ 246a^3bd^4fg^2$	$- 12ab^4dfg^3$	$- 12ab^3d^2e^4g$
$- 84a^5d^3efg^2$.	$- 372a^3bd^3e^2g^2$	$- 3ab^4e^2g^3$	$+ 288ab^3d^2e^3f^2$
$+ 66a^5d^2e^3g^2$	$- 816a^3bd^3ef^2g$	$+ 24ab^4ef^2g^2$	$- 288abc^4dg^3$
$+ 228a^5d^2e^2f^2g$	$+ 480a^3bd^3f^4$	$- 12ab^4f^4g$	$+ 48abc^4efg^2$
$- 120a^5d^2ef^4$	$+ 1536a^3bd^2e^3fg$	$- 24ab^3c^2fg^3$	$+ 1536abc^3d^2fg^2$
$- 288a^5de^4fg$	$- 888a^3bd^2e^2f^3$	$- 108ab^3cdeg^3$	$+ 1200abc^3de^2g^2$
$+ 160a^5de^3f^3$	$- 480a^3bde^5g$	$+ 180ab^3cdf^2g^2$	$- 4188abc^3def^2g$
$+ 80a^5e^6g$	$+ 288a^3bde^4f^2$	$+ 312ab^3ce^2fg^2$	$+ 2268abc^3df^4$
$- 48a^5e^5f^2$	$- 27a^3c^4eg^3$	$- 564ab^3cef^3g$	$+ 1200abc^3e^3fg$
$- 3a^2b^5c^3g^4$	$- 3a^2c^4f^2g^2$	$+ 216ab^3cf^5$	$- 1296abc^3e^2f^3$
$+ 24a^2b^5cdfg^3$	$+ 66a^2c^3d^2g^3$	$- 72ab^3d^2efg^2$	$- 2664abc^2d^3eg^2$
$+ 6a^2b^2ce^2g^3$	$+ 156a^2c^3defg^2$	$+ 168ab^3d^2f^2g$	$- 240abc^2d^3f^2g$
$- 48a^2b^2cef^2g^2$	$+ 12a^2c^3df^3g$	$+ 12ab^3de^2g^2$	$+ 5736abc^2d^2e^2fg$
$+ 24a^2b^2cf^4g$	$+ 270a^2c^2e^3g^2$	$- 84ab^3de^2f^2g$	$- 3048abc^2d^2ef^3$
$+ 42a^2b^2d^2eg^3$	$- 804a^2c^2e^2f^2g$	$- 144ab^3def^4$	$- 3252abc^2de^4g$
$- 90a^2b^2d^2f^2g^2$	$+ 405a^2c^2ef^4$	$+ 108ab^2e^3f^3$	$+ 2340abc^2de^3f^2$
$- 60a^2b^2de^2fg^2$	$- 372a^2c^2d^3fg^2$	$+ 114ab^2c^2eg^3$	$+ 324abc^2e^5f$
$+ 180a^2b^2def^3g$	$- 870a^2c^2d^2e^2g^2$	$- 6ab^2c^2f^2g^2$	$+ 816abcd^5g^2$
$- 48a^2b^2df^5$	$+ 1428a^2c^2d^2ef^2g$	$+ 228ab^2c^2d^2g^3$	$- 1572abcd^4efg$
$- 3a^2b^2e^4g^2$	$- 660a^2c^2d^2f^4$	$- 828ab^2c^2defg^2$	$+ 1224abcd^4f^3$
$- 6a^2b^2e^3f^2g$	$+ 1200a^2c^2de^2fg$	$- 84ab^2c^2df^3g$	$+ 1656abcd^3e^2g$
$- 18a^2b^2e^2f^4$	$- 660a^2c^2def^3$	$- 804ab^2c^2e^3g^2$	$- 1296abcd^3e^2f^2$
$+ 12a^2bc^3fg^3$	$- 429a^2c^2e^4g$	$+ 1890ab^2c^2e^2f^2g$	$- 432abcd^2e^2f$
$- 24a^2bc^3deg^3$	$+ 225a^2c^2e^3f^2$	$- 648ab^2c^2ef^4$	$- 168abd^6fg$
$- 60a^2bc^2df^2g^2$	$+ 957a^2cd^4eg^2$	$- 816ab^2cd^3fg^2$	$- 192abd^5e^2g$
$- 96a^2bc^2e^2fg^2$	$- 84a^2cd^4f^2g$	$+ 1428ab^2cd^2e^2g^2$	$+ 144abd^5ef^2$
$+ 312a^2bc^2ef^3g$	$- 2664a^2cd^3e^2fg$	$+ 2148ab^2cd^2ef^2g$	$+ 144abd^4e^3f$
$- 162a^2bc^2f^5$	$+ 1440a^2cd^3ef^3$	$- 1320ab^2cd^2f^4$	$+ 80ac^6g^3$
$- 84a^2bcd^3g^3$	$+ 972a^2cd^2e^3g$	$- 4188ab^2cde^3fg$	$- 480ac^5dfg^2$

APPLICATIONS TO BINARY QUANTICS. 209

$- 429ac^5e^2g^u$	$- 60b^4c^4eg^3$	$- 640b^3d^4e^3g$	$- 648bc^4de^2f^2$
$+ 1458ac^5ef^2g$	$- 18b^4c^2f^2g^2$	$+ 1392b^3d^3e^2f^2$	$- 1872bc^4e^4f$
$- 729ac^5f^4$	$- 120b^4cd^2g^3$	$- 648b^3d^2e^4f$	$+ 144bc^3d^4fg$
$+ 972ac^4d^2eg^u$	$+ 348b^4cdefg^2$	$- 48b^2c^5g^3$	$+ 1980bc^3d^3e^2g$
$- 12ac^4d^2f^2g$	$- 144b^4cdf^3g$	$+ 288b^2c^4dfg^u$	$+ 144bc^3d^3ef^u$
$- 3252ac^4de^2fg$	$+ 405b^4ce^3g^u$	$+ 225b^2c^4e^2g^2$	$+ 852bc^3d^2e^2f$
$+ 1620ac^4def^3$	$- 648b^4ce^2f^2g$	$- 1296b^2c^4ef^2g$	$+ 3780bc^3de^5$
$+ 402ac^4e^4g$	$+ 252b^4cef^4$	$+ 243b^2c^4f^4$	$- 1200bc^3d^5eg$
$+ 270ac^4e^3f^u$	$+ 480b^4d^3fg^u$	$- 540b^2c^3d^2eg^u$	$+ 192bc^3d^5f^2$
$- 324ac^3d^4g^u$	$- 660b^4d^2e^2g^u$	$+ 288b^2c^3d^2f^2g$	$+ 2478bc^2d^4e^2f$
$+ 1656ac^3d^3efg$	$- 1320b^4d^2ef^2g$	$+ 2340b^2c^3de^2fg$	$- 6660bc^2d^6e^4$
$- 640ac^3d^3f^3$	$+ 552b^4d^2f^4$	$- 2268b^2c^3def^3$	$+ 240bcd^7g$
$+ 1818ac^3d^2e^3g$	$+ 2268b^4de^3fg$	$+ 270b^2c^3e^3g$	$- 2088bcd^6ef$
$- 1230ac^3d^2e^2f^2$	$- 864b^4de^2f^3$	$+ 1530b^2c^3e^8f^2$	$+ 3840bcd^5e^3$
$- 1080ac^3de^4f$	$- 729b^4e^5g$	$+ 180b^2c^2d^4g^u$	$+ 408bd^8f$
$+ 135ac^3e^6$	$+ 243b^4e^4f^2$	$- 1296b^2c^2d^3efg$	$- 720bd^7e^2$
$- 192ac^2d^5fg$	$+ 160b^3c^3dg^3$	$+ 1392b^2c^2d^3f^3$	$+ 135c^8e^3g$
$- 2673ac^2d^4e^2g$	$+ 36b^3c^3efg^u$	$- 1230b^2c^2d^2e^3g$	$- 1215c^6e^2f^2$
$+ 780ac^2d^4ef^u$	$+ 108b^3c^3f^3g$	$+ 1134b^2c^2d^2e^2f^2$	$- 270c^5d^3e^3g$
$+ 1980ac^2d^3e^2f$	$- 888b^3c^2d^2fg^u$	$- 648b^2c^2de^4f$	$+ 1620c^5d^u ef^u$
$- 270ac^2d^2e^5$	$- 660b^3c^2de^2g^2$	$- 1215b^2c^2e^6$	$+ 3780c^5de^3f$
$+ 1308acd^6eg$	$+ 2772b^3c^2def^u g$	$+ 144b^2cd^5fg$	$- 198c^5e^6$
$- 120acd^5f^2$	$- 864b^3c^2df^4$	$+ 780b^2cd^4e^2g$	$+ 180c^4d^4eg$
$- 1200acd^5e^2f$	$- 1296b^3c^2e^3fg$	$- 1920b^2cd^4ef^2$	$- 540c^4d^4f^3$
$+ 180acd^4e^4$	$+ 360b^3c^2e^2f^3$	$+ 144b^2cd^3e^3f$	$- 6660c^4d^3e^2f$
$- 228ad^8g$	$+ 1440b^3cd^3eg^u$	$+ 1620b^2cd^2e^5$	$- 2490c^6d^2e^4$
$+ 240ad^7ef$	$- 240b^3cd^8f^u g$	$- 120b^2d^5eg$	$- 40c^3d^5g$
$- 40ad^6e^3$	$- 3048b^3cd^u e^2fg$	$+ 336b^2d^6f^2$	$+ 3840c^3d^3ef$
$- b^6g^4$	$+ 1896b^3cd^u ef^3$	$+ 192b^2d^3e^2f$	$+ 6095c^3d^4e^6$
$+ 12b^5cfg^8$	$+ 1620b^3cde^4g$	$- 540b^2d^4e^4$	$- 720c^2d^7f$
$+ 60b^5deg^3$	$- 2268b^3cde^3f^u$	$+ 324bc^5e^2fg$	$- 5130c^2d^6e^u$
$- 48b^5df^2g^2$	$+ 972b^3ce^5f$	$+ 972bc^5ef^3$	$+ 1860cd^3e$
$- 162b^5e^2fg^2$	$- 432b^3d^5g^u$	$- 432bc^4d^2efg$	$- 248d^{10}$
$+ 216b^5ef^3g$	$+ 1224b^3d^4efg$	$- 648bc^4d^2f^3$	
$- 72b^5f^5$	$- 976b^3d^4f^3$	$- 1080bc^4de^3g$	

E E

·242. Mr. Cayley's theory of the number of invariants of a quantic, shows that, in addition to the four invariants already mentioned, sextics have a skew invariant E of the fifteenth degree. I do not know whether in the same manner as the skew invariant of a quintic is the resultant of two covariants, this invariant also can be got by operating with one of two known covariants on the other. The invariant however can be formed by means of the differential equation. The highest power of a which occurs in it is a^6, and the factor multiplying it is

$$(dg^2 - 3efg + 2f^3)(eg - f^2)^3.*$$

The expression for E in terms of the other invariants may be got from the following considerations. If in the sextic b, d, f vanish, E necessarily vanishes. For, since the weight of E is forty-five (Art. 139), the weight of some one of the constituent coefficients in each term must be expressed by an odd number; and when we make in the equation all the terms vanish whose weight is odd, E vanishes. $E=0$ is therefore the condition that the roots of the sextic should form a system in involution. If then we make b, d, $f=0$ in A, B, C, D, and eliminate a, c, e, g from the results, the relation thus obtained between A, B, C, D must be satisfied when E vanishes, and must therefore contain it as a factor.

If we write $ag = \lambda$, $ce = \mu$, $ae^3 + gc^3 = \nu$, the values of the invariants got by making b, d, $f = 0$, may be written

$$A = \lambda + 15\mu, \quad B = \lambda\mu + \mu^2 - \nu,$$
$$C = -24\lambda\mu^2 - 8\mu^3 + 4(\lambda + 3\mu)\nu,$$
$$\Delta = \lambda\{\lambda^2 - 150\lambda\mu - 1875\mu^2 + 500\nu\}^2.$$

Eliminating ν in the first place, the last two equations become

$$C = 4\mu(\lambda-\mu)^2 - 4(\lambda+3\mu)B, \quad \Delta = \lambda(\lambda^2 + 350\lambda\mu - 1375\mu^2 - 500B)^2.$$

Then eliminating μ by the help of the first equation, we get

$$1024\lambda^3 - 1152\lambda^2 A + (132A^2 - 10800B)\lambda + 3375C + 2700AB - 4A^3 = 0,$$
$$\lambda(256\lambda^2 - 320A\lambda + 55A^2 + 4500B)^2 - \Delta = 0.$$

---

* I have partly calculated the value of E. If I find leisure to complete the calculation I will give the result in an appendix.

The resultant of these two equations is of the thirtieth degree in the coefficients; and therefore from what we have seen can only differ by a constant multiplier from E^2.

243. By Art. 225, when the discriminant is positive the sextic has either six or two real roots; and when it is negative, has either four or none. We can readily anticipate that the discussion of this expression for E is likely to lead to the same results in affording criteria for further distinguishing these cases, as the corresponding discussion of the expression G in the case of the quintic. Analogy also leads us to expect that what will be important to examine will be the result of making $\Delta = 0$ in the expression for E. Now, although the calculation of the general expression for E may be a little laborious, that part of it which is independent of Δ is easily obtained. It will evidently be the product of $3375C + 2700AB - 4A^3$ by the square of the resultant of the cubic and of the quadratic

$$256\lambda^3 - 320A\lambda + 55A^2 + 4500B.$$

And again, analogy leads us to believe that the first of these factors is not important in the question of the criteria for real roots, and that it is the square factor alone which needs to be attended to.

The result I find is that, writing for convenience B' for $100B$, C' for $125C$, the quantity squared differs only by a constant multiplier from

$$4A^6 - 19A^4B' - 49A^2B'^2 - 4A^3C' - 80B'^3 + 52AB'C' - 4C'^2.$$

Analogy then leads me to suppose that the criteria for the number of real roots of a sextic depend on the signs of this quantity, and of $A^3 - 100B^2$, $A^3 - 125C$,

$$(A^3 - 300AB + 250C^2)^2 - 5(A^2 - 100B)^3,$$

which, as we saw, vanish when three roots are all equal.

LESSON XVIII.

ON THE ORDER OF RESTRICTED SYSTEMS OF EQUATIONS.*

244. THE problems discussed in this lesson are purely algebraical, and in the investigation of them I do not make use of any geometrical principles. But I find it convenient to borrow one or two *terms* from geometry, because we can thus avoid circumlocution, and also can more readily see how to extend to quantics in general, theorems already known for ternary and quaternary quantics.

We saw (Art. 74) that if we are given k equations in k† independent variables, the number of systems of common values of the variables which can be found to satisfy all the equations, will be equal to the product of the orders of the equations. Now, in the geometry of two and three dimensions respectively, the system of values $x = a$, $y = b$; or $x = a$, $y = b$, $z = c$, denotes a point. I find it convenient therefore to use the word "point" in general instead of "system of values of the variables"; and the theorem already stated may be enunciated. "A system of k equations in k variables of degrees l, m, n, p, q, &c. respectively, represents $lmnpq$ &c. *points*," by which we mean that so many "systems of values of the variables" can be found to satisfy all the equations. This number $lmnpq$ &c. will be called the *order* of the system of equations.

245. If we have a system of $k-1$ equations in k independent variables, we have not data enough to *determine* any

---

* In this place in the former edition followed a Lesson on the application of this theory to ternary quantics. This would now require much development in order to bring it up to the present state of the subject; and as the results all admit of geometrical interpretation, I have thought that they would be placed more properly in a geometrical treatise. Instead, I have inserted here what has in substance already been published as an appendix to my work on the *Geometry of Three Dimensions*, but which seems properly to belong to the algebraical theory explained in this volume.

† If as is usual we employ homogeneous equations, the number of variables will of course be $k + 1$.

system of common values of the variables, and the system of equations denotes an infinity of "points." Such a system of equations we shall speak of as denoting a *curve*. If with the given system of $k-1$ equations, we combine any arbitrary equation of the first degree, we have then data enough to determine points, which will be equal in number to the product of the degrees of the equations. We shall define the *order* of a curve as the number of points which are obtained when with the equations which denote the curve, we combine an arbitrary equation of the first degree.

When we are given a system of $k-2$ equations, these denote a doubly infinite series of points, since we cannot determine any points unless we are given two other equations. Such a system we shall speak of as denoting a *surface*. If with the system of $k-2$ equations we combine an arbitrary equation of the first degree, we shall have a "curve" whose order is the product of the degrees of the $k-2$ equations. In general by the order of a surface, we mean either the order of the curve obtained by combining with the given equations an equation of the first degree, or, what comes to the same thing, the number of points obtained by combining with the given equations two equations of the first degree.

And so more generally, if we have any system of fewer than k equations, by the order of the system we mean the number of points that are obtained, when with the given equations we combine as many equations of the first degree as are wanting to make the entire number of equations up to k, and so afford data enough to determine systems of values of the variables. It is evident that in the case under consideration, the order of the system is the product of the degrees of the equations which compose it.

246. If we have $k+1$ equations in k independent variables, whose degrees are l, m, n, &c., we can eliminate the variables; and we have seen (Arts. 72, 74) that the order in which the coefficients of each equation enter into the resultant, will be equal to the product of the degrees of the remaining equations. Taking then, to fix the ideas, the case of four equations: let their orders be l, m, n, r, and let any quantity enter into the

coefficients of the equations in the degrees λ, μ, ν, ρ respectively, this quantity will enter into the resultant in the degree

$$\lambda mnr + \mu nrl + \nu rlm + \rho lmn.$$

We shall use the word 'order' to denote the degrees l, m, n, r, in which the equations contain the variables which are to be eliminated, and *weight* to denote the degrees λ, μ, ν, ρ in which they contain the quantity not eliminated: and the result just written may be stated, that the weight of the resultant, or the weight of the system, is equal to the sum of the weights of each equation multiplied by the order of the system formed by the remaining equations.

And this is still true, if we break the given system up into partial systems. Thus, the first two equations form a system whose order is lm and weight $\lambda m + \mu l$, and the second two equations a system whose order is nr and weight $\nu r + \rho n$; and the value just given for the weight of the entire system, is

$$nr(\lambda m + \mu l) + lm(\nu r + \rho n),$$

that is, it is the sum of the weights of each component system multiplied by the order of the other. The advantage of so stating the matter will appear presently.

247. What has been hitherto said in this lesson, is but a re-statement in other words of principles already laid down in the Lesson on Elimination: but my object has been to make more intelligible the object of investigations, on which we shall now enter as to the order and weight of systems of a somewhat different kind. We have seen that k equations in k variables represent $lmnp$ &c. points. But now we may combine with these k equations an additional equation, which is satisfied for some of the points but not for others of them. We have then a system of $k + 1$ equations representing points, that is to say, all satisfied by a number of systems of common values of the variables, that number being now however generally smaller than the product of the degrees of any k of the equations. Cases are of constant occurrence where a number of points can be expressed in no other way than that here described. A simple geometrical example will suffice. Consider p points in a plane where p is a prime number, and where the points do not lie in a

right line, then these points cannot be represented as the complete intersection of any two curves, and if we have any two curves going through the points, their intersection includes not only these points but others besides. To define the points completely, we must add a third curve going through the p given points, but not through the remaining points of intersection of the first two curves. The points are then completely defined as the only points common to all three curves. Our object then is, in some important cases where a system of points is defined by more than k equations, to lay down rules for ascertaining the order of the system; that is to say, how many systems of common values satisfy all the equations.

In like manner a system of $k-1$ equations is satisfied by an infinity of common values. But it may happen that we can write down an additional equation satisfied by part of this series of common values, but not by the remaining part. In such a case, the system of $k-1$ equations denotes a complex curve, and it requires the system of k equations to define that part of it for which all the equations are satisfied. It will be the object of this lesson to ascertain the order and weight of what we may call restricted systems: that is to say, where to a number of equations sufficient to define points, curves, &c., is added one or more others which exclude from consideration those values of the variables which satisfy the first set of equations, but do not satisfy the additional equations.

248. The simplest example of such a system is the set of determinants

$$\left\| \begin{array}{ccc} u, & v, & w \\ u', & v', & w' \end{array} \right\| = 0,$$

or, at full length,

$$vw' - v'w = 0, \quad wu' - w'u = 0, \quad uv' - vu' = 0.$$

By writing these equations in the form

$$\frac{u}{u'} = \frac{v}{v'} = \frac{w}{w'},$$

it is evident that in general values of the variables which satisfy two of the equations must satisfy the third. But there is an exception for the case of values which make either u and u', v and v', or w and $w' = 0$. In any of these cases it is easy to see that two of the equations will be satisfied, but not the third. And now it is easy to see how to calculate the order of the system common to all three. Let the orders of u and u', of v and v', of w and w', be l, m, n respectively; then the orders of the first two equations are $m+n$, $n+l$, and of the system formed by them is $(m+n)(n+l)$. But in this system will be included values which satisfy both w and w', these values not satisfying the third equation. Excluding then this system, the order of which is n^2, the order of the system common to the three determinants is $mn + nl + lm$.

In like manner suppose we have a system with three rows and four columns,

$$\begin{Vmatrix} u, & u', & u'', & u''' \\ v, & v', & v'', & v''' \\ w, & w', & w'', & w''' \end{Vmatrix} = 0.$$

Let us write at full length the determinants formed by the omission of the third and fourth columns

$$u''(vw' - v'w) + v''(wu' - w'u) + w''(uv' - u'v) = 0,$$

$$u'''(vw' - v'w) + v'''(wu' - w'u) + w'''(uv' - u'v) = 0,$$

then these two equations are obviously satisfied for all values which satisfy the three $vw' = v'w$, $wu' = w'u$, $uv' = u'v$. But these values will not satisfy the other determinants of the given system. From $(l+m+n)^2$, then, which is the order of the system formed by the two equations written at length, we must subtract $mn + nl + lm$ which has just been found to be the order of the system special to these two equations, and the remainder $l^2 + m^2 + n^2 + mn + nl + lm$ is the order of the system common to all the determinants. Having thus determined the order of a system with three rows and four columns, we have, in like manner, thence derived the order of a system with four rows and five columns. Proceeding thus step by step I arrive by induction at a general formula which I establish by showing that if it is true for a system with k rows, it is true for a system with $k+1$.

249. The formula at which I arrive may be written most simply and generally as follows: Let the orders of the several functions be represented by the letters which denote them,

$$\begin{Vmatrix} a+\alpha, & b+\alpha, & c+\alpha, & d+\alpha, & \&c. \\ a+\beta, & b+\beta, & c+\beta, & d+\beta, & \&c. \\ a+\gamma, & b+\gamma, & c+\gamma, & d+\gamma, & \&c. \\ \&c. & & & & \end{Vmatrix};$$

let p denote the sum of the quantities α, β, γ, &c., and let q denote the sum of their products in pairs; and let P and Q denote the corresponding sums for the quantities a, b, c, &c.; then I say that the order of the system is $Q+pP+p^2-q$. For if we consider the two determinants obtained by omitting the first and second columns alternately, these will be of the orders $p+P-a$, $p+P-b$, forming a system whose order is $(p+P-a)(p+P-b)$. But these two determinants will be satisfied for all values which satisfy all the minor determinants of the inferior system formed by striking out the first two columns; and yet these values will not satisfy the other determinants of the given system. To find then the order of the system common to all, we must subtract from $(p+P-a)(p+P-b)$ the order of the inferior system; which we can calculate if the truth of the formula given is assumed. Since in the inferior system the number of rows is greater than the number of columns, we must, in order to apply the formula, write the rows as columns and the columns as rows; and thus the new P and Q will be the old p and q. The new p will evidently be the old $P-a-b$; and the new q, which is the sum of the products in pairs of c, d, &c., can easily be seen to be

$$Q-(a+b)P+a^2+ab+b^2.$$

Hence by the formula the order of the lower system is

$$P^2-Q-P(a+b)+ab+p(P-a-b)+q.$$

Subtracting this number from $(p+P-a)(p+P-b)$ we obtain as we ought $Q+pP+p^2-q$. Thus it is proved that the formula, if true for an inferior system, is true for that next above it; and since it can easily be verified for the case of two rows and three columns it is true generally. In applying it we have generally either a or $\alpha = 0$. When all the rows are of the same order we have α, β, &c. all $= 0$, and therefore p, q both $= 0$, and the order of the system is Q.

250. We may proceed in like manner to calculate the weight of the system of determinants considered in the last article. Beginning again with the simplest case, let us suppose that the system $\begin{Vmatrix} u, & v, & w \\ u', & v', & w' \end{Vmatrix}$ is to be combined with one or more other equations and the variables eliminated. Now the result of elimination between $uv' - u'v$, $uw' - u'w$, and any other equations will contain as a factor the resultant of u, u', and the other equations. If we reject this factor we get the same result as if we had eliminated between $uv' - u'v$, $vw' - vw'$ and the other equations, and then rejected the factor got by eliminating between v, v', and the other equations. To illustrate the method employed, let us suppose that u, u'; v, v'; w, w' respectively contain any quantity not eliminated in the degrees λ, μ, ν; and that we are to combine with the determinants of the given system another equation $R = 0$, whose order is r, and containing the uneliminated quantity in the degree ρ. This quantity then will enter into the resultant of R, $uv' - u'v$, $uw' - u'w$, in the degree

$$\rho(l+m)(l+n) + r\{(l+m)(\lambda+\nu) + (l+n)(\lambda+\mu)\}.$$

But the resultant of R, u, u', will contain the same quantity in the degree

$$\rho l^2 + 2rl\lambda.$$

When then this factor is rejected from the former result, the remainder is

$$\rho(mn + nl + lm) + r\{\lambda(m+n) + \mu(n+l) + \nu(l+m)\}.$$

The order then of the system of three determinants is the quantity multiplying ρ, and the weight is the quantity multiplying r.

251. Finding in this way the weight of a system with two rows and three columns, we find, step by step, as in Art. 249, the weight of a system with k rows and $k+1$ columns. The result is that if the orders of the several functions be as written

in Art. 249, and if their weights (that is to say, the degrees in which they contain the variable not eliminated) be $a'+a'$, $b'+a'$, &c., $a'+\beta'$, &c., then the formula for the weight is derived from that for the order, by performing on it the operation $a'\dfrac{d}{da}+b'\dfrac{d}{db}+$&c.$+a'\dfrac{d}{da}+\beta'\dfrac{d}{d\beta}+$&c. That is to say, the weight is

$$\Sigma(ab')+pP'+p'P+2\Sigma(\alpha a')+\Sigma(\alpha\beta'),$$

which may also be written

$$(P+p)(P'+p')+\Sigma(\alpha a')-\Sigma(aa');$$

this being the result of performing the operation $a'\dfrac{d}{da}+$&c. on $Q+pP+p^2-q$.

252. If we form the condition that the two equations

$$at^m+bt^{m-1}+ct^{m-2}+\&c.=0,\quad a't^n+b't^{n-1}+c't^{n-2}+\&c.=0,$$

should have a common root, we obtain a single equation, namely the resultant of the equations. But if we form the conditions that they should have two common roots, we obtain (Art. 78) not two equations, but a whole system, no doubt equivalent to two conditions, yet such that two equations of the system would not adequately define the conditions in question. Now we may suppose that t is a parameter eliminated, and that a, b, &c. contain variables, and we may propose to investigate the order of the system of conditions in question. Now, Art. 78, these conditions are the determinants of the system

$$\left\|\begin{array}{l} a,\ b,\ c,\ d\ \ldots \\ \ \ \ a,\ b,\ c\ \ldots \\ \ \ \ \ \ \ a,\ b\ \ldots \\ \ldots\ldots\ldots\ldots\ldots \\ a',\ b',\ c',\ d'\ \ldots \\ \ \ \ a',\ b',\ c'\ \ldots \\ \ldots\ldots\ldots\ldots\ldots \end{array}\right\|,$$

where the first line is repeated $n-1$ times, and the second $m-1$ times; there are $m+n-2$ rows and $m+n-1$ columns. The problem is then a particular case of that of Art. 249. We suppose the degrees of the functions introduced to be equidifferent: that is to say, if the degrees of a, a' be λ, μ, we suppose those of b, b'' to be $\lambda+\alpha$, $\mu+\alpha$; of c, c' to be $\lambda+2\alpha$ $\mu+2\alpha$, &c. We may write the formula of Art. 249 in the somewhat more convenient form $Q+pP+\frac{1}{2}(p^2+s_2)$, where s_2 is the sum of the squares of α, β, &c. To apply it to the present case we may take for the quantities a, b, c, &c. 0, α, 2α, &c.; and for the quantities α, β, γ, &c. of Art. 249, λ, $\lambda-\alpha$, $\lambda-2\alpha$, &c. P is then the sum of $m+n-2$ terms of the series α, 2α, 3α, &c., and is therefore, if we write $m+n=k$, $\frac{1}{2}(k-1)(k-2)\alpha$. In the same case Q is the sum of products in pairs of these quantities, and is therefore

$$= \frac{(k-3)(k-2)(k-1)(3k-4)}{1.2.3.4}\alpha^2.$$

Again p is the sum of $n-1$ terms of the series λ, $\lambda-\alpha$, $\lambda-2\alpha$, &c., and of $m-1$ terms of the series μ, $\mu-\alpha$, $\mu-2\alpha$, &c. We have then

$$p = (n-1)\lambda + (m-1)\mu - \tfrac{1}{2}\alpha\{(n-1)(n-2)+(m-1)(m-2)\}.$$

In like manner s_2 is the sum of the squares of the same quantities, and is

$$(n-1)\lambda^2 + (m-1)\mu^2 - \lambda\alpha(n-1)(n-2) - \mu\alpha(m-1)(m-2)$$
$$+ \alpha^2\left\{\frac{(n-1)(n-2)(2n-3)}{1.2.3} + \frac{(m-1)(m-2)(2m-3)}{1.2.3}\right\}.$$

Collecting all the terms, the order of the required system is found to be

$$\tfrac{1}{2}n(n-1)\lambda^2 + \tfrac{1}{2}m(m-1)\mu^2 + (m-1)(n-1)\lambda\mu$$
$$+ \tfrac{1}{2}n(n-1)(2m-1)\lambda\alpha + \tfrac{1}{2}m(m-1)(2n-1)\mu\alpha$$
$$+ \tfrac{1}{2}mn(m-1)(n-1)\alpha^2.$$

If the two equations considered are of the same degree, that is to say, if $m=n$, we may write $\lambda+\mu=p$, $\lambda\mu=q$, and the order becomes

$$\tfrac{1}{2}n(n-1)(p+n\alpha)\{p+(n-1)\alpha\} - (n-1)q.$$

If all the functions a, b, &c. are of the first degree, writing $\lambda=\mu=1$, and $\alpha=0$, in the preceding formula, the order is found to be $\frac{1}{2}(m+n-1)(m+n-2)$.

253. If the degrees in which the uneliminated variables occur in any terms be denoted by the accented letters corresponding

ORDER OF RESTRICTED SYSTEMS OF EQUATIONS. 221

to those which express their degrees in the variables to be eliminated, then the formula for the weight of the system is obtained from that for the order by performing on it the operation $\lambda' \dfrac{d}{d\lambda} + \mu' \dfrac{d}{d\mu} + a' \dfrac{d}{da}$. In other words the weight is

$n(n-1)\lambda\lambda' + m(m-1)\mu\mu' + (m-1)(n-1)(\lambda\mu' + \lambda'\mu)$
$+ \tfrac{1}{2}n(n-1)(2m-1)(\lambda a' + \lambda' a) + \tfrac{1}{2}m(m-1)(2n-1)(\mu a' + \mu' a)$
$+ mn(m-1)(n-1)aa'.$

254. The next system we discuss is that formed by the system of conditions that the three equations

$at^l + bt^{l-1} + \&c. = 0, \quad a't^m + b't^{m-1} + \&c. = 0, \quad a''t^n + b''t^{n-1} + \&c. = 0,$

may have a common factor. The system may be expressed by the three equations obtained by eliminating t in turn between every pair of these equations, a system equivalent to two conditions. The order of the system may be found by eliminating from the equations the variables which enter implicitly into a, b, c, &c., when the order of the resulting equation in t determines the order of the system.

Let us suppose that a, a', a'' are homogeneous functions in x, y, z of the degrees λ, μ, ν respectively; that b, b', b'' are of the degrees $\lambda - 1$, $\mu - 1$, $\nu - 1$, &c., and if we take the reciprocal of t as a fourth variable, the equations are of the orders respectively λ, μ, ν, forming a system of the order $\lambda\mu\nu$. But the system of values $x = 0$, $y = 0$, $z = 0$ is a multiple point in the three equations of the orders $\lambda - l$, $\mu - m$, $\nu - n$ respectively. The order then is to be reduced by $(\lambda - l)(\mu - m)(\nu - n)$. It is therefore

$$l\mu\nu + m\nu\lambda + m\lambda\mu - \lambda mn - \mu nl - \nu lm + lmn.$$

This then is the order of the system we are investigating. If the orders of b, b', c, c', &c. had been $\lambda + a$, $\mu + a$, $\lambda + 2a$, $\mu + 2a$, &c., then the order of the system would have been

$$l\mu\nu + m\nu\lambda + n\lambda\mu + a(mn\lambda + nl\mu + lm\nu) + a^2 lmn.$$

The weight is found by operating on this with $\lambda' \dfrac{d}{d\lambda} + \&c.$, and is

$l(\mu\nu' + \mu'\nu) + m(\nu\lambda' + \nu'\lambda) + n(\lambda\mu' + \lambda'\mu) + mn(a\lambda' + a'\lambda)$
$\qquad + nl(a\mu' + a'\mu) + lm(a\nu' + a'\nu) + 2lmn aa'.$

255. It is a particular case of the preceding to find the order and weight of the system of conditions that an equation $at^n + bt^{n-1} + \&c.$ may have three equal roots; because these conditions are found by expressing that the three second differentials may have a common factor. Writing in the preceding for l, m, and n, $n-2$; for μ, $\lambda + \alpha$; and for ν, $\lambda + 2\alpha$; we find, for the order of the system,

$$3(n-2)\lambda(\lambda + n\alpha) + n(n-1)(n-2)\alpha^2;$$

and in like manner for its weight

$$6(n-2)\lambda\lambda' + 3n(n-2)(\alpha'\lambda + \alpha\lambda') + 2n(n-1)(n-2)\alpha\alpha'.$$

Again, to find the order and weight of the system of conditions that the same equation may have two distinct pairs of equal roots; we form first, by Art. 249, the order and weight of the system of conditions that the two first differentials $at^{n-1} + \&c.$, $bt^{n-1} + \&c.$ may have two common factors. We subtract then the order and weight of the system found in the first part of this article. The result is that the order is

$$2(n-2)(n-3)\lambda(\lambda + n\alpha) + \tfrac{1}{2}n(n-1)(n-2)(n-3)\alpha^2,$$

and the weight is

$$4(n-2)(n-3)\lambda\lambda' + 2n(n-2)(n-3)(\alpha'\lambda + \alpha\lambda')$$
$$+ n(n-1)(n-2)(n-3)\alpha\alpha'.$$

Before proceeding further in investigating the order of other systems, it is necessary to discuss a different problem, and I commence by explaining the use of one or two other terms which I borrow from geometry.

256. *Intersection of quantics having common curves.* Two systems of quantics are said to *intersect* if they have one or more "points" common, that is to say, if they are both capable of being satisfied by the same system of values of the variables. A "surface" is said to *contain* a "curve" if every system of values which satisfies the $k-1$ equations constituting the curve, satisfies also the $k-2$ equations constituting the surface. Thus, in the case of four variables, three equations $U=0$, $V=0$, $W=0$ constitute a curve, and the two equations $U=0$, $V=0$ constitute a surface which evidently contains that curve.

ORDER OF RESTRICTED SYSTEMS OF EQUATIONS. 223

Now a system of k quantics in k variables, in general, as we have seen, intersect in a definite number of points, that number being the product of the orders of the quantics. But it may happen that they may have an infinity of points common, these points forming a "curve" in the sense in which we have already defined that word. Besides that curve they will have ordinarily a finite number of points common, which it is our object now to determine. Let us take, for example, to fix the ideas, the case of four independent variables; and suppose that we have four equations of the form

$$U = Au\ +Bv\ +Cw\ =0,$$
$$V = A'u\ +B'v\ +C'w\ =0,$$
$$W = A''u\ +B''v\ +C''w = 0,$$
$$Z = A'''u + B'''v + C'''w = 0.$$

We suppose the degrees of U, V, W, Z to be l, m, n, p; of u, v, w to be λ, μ, ν; and A, B; A', B', &c. are therefore functions of the degrees $l-\lambda$, $l-\mu$; $m-\lambda$, $m-\mu$, &c. Now, evidently, these equations will be all satisfied by every system of values which make $u=0, v=0, w=0$; and these equations not being sufficient to determine "points," will be satisfied by an infinity of values of variables. In other words, the four quantics U, V, W, Z have a common curve uvw. And yet U, V, W, Z may be satisfied by a number of values which do not make u, v, w all $= 0$. It is our object to determine this latter number; and our problem is, When a system of quantics has a common curve, it is required to find how many of their $lmnp$ &c. points of intersection are absorbed by that curve, and in how many points they intersect not on that curve.

257. Let us first consider the curve formed by $k-1$ of the quantics; for instance, in the example we have chosen for illustration, the curve UVW. Now evidently a portion of this curve is the curve uvw, but there are besides an infinity of points satisfying UVW which do not satisfy u, v, w. We speak then of the curve UVW, as a complex curve consisting of the curve uvw and a complementary curve. Now the order of a complex curve is always equal to the sum of the orders of its components. For by definition the order of the complex curve

224 ORDER OF RESTRICTED SYSTEMS OF EQUATIONS.

UVW is the number of points obtained by combining with the equations of the system an additional one of the first degree: that order being in the present case lmn. And evidently, since of those lmn points $\lambda\mu\nu$ lie on the curve uvw, there must be $lmn - \lambda\mu\nu$ on the complementary curve.

The two curves intersect in points whose number i is easily obtained. For evidently all points which satisfy the three equations

$$Au + Bv + Cw = 0, \quad A'u + B'v + C'w = 0, \quad A''u + B''v + C''w = 0,$$

and which do not satisfy u, v, w; must satisfy the determinant

$$\begin{vmatrix} A, & B, & C \\ A', & B', & C' \\ A'', & B'', & C'' \end{vmatrix} = 0,$$

the degree of which is $l + m + n - \lambda - \mu - \nu$. The intersection of this new quantic with uvw gives all the points in which uvw meets the complementary curve. We have therefore

$$i = \lambda\mu\nu (l + m + n - \lambda - \mu - \nu).$$

258. To find now the number of points common to $UVWZ$, we have to consider the points in which the curve UVW meets Z; and it is required to find how many of these are not on the curve uvw. But since uvw is itself a part of the curve UVW, it is evident that the points required are contained among the $p(lmn - \lambda\mu\nu)$ points in which the complementary curve meets Z. And from these points must be excluded the i points in which the complementary curve meets uvw. Using then the value given in the last article for i, we find, for the number which we seek to determine,

$$lmnr - \lambda\mu\nu (l + m + n + r) + \lambda\mu\nu (\lambda + \mu + \nu).$$

We shall state this result thus, that if k quantics of orders l, m, n, p, &c. have common a curve of order α; then the number of points which they will have common in addition to this curve is less than the product of the orders of the quantics by $\alpha(l + m + n + \&c.) - \beta$, where β is a constant depending only on the nature of the curve and not involving the orders of the quantics. We shall call this constant the rank of the curve. We have seen that when the curve is given as the

ORDER OF RESTRICTED SYSTEMS OF EQUATIONS. 225

intersection of quantics u, v, w, the order is $\lambda\mu\nu$ and the rank $\lambda\mu\nu(\lambda + \mu + \nu)$.

We saw, in the last article, that if the intersection UVW consists of two complementary curves whose orders are α, α', and whose ranks are β, β', the number of points in which the two curves intersect is $\alpha(l+m+n)-\beta$; and by parity of reasoning it is $\alpha'(l+m+n)-\beta'$. Hence the orders and ranks of the two complementary curves are connected by the equations $\alpha + \alpha' = lmn$, $\beta - \beta' = (\alpha - \alpha')(l+m+n)$.

259. Next, let us consider the case where the quantics have common two or more distinct curves uvw, $u'v'w'$, &c. Let the intersection for instance of UVW consist of the two curves uvw, $u'v'w'$, and of a complementary curve α''; then, in the first place, the order of α'' is evidently $lmn - \lambda\mu\nu - \lambda'\mu'\nu'$. Secondly, we have seen that uvw meets the remaining intersection of UVW in points whose number is

$$\lambda\mu\nu(l+m+n-\lambda-\mu-\nu).$$

If then i of these lie on $u'v'w'$ (that is to say, if uvw, $u'v'w'$ intersect in i points) there must be on the complementary curve α''

$$\lambda\mu\nu(l+m+n-\lambda-\mu-\nu)-i.$$

And in like manner α'' meets $u'v'w'$ in

$$\lambda'\mu'\nu'(l+m+n-\lambda'-\mu'-\nu')-i \text{ points.}$$

As before, then, the number of points on neither curve in which α'' meets any other quantic Z is

$$(lmn - \lambda\mu\nu - \lambda'\mu'\nu')p - \lambda\mu\nu(l+m+n-\lambda-\mu+\nu)$$
$$-\lambda'\mu'\nu'(l+m+n-\lambda'-\mu'-\nu')+2i,$$

or $lmnp - (\lambda\mu\nu + \lambda'\mu'\nu')(l+m+n+p) + \lambda\mu\nu(\lambda+\mu+\nu)$
$$+\lambda'\mu'\nu'(\lambda'+\mu'+\nu')+2i.$$

Thus, then, the diminution from the number $lmnp$ effected by a complex curve is equal to the sum of the diminutions effected by the simple curves less double the number of their points of intersection. The same holds no matter how many be the curves common to the quantics: and we may say that when a complex curve consists of several simple curves the order of the complex is equal to the sum of the orders of its

G G

components; and the rank of the complex is equal to the sum of their ranks increased by double the number of points common to every pair of curves.

260. We give, as an illustration of the application of these principles, the problem to determine how many surfaces of the second degree can be described through five points to touch four planes. Let S, T, U, V, W be five surfaces passing through the five points, then any other will be of the form $aS + \beta T + \gamma U + \delta V + \varepsilon W$; and the condition that this should touch a plane will be a cubic function of the five quantities a, β, γ, δ, ε. We are given four such equations, and it is required to find how many systems of values can be got to satisfy them all. If the four equations had no common "curves," the number of their common "points" would be 3^4 or 81. But the existence of common curves may be seen in this way. The condition that a surface of the second order should touch a plane vanishes identically when the surface consists of two planes. Let us take then for S and T two pairs of planes passing through the five given points, $S = (123)(145)$, $T = (123)(245)$; then, evidently, the condition that $aS + \beta T + \gamma U + \delta V + \varepsilon W$ should touch any plane whatever, must be satisfied by the supposition $\gamma = 0$, $\delta = 0$, $\varepsilon = 0$. This "curve," then, which is of the first degree, will be common to all four quantics. And, if if we call this the line $(123)(45)$, it is evident, by parity of reasoning, that the quantics have common ten such lines $(124)(35)$, &c. Now if, as before, we take S as the system of two planes $(123)(145)$, $T = (123)(245)$, and take $U = (145)(234)$; then, while the line $(123)(45)$ is denoted by $\gamma = 0$, $\delta = 0$, $\varepsilon = 0$, the line $(145)(23)$ is denoted by $\beta = 0$, $\delta = 0$, $\varepsilon = 0$; and these two lines intersect, being both satisfied by the common values $\beta = 0$, $\gamma = 0$, $\delta = 0$, $\varepsilon = 0$. And, in like manner, $(123)(45)$ is intersected by $(245)(13)$, $(345)(12)$. Thus then the ten lines have fifteen points of mutual intersection. The rank of a single curve of the first degree being got by making $\lambda = \mu = \nu = 1$ in the formula $\lambda \mu \nu (\lambda + \mu + \nu)$ is three. Hence the rank of the entire system is ten times three increased by twice fifteen or is 60. And the number of points which satisfy the four quantics is $81 - 10(3 + 3 + 3 + 3) + 60$ or is 21.

ORDER OF RESTRICTED SYSTEMS OF EQUATIONS. 227

261. We have shown, Art. 249, how to determine the order of a system of determinants, the number of rows and columns in whose matrix differ by one. We shall now show how, when such a system represents a curve, to determine the rank of the curve. Commence, as before, with the simple case

$$\left\| \begin{array}{ccc} u, & v, & w \\ u', & v', & w' \end{array} \right\|,$$

and we see that the intersection of $uv' - u'v$, $uw' - u'w$ is a complex curve, consisting of the curve uu' and of the curve with which we are concerned. And knowing the order and rank of uu', we find the order and rank of the other curve. Knowing these again, we obtain, in like manner, the characteristics of the curve represented by a system with four columns and three rows, and so on, step by step, until we arrive, by induction, at the result that the general expression for the rank is

$$(p^2 - q + pP + Q)(P + 2p) - q(p + P) + R + r,$$

where p, q, P, Q have the same meaning as in Art. 249, and R, r are the sums of the products in threes of the quantities $a, b, \&c., \alpha, \beta, \&c.$ For we have seen, Art. 258, that the ranks of two complementary curves which together make up the intersection of two quantics whose orders are μ, ν, are connected by the equation $\beta - \beta' = (\alpha - \alpha')(\mu + \nu)$. And we saw (Art. 249) that the intersection of two quantics whose orders are $p + P - a$, $p + P - b$, is made up of the curve whose order is $\alpha = p^2 - q + pP + Q$, and of the curve of lower order for which $\alpha' = P^2 - Q + pP + q - p'(p + P) + q'$, where we have written $p' = a + b$, $q' = ab$. Now to find, by the formula of this article assumed true, the rank of the lower system, we are to write for the new R, the old r; and for the new r,

$$R - p'Q + (p'^2 - q')P - (p'^3 - 2p'q').$$

We have, then, after a little reduction,

$$\beta' = (P^2 + pP + q)(2P + p) - Q(3P + 2p) + R + r$$
$$- p'(P + p)(3P + p) + 2p'(Q - q) + p'^2(P + p) + 2q'(P + p) - p'q'.$$

To this value of β' add $(\alpha - \alpha')(\mu + \nu)$; that is to say,

$$\{p^2 - P^2 + 2(Q - q) + p'(P + p) - q'\}(2P + 2p - p'),$$

228 ORDER OF RESTRICTED SYSTEMS OF EQUATIONS.

and the result is the value already given for β. As then the formula can easily be verified for the case of two rows, it is true generally.

262. Let us next consider such a system as

$$\left\| \begin{array}{cccccc} a, & b, & c, & d, & e, & f \\ a', & b', & c', & d', & e', & f' \\ a'', & b'', & c'', & d'', & e'', & f'' \\ a''', & b''', & c''', & d''', & e''', & f''' \end{array} \right\|,$$

where the number of columns exceeds the number of rows by two; and let us examine the order of the restricted system common to all these determinants. Now any three of these $(ad'e''f''')$, $(bd'e''f''')$, $(cd'e''f''')$ have common the curve

$$\left\| \begin{array}{cccc} d, & d', & d'', & d''' \\ e, & e', & e'', & e''' \\ f, & f', & f'', & f''' \end{array} \right\|.$$

If then l, m, n be the order of the three determinants; α, β, the order and weight of this curve, the determinants intersect in points not on this curve, in number

$$lmn - \alpha(l+m+n) + \beta.$$

But if we represent the orders of the several functions in the same way as in Art. 249, it is easy to see that the degrees of the three determinants are $P+p-b-c$, $P+p-c-a$, $P+p-a-b$; so that if we write $a+b+c=p'$, $bc+ca+ab=q'$, $abc=r'$, we are to substitute for lmn,

$$(P+p)^3 - 2p'(P+p)^2 + (P+p)(p'^2+q') - (p'q'-r');$$

and for $l+m+n$, $3P+3p-2p'$. The order and rank of the curve, α and β, are found from the formulæ of Arts. 249, 261, by writing for P, Q, and R, p, q, and r; and for p, q, r respectively

$$P-p', \quad Q-Pp'+p'^2-q', \quad R-p'Q+(p'^2-q')P-(p'^3-2p'q'+r').$$

We find thus, as before,

$$\alpha = P^2 - Q + pP + q - p'(P+p) + q',$$
$$\beta = (P^2+pP+q)(2P+p) - Q(3P+2p) + R + r$$
$$\quad - p'(P+p)(3P+p) + 2p'(Q-q) + p'^2(P+p) + 2q'(P+p) - p'q' - r'.$$

ORDER OF RESTRICTED SYSTEMS OF EQUATIONS. 229

Add, then, to the value of β just found the value given for $lmn - \alpha(l+m+n)$, and we find the required result, viz.

$$R + pQ + (p^2 - q)P + p^3 - 2pq + r.$$

If the several rows are of the same degrees, that is, if α, β, &c. all $= 0$, then the order of the system is R.

The correspondence of this result with that of Art. 249, may be made more manifest by writing the symmetric function $\alpha^2 + \alpha\beta + \beta^2 +$ &c. p_2; $\alpha^3 + \alpha^2\beta + \alpha\beta^2 + \beta^3 +$ &c. p_3, &c. then the result of Art. 249 is $Q + pP + p_2$, and of this article is $R + pQ + p_2 P + p_3$. And we are led by induction to the conclusion, that the order of a system where the columns exceed the rows by three will be $S + pR + p_2 Q + p_3 P + p_4$, and so on generally. I have not actually carried the calculation further than the case considered in the present article, but the following articles will show how the process here used can be extended to the higher cases.

In all cases the weight of the system may be deduced from its order by the rule already given.

263. The formula of the last article may be applied to calculate the order of the system of conditions, that the equations $at^m +$ &c., $a't^n +$ &c. may have three common roots. The conditions are formed by a system of determinants, the matrix for which is formed as in Art. 252; save that the line a, b, c is repeated $n-2$ times, and the line a', b', c', $m-2$ times. The matrix consists of $m + n - 2$ columns and $m + n - 4$ rows. The order of the system then calculated by the last article is found to be

$$\frac{n(n-1)(n-2)}{1.2.3}\lambda^3 + \frac{m(m-1)(m-2)}{1.2.3}\mu^3 + \tfrac{1}{2}(n-1)(n-2)(m-2)\lambda^2\mu$$

$$+ \tfrac{1}{2}(m-1)(m-2)(n-2)\lambda\mu^2 + \tfrac{1}{3}(m-1)n(n-1)(n-2)\lambda^2\alpha$$

$$+ \tfrac{1}{3}(n-1)m(m-1)(m-2)\mu^2\alpha + \tfrac{1}{2}(m-2)(n-2)\{m(n-1)+n(m-1)\}\lambda\mu\alpha$$

$$+ \{\tfrac{1}{2}n(n-1)(n-2)m(m-2) + \tfrac{1}{3}n(n-1)(n-2)\}\alpha^2\lambda$$

$$+ \{\tfrac{1}{2}m(m-1)(m-2)n(n-2) + \tfrac{1}{3}n(m-1)(m-2)\}\alpha^2\mu$$

$$+ \tfrac{1}{6}m(m-1)(m-2)n(n-1)(n-2)\alpha^3.$$

230   ORDER OF RESTRICTED SYSTEMS OF EQUATIONS.

In the case where we have $\alpha = 0$, $\lambda = \mu = 1$, this reduces to

$$\tfrac{1}{6}(m+n-2)(m+n-3)(m+n-4).$$

The weight of the system, found by the same process as before, is

$\tfrac{1}{2}n(n-1)(n-2)\lambda^2\lambda' + \tfrac{1}{2}m(m-1)(m-2)\mu^2\mu'$

$+ (n-1)(n-2)(m-2)(\lambda\mu\lambda'+\tfrac{1}{2}\lambda^2\mu')+(m-1)(m-2)(n-2)(\lambda\mu\mu'+\tfrac{1}{2}\mu^2\lambda')$

$+ (m-1)n(n-1)(n-2)(\lambda\lambda'\alpha + \tfrac{1}{2}\lambda^2\alpha')$

$+ (n-1)m(m-1)(m-2)(\mu\mu'\alpha + \tfrac{1}{2}\mu^2\alpha')$

$+ \tfrac{1}{2}(m-2)(n-2)\{2mn - m - n\}(\lambda\mu'\alpha + \lambda'\mu\alpha + \lambda\mu\alpha')$

$+ \{\tfrac{1}{2}n(n-1)(n-2)m(m-2) + \tfrac{1}{3}n(n-1)(n-2)\}(\alpha^2\lambda' + 2\alpha\alpha'\lambda)$

$+ \{\tfrac{1}{2}m(m-1)(m-2)n(n-2) + \tfrac{1}{3}m(m-1)(m-2)\}(\alpha^2\mu' + 2\alpha\alpha'\mu)$

$+ \tfrac{1}{2}m(m-1)(m-2)n(n-1)(n-2)\alpha^2\alpha'.$

264. The next problem we investigate is when a system of quantics have a "surface" common, to find how many of their points of intersection are absorbed by the common surface. We mean by the *order* and *rank* of a surface, the order and rank of the curve which is the section of the surface by any quantic of the first degree. Thus, consider the case of five independent variables, then a system of three equations constitutes a surface, and if their orders be λ, μ, ν, the order of the surface will be $\lambda\mu\nu$, and its rank $\lambda\mu\nu(\lambda+\mu+\nu)$; these being the order and rank of the curve got by uniting with the given equations an additional one of the first degree.

Now, first let $k-1$ quantics have a surface in common, whose order and rank are α, β; they will also in general have common besides a complementary curve whose order is readily found. For join with the given quantics another of the first degree, and we then have a system of k* quantics, having a curve common, and therefore by Art. 258 intersecting in $lmnr - \alpha(l+m+n+r) + \beta$ points besides. But these are the points in which the quantic of the first degree meets the complementary curve, and therefore this is the order of that curve.

---

* To fix the ideas we take $k = 5$ in the next article.

265. Next let us investigate the number of points in which the surface and complementary curve intersect each other. Let U, V, W, Y be respectively of the forms

$$Au + Bv + Cw = 0,$$
$$A'u + B'v + C'w = 0,$$
$$A''u + B''v + C''w = 0,$$
$$A'''u + B'''v + C'''w = 0.$$

Then the points common to U, V, W, Y which do not make u, v, $w = 0$, will satisfy the system of determinants

$$\left\| \begin{array}{cccc} A, & A', & A'', & A''' \\ B, & B', & B'', & B''' \\ C, & C', & C'', & C''' \end{array} \right\| = 0.$$

But since A is of the order $l - \lambda$, B of the order $l - \mu$, A' of the order $m - \lambda$, &c., it follows (Art. 249) that the order of the set of determinants is $Q - pP + p_2$, where P and Q are the sum and sum of products in pairs of l, m, n, r, where $p = \lambda + \mu + \nu$, and $p_2 = \lambda^2 + \mu^2 + \nu^2 + \lambda\mu + \mu\nu + \nu\lambda$. If now we combine this system of determinants (equivalent to two conditions), with the $k - 2$ conditions which constitute the surface, we determine the points common to the surface and complementary curve. And their number is the order of the system of determinants, multiplied by $\lambda\mu\nu$. Writing then α and β for the order and rank of the surface $\lambda\mu\nu$, $\lambda\mu\nu(\lambda + \mu + \nu)$, and denoting by γ the new characteristic $\lambda\mu\nu(\lambda^2 + \mu^2 + \nu^2 + \lambda\mu + \mu\nu + \nu\lambda)$, which we may call the class of the surface, we find

$$i = \alpha Q - \beta P + \gamma.$$

266. If then we have an additional quantic Z also containing the given surface, and if it be required to find how many points not on the surface are common to all k quantics, these will be evidently the points of intersection of the complementary curve with Z, less the number of points of intersection of the complementary curve with the surface. If then l, m, n, r, s be the orders of the quantics, the number sought will be got by subtracting from

$$s \{lmnr - \alpha (l + m + n + r) + \beta\},$$

the number
$$\alpha(lm + ln + mn + lr + mr + nr) - \beta(l + m + n + r) + \gamma.$$
And the difference is
$$lmnrs - \alpha Q + \beta P - \gamma,$$
which is the formula required.

267. Next let us consider the case where a system of quantics have common not only a surface, whose characteristics are α, β, γ, but also a curve, whose characteristics are α', β', intersecting the surface in i points. As before, consider first $k-1$ of the quantics. Their intersection we have seen consists of the surface and of a complementary curve, whose order is
$$lmnr - \alpha(l + m + n + r) + \beta.$$
And if the complementary curve be itself complex, consisting in part of the curve α', and also of another curve, whose order is α'', we have evidently
$$\alpha'' = lmnr - \alpha(l + m + n + r) + \beta - \alpha'.$$
The points therefore which we desire to determine are got by subtracting from the α''s points of intersection of the curve α'' with the remaining one of the given system of k quantics, $\delta + \delta'$ where δ is the number of points in which the curve α'' meets the surface (α, β, γ), and δ' is the number of points where it meets the curve. But we know δ, since we know, by Art. 265, the number of points where the surface is met by the entire curve complementary to it; and therefore have
$$\delta + i = \alpha(lm + ln + \&c.) - \beta(l + m + n + r) + \gamma;$$
and we know δ', knowing, by Art. 257, the number of points in which the curve α' is met by the entire curve complementary to it, and therefore have
$$\delta' + 2i = \alpha'(l + m + n + r) - \beta'.$$
Substituting the values thence derived for δ and δ' in $\alpha''s - \delta - \delta'$, we get
$$lmnrs - \alpha Q + \beta P - \gamma - \alpha' P + \beta' + 3i.$$
In other words, the diminution from the number $lmnrs$ produced by curve and surface together is equal to the sum of their separate diminutions lessened by three times the number of their common points.

ORDER OF RESTRICTED SYSTEMS OF EQUATIONS. 233

268. This result may be confirmed by supposing one of the quantics to be a complex one $Z'Z''$, where Z' contains the common surface, and Z'' the common curve; and the degrees of Z', Z'' are s', s''. Then the quantics U, V, W, Y, Z', by Art. 266, have common points not on the common surface

$$lmnrs' - \alpha\{s'(l+m+\&c.) + lm + mn + \&c.\} + \beta(s'+l+m+\&c.) - \gamma.$$

But among these will be reckoned the $\alpha's'$ points in which the common curve meets Z', deducting however the i points common to the curve and surface. To find then the number of points $UVWYZ'$ which lie on neither curve nor surface, we must deduct from the number last written $\alpha's' - i$.

Consider now the intersections of U, V, W, Y, Z''; these are a system of quantics having common two curves intersecting in i points; viz. the given curve α', and the curve of intersection of the common surface by Z'', whose order will be $\alpha s''$, and whose rank will be $\alpha s''(\lambda + \mu + \nu + s'')$. The number of points $UVWYZ''$ which lie on neither curve nor surface will be

$$lmnrs'' - (\alpha' + \alpha s'')(l+m+n+r+s'') + \beta' + \alpha s''(\lambda + \mu + \nu + s'') + 2i.$$

Adding, and writing s for $s' + s''$, we get

$$lmnrs - \alpha Q + \beta P - \gamma - \alpha' P + \beta' + 3i,$$

as in the last article.

269. We next suppose the quantics to have common two surfaces having i points of intersection. The method would be the same if there were several surfaces. Let the last quantic be a complex one, consisting of Z' which passes through the first surface and Z'' which passes through the second. Then the system U, V, W, Y, Z', have common the surface $\lambda\mu\nu$ and the curve $\lambda'\mu'\nu's'$, which have i points common, and they will therefore intersect in points which lie on neither

$$lmnrs' - \lambda\mu\nu\{(l+m+n+r)s' + lm + \&c.\} + \beta(l+m+n+r+s')$$
$$- \gamma - \lambda'\mu'\nu's'(l+m+n+r+s') + \lambda'\mu'\nu's'(\lambda' + \mu' + \nu' + s') + 3i.$$

In like manner the points common with Z'' are

$$lmnrs'' - \lambda\mu\nu s''(l+m+n+r+s'') + \lambda\mu\nu s''(\lambda + \mu + \nu + s'')$$
$$- \lambda'\mu'\nu'\{(l+m+n+r)s'' + lm + \&c.\} + \beta'(l+m+n+r+s'') + 3i.$$

H H

234   ORDER OF RESTRICTED SYSTEMS OF EQUATIONS.

Adding which we have
$$lmnr(s'+s'') - (\lambda\mu\nu + \lambda'\mu'\nu')\{(l+m+n+r)(s'+s'') + lm + \&c.\}$$
$$+ (\beta+\beta')(l+m+n+r+s'+s'') - \gamma + 6i - \gamma'.$$
In other words, the combined effect of the two surfaces is equal to the sum of the effects of the surfaces separately considered, diminished by six times the number of their common points. When there are only four variables, two surfaces always *must* have common points of intersection.

270. Lastly, let the two surfaces have a common curve whose order and rank are α'', β''. Proceeding, as in the last article, we find that the system $UVWYZ'$ have common indeed the surface $\lambda\mu\nu$; and the curve $\lambda'\mu'\nu's$. But since this curve is a complex one, consisting in part of the curve α'', β'' which lies on $\lambda\mu\nu$, we are only to take into account the complementary curve which, by Art. 259, has for its order $\lambda'\mu'\nu's - \alpha''$, while its rank is
$$\beta'' + (\lambda'\mu'\nu's' - 2\alpha'')(\lambda'+\mu'+\nu'+s');$$
and the complementary curve intersects $\alpha''\beta''$ in
$$\alpha''(\lambda'+\mu'+\nu'+s') - \beta'' \text{ points.}$$
The number of intersections is therefore
$$lmnrs' - \lambda\mu\nu\{(l+m+n+r)s' + lm + \&c.\} + \beta(l+m+n+r+s') - \gamma$$
$$- (\lambda'\mu'\nu's' - \alpha'')(l+m+n+r+s')$$
$$+ \beta'' + (\lambda'\mu'\nu's' - 2\alpha'')(\lambda'+\mu'+\nu'+s') + 3\alpha''(\lambda'+\mu'+\nu'+s') - 3\beta''.$$
Similarly the intersections for $UVWYZ'$ are
$$lmnrs'' - \lambda'\mu'\nu'\{(l+m+n+r)s'' + lm + \&c.\} + \beta'(l+m+n+r+s'') - \gamma'$$
$$- (\lambda\mu\nu s'' - \alpha'')(l+m+n+r+s'')$$
$$+ \beta'' + (\lambda\mu\nu s'' - 2\alpha'')(\lambda+\mu+\nu+s'') + 3\alpha''(\lambda+\mu+\nu+s'') - 3\beta''.$$
Adding, we have
$$lmnr(s'+s'') - (\lambda\mu\nu + \lambda'\mu'\nu')\{(l+m+n+r)(s'+s'') + lm + \&c.\}$$
$$+ (\beta+\beta'+2\alpha'')(l+m+n+r+s'+s'')$$
$$- \gamma - \gamma' + \alpha''(\lambda+\mu+\nu+\lambda'+\mu'+\nu') - 4\beta''.$$

In other words, the two surfaces make up a complex surface, whose order is the sum of their orders, whose rank is the sum of their ranks increased by twice the order of the common curve, and whose class is the sum of their classes increased by four times the rank of the common curve and diminished by $a''(\lambda + \mu + \nu + \lambda' + \mu' + \nu')$.

We must leave untouched some other cases which ought to be discussed in order to complete the subject; in particular the case where the surfaces touch in points or along a curve. The cases examined suffice to enable us to ascertain the order of a set of determinants in whose matrix the number of columns exceeds that of the rows by three. Taking four determinants of the system we find that they have a surface common; and by determining the three characteristics of that surface and applying the formula of Art. 265, the order of the system can be obtained.

271. We come now to the problem of finding the order of the system of conditions that three ternary quantics should have two common points. The method followed is the same as that given by Mr. Cayley for eliminating between three homogeneous equations in three variables, and which we have explained (Art. 90). Let the three equations be of the degrees l, m, n. Multiply the first by all the terms x^{m+n-3}, yx^{m+n-4}, &c. of an equation of the degree $m+n-3$, the second in like manner by all the terms of an equation of the degree $n+l-3$, and the third by all the terms of an equation of the degree $l+m-3$. We have thus in all

$$\tfrac{1}{2}(m+n-1)(m+n-2) + \tfrac{1}{2}(n+l-1)(n+l-2) + \tfrac{1}{2}(l+m-1)(l+m-2),$$

equations of the degree $l+m+n-3$; from which we are to eliminate the $\tfrac{1}{2}(l+m+n-1)(l+m+n-2)$ terms $x^{l+m+n-3}$, &c. But as it has been shewn, in the place referred to, the equations we use are not independent but are connected by

$$\tfrac{1}{2}(l-1)(l-2) + \tfrac{1}{2}(m-1)(m-2) + \tfrac{1}{2}(n-1)(n-2)$$

relations. Subtracting then the number of relations, the number of independent equations is found to be one less than the number of quantities to be eliminated; and we have a matrix in which the number of columns is one more than the number of rows, the case considered in Art. 249. But, as was shewn, Art. 89, when we are given a number of equations connected by relations, the determinants formed by taking a sufficient number of

236. ORDER OF RESTRICTED SYSTEMS OF EQUATIONS.

the equations, require to be reduced by dividing out extraneous factors, these factors being determinants formed with the co-efficients of the equations of relation. If then, in the present case, we took a sufficient number of the equations and determined the order by the rule of Art. 249, our result would require to be reduced by a number which we proceed to determine.

272. Let us commence with the simplest case where we have k equations in k variables, the equations being connected by a single relation. To fix the ideas we write down the system with three rows

$$\begin{vmatrix} a, & b, & c & | & \lambda \\ a', & b', & c' & | & \lambda' \\ a'', & b'', & c'' & | & \lambda'' \end{vmatrix},$$

where we mean to imply that the quantities involved are connected by the relations

$$\lambda a + \lambda' a' + \lambda'' a'' = 0, \quad \lambda b + \lambda' b' + \lambda'' b'' = 0, \quad \lambda c + \lambda' c' + \lambda'' c'' = 0.$$

We also suppose that $\lambda a, \lambda' a', \lambda'' a''$ are of the same order, so that the orders $\lambda + a = \lambda' + a' = \lambda'' + a''$. Now let us, in the first place, suppose that the two first equations are in the simplest form, and that $\lambda'' = -1$. The true order then is that determined from the first two equations; that is to say, if we indicate the orders, as in Art. 249, $Q + P(\alpha + \beta) + \alpha^2 + \beta^2 + \alpha\beta$. Now suppose that we had omitted the first row, the order deduced from the second and third, would be $Q + P(\beta + \gamma) + \beta^2 + \gamma^2 + \beta\gamma$; which we see, in order to give the true order, requires to be reduced by $(\gamma - \alpha)(P + \alpha + \beta + \gamma)$; in other words, by the order of λ multiplied by the order of the determinant obtained from the three equations. And the general rule to which we are thus led is: Leave out one of the rows and determine the order of the remaining system by the rule of Art. 249; from the number so found, subtract the order of the determinant formed from all the equations, multiplied by the order of the term in the relation column belonging to the omitted row. It is easy to verify, that we are thus led to the same result whatever be the omitted row. Thus

$$Q + P(\alpha + \beta) + \alpha^2 + \beta^2 + \alpha\beta - \lambda''(P + \alpha + \beta + \gamma)$$
$$= Q + P(\beta + \gamma) + \beta^2 + \gamma^2 + \beta\gamma - \lambda(P + \alpha + \beta + \gamma),$$

since the orders $\quad\lambda-\lambda''=\gamma-\alpha.$

And our result may be written in a symmetrical form if we write A for the common value of $\lambda+\alpha$, $\lambda'+\beta$, $\lambda''+\gamma$, when it becomes

$$Q+P(\alpha+\beta+\gamma)+\alpha^2+\beta^2+\gamma^2+\beta\gamma+\gamma\alpha+\alpha\beta-A(P+\alpha+\beta+\gamma),$$
or $\quad\quad\quad Q+Pp+p^2-q-A(P+p).$

273. And, generally, if there be any number of relation columns, I have been led by a similar process to the following result: Let the terms in the relation columns be λ, λ', λ'', &c. μ, μ', μ'', &c. ν, ν', ν'', &c.; then we must have

$$\lambda+\alpha=\lambda'+\beta', \&c., \quad \mu+\alpha=\mu'+\beta', \&c., \quad \nu+\alpha=\nu'+\beta', \&c.$$

Let A, B, C denote the common values of these sums, and let P', Q' denote the sum and sums of products in pairs of the quantities A, B, C; then the order of the system is

$$Q+Pp+p^2-q-P'(P+p)+Q'.$$

This result may be stated as follows, in a way which leads us at once to foresee the answer to some other questions that may be proposed as to the order of systems of these equations. In the case we are considering, the entire number of columns, counting the relation columns, is one more than the number of rows; *and the order of the system is that given by the rule of* Art. 249, *if we give a negative sign to the orders in the relation columns.* In like manner, when the number of columns, counting the relation columns, is equal to the number of rows, the system, by Mr. Cayley's theorem, represents a determinant whose order is that which we should obtain by calculating the order of the entire system considered as a determinant, the orders in the relation columns being taken negatively. And so no doubt if the entire number of columns exceeded the number of rows by two, the order of the system would be found by the same modification from the rule of Art. 262.

274. Let us now apply the rule just arrived at to the problem proposed in Art. 271. We suppose that in the three ternary quantics the coefficients of the highest powers of x, viz. x^l, x^m, x^n are of the orders λ, μ, ν; those of $x^{l-1}y$, $x^{l-1}z$

are of the orders $\lambda + \alpha, \lambda + \alpha'$, and so on, the orders of the coefficients increasing by α for every power of y, and by α' for every power of z. Then the terms in the first column consist first of $\frac{1}{2}(m+n-1)(m+n-2)$ terms whose orders are $\lambda; \lambda - \alpha, \lambda - \alpha'; \lambda - 2\alpha, \lambda - \alpha - \alpha', \lambda - 2\alpha'$, &c.; secondly, of $\frac{1}{2}(n+l-1)(n+l-2)$ terms whose orders are $\mu, \mu - \alpha, \mu - \alpha'$, &c., and thirdly of $\frac{1}{2}(l+m-1)(l+m-2)$ similar terms in ν. These may be taken for the numbers α, β, γ, &c. of Art. 249. The numbers a, b, c, &c. of that article are $0, \alpha, \alpha' : 2\alpha, \alpha + \alpha'$, $2\alpha'$, &c., there being in all $\frac{1}{2}(l+m+n-1)(l+m+n-2)$ such terms. Lastly, the numbers A, B, C, &c. of the last article are found to consist of $\frac{1}{2}(l-1)(l-2)$ terms, $\mu+\nu, \mu+\nu-\alpha, \mu+\nu-\alpha'$; together with $\frac{1}{2}(m-1)(m-2)$ and $\frac{1}{2}(n-1)(n-2)$ corresponding terms in $\nu + \lambda$ and $\lambda + \mu$. In calculating I have found it convenient to throw the formula of the last article into the shape

$$\tfrac{1}{2}\{(P+p-P')^2 - (S_2 + S_2' - s_2)\},$$

where s_2 denotes the sum of the squares of the terms a, b, c, &c. Also if $\phi(\lambda) = Al^4 + Bl^3 + Cl^2 + Dl + E$, it is convenient to take notice that

$\phi(l+m+n) + \phi(l) + \phi(m) + \phi(n) - \phi(l+m) - \phi(m+n) - \phi(n+l)$
$\qquad = 12Almn(l+m+n) + 6Blmn + E.$

I have thus arrived at the result, that the order of the system is

$\frac{1}{2}mn(mn-1)\lambda^2 + \frac{1}{2}nl(nl-1)\mu^2 + \frac{1}{2}lm(lm-1)\nu^2$

$+ \{(nl-1)(lm-1) - \frac{1}{2}(l-1)(l-2)\}\mu\nu$

$+ \{(lm-1)(mn-1) - \frac{1}{2}(m-1)(m-2)\}\nu\lambda$

$+ \{(nl-1)(nm-1) - \frac{1}{2}(n-1)(n-2)\}\lambda\mu$

$+ mn\lambda\{lmn - l + 1 - \frac{1}{2}(m+n)\}(\alpha + \alpha')$

$+ nl\mu\{lmn - m + 1 - \frac{1}{2}(n+l)\}(\alpha + \alpha')$

$+ lm\nu\{lmn - n + 1 - \frac{1}{2}(l+m)\}(\alpha + \alpha')$

$+ \frac{1}{2}lmn(lmn - l - m - n + 2)(\alpha^2 + \alpha'^2) + \frac{1}{2}lmn(2lmn - l - m - n + 1)\alpha\alpha'.$

If the order of all the terms in the first equation be λ, in the second μ, in the third ν, we have only to make α and $\alpha' = 0$ in the preceding formula. In this case, supposing $\lambda = \mu = \nu = 1$, and $l = m = n$, the order becomes $\frac{3}{2}n(n-1)(n^2+n-1)$.

(239)

LESSON XIX.

APPLICATIONS OF SYMBOLICAL METHODS.

275. IN this Lesson, which is supplementary to Lesson XIV., we wish to show how the symbolical notation there explained affords a calculus by means of which invariants and covariants can be transformed, and the identity of different expressions ascertained. The basis of the whole, as applied to binary quantics, is the following identical equation: Let D_1 denote $x\frac{d}{dx_1} + y\frac{d}{dy_1}$, and as before let $\overline{12} = \frac{d}{dx_1}\frac{d}{dy_2} - \frac{d}{dx_2}\frac{d}{dy_1}$, then it is easy to verify that, identically

$$D_1\overline{23} + D_2\overline{31} + D_3\overline{12} = 0, \quad \ldots\ldots\ldots\ldots(A),$$

for on expansion the coefficients of x and y separately vanish. To illustrate the use to be made of this equation, we write out the first application we make of it in greater fulness of detail than we shall think it necessary to do afterwards. Squaring equation A, which may be written $D_1\overline{23} = D_3\overline{13} - D_2\overline{12}$, we get

$$2D_2D_3\overline{12}.\overline{13} = D_3^2\overline{13}^2 + D_2^2\overline{12}^2 - D_1^2\overline{23}^2 \ldots\ldots (B).$$

In this form the equation is true, even if the functions $U_1, U_2, U_3,$ supposed to be operated on, are all different in form. But the case which we shall exclusively consider in this Lesson is when we are forming derivatives of a single function U, with which U_1, U_2, U_3 all become identical when their suffixes are suppressed. And in this case we have seen (Art. 149) that $D_1^2\overline{23}^2, D_2^2\overline{31}^2, D_3^2\overline{12}^2$ are all only different expressions for the same thing. Hence, equation B becomes

$$D_3^2\overline{12}^2 = 2D_2D_3\overline{12}.\overline{13}.$$

Now, by the theorem of homogeneous functions, the operation D performed on any function only affects it with a numerical multiplier. On the left-hand side of the equation, since D_3^2 only affects U_3 which is not there elsewhere differentiated, that multiplier is $n(n-1)$; on the right-hand side $D_2, D_3,$ each affect functions which have been once differentiated besides; and therefore, introduces the numerical multiplier $n-1$; consequently, if we expand the equation $2D_2D_3\overline{12}.\overline{13} = D_3^2\overline{12}^2$, drop suffixes, and divide both sides by $2(n-1)$, we get

$$(n-1)\left\{\frac{d^2U}{dx^2}\left(\frac{dU}{dy}\right)^2 - 2\frac{d^2U}{dxdy}\frac{dU}{dx}\frac{dU}{dy} + \frac{d^2U}{dy^2}\left(\frac{dU}{dx}\right)^2\right\}$$
$$= nU\left\{\frac{d^2U}{dx^2}\frac{d^2U}{dy^2} - \left(\frac{d^2U}{dxdy}\right)^2\right\}.$$

276. It will be observed, that whenever by transformation the number of figures in the symbol is diminished, it shows that the quantic U is itself a factor in the derivative. Thus, in the example chosen, we proved that $\overline{12}.\overline{13}$ differs only by a numerical factor from $\overline{12}^2$. But as what we operate on is the product $U_1U_2U_3$, and as the symbol $\overline{12}^2$ does not affect U_3 at all; when we drop the suffixes, U remains as a factor.

277. And in general any symbol may be so transformed that the highest power of any factor $\overline{12}$ may be even. For the signification of the symbol is not altered if we interchange the figures 1 and 2; therefore,

$$\overline{12}^{2m+1}\phi_1 = -\overline{12}^{2m+1}\phi_2 = \tfrac{1}{2}\overline{12}^{2m+1}(\phi_1 - \phi_2),$$

and by the help of equation A the quantity $\phi_1 - \phi_2$ can be so transformed as to be divisible by $\overline{12}$. Thus, for example, the derivative $\overline{12}^m.\overline{13}$, where m is odd, differs only by a numerical multiplier from $\overline{12}^{m+1}$. For

$$2D_2\overline{12}^m.\overline{13} = \overline{12}^m\{D_2\overline{13} - D_1\overline{23}\} = D_3\overline{12}^{m+1},$$

or if n be the degree of U the function operated on,

$$2(n-m)\overline{12}^m.\overline{13} = n(12)^{m+1}.$$

278. In addition to the identical equation already mentioned, we employ another which can be easily verified, viz.:

$$\overline{12}.\overline{34} + \overline{13}.\overline{42} + \overline{14}.\overline{23} = 0 \ldots\ldots\ldots\ldots(C).$$

By the help of these equations we can reduce all symbols to certain standard forms, which we denote by special letters, as follows: For two factors we take as our standard form the

APPLICATIONS OF SYMBOLICAL METHODS. 241

Hessian $\overline{12}^2$ which we call H; for three factors $\overline{12}^2.\overline{13} = G$, which is given at length (Art. 152); for four factors $\overline{12}^4 = S$ (Art. 201), or $\overline{12}^2.\overline{34}^2 = H^2$; for five factors $\overline{12}^4.\overline{13} = F$, or $\overline{12}^2.\overline{13}.\overline{45}^2 = GH$; for six factors $\overline{12}^6 = A$ (Art. 236), or $\overline{12}^2.\overline{23}^2.\overline{31}^2 = T$ (Art. 201), or $\overline{12}^2.\overline{34}^2.\overline{56}^2 = H^3$, and so on. The following examples will sufficiently illustrate how these reductions are effected.

279. The first examples which are given, are chosen to illustrate the following theorem: "The result of substituting in any function U, $\dfrac{dU}{dy}$ for x, and $-\dfrac{dU}{dx}$ for y, is a covariant (Art. 157), and is always divisible by U." Let us first see how to express symbolically the covariant in question. The given function may be written $\left(x\dfrac{d}{dx} + y\dfrac{d}{dy}\right)^n U$. When we make the proposed change, each of the n factors which operate on U will assume the form $\dfrac{d}{dx}\dfrac{d}{dy_2} - \dfrac{d}{dy}\dfrac{d}{dx_2}$. The required covariant then $\dfrac{d^n U}{dx^n}\left(\dfrac{dU}{dy}\right)^n - \&c.$ will be expressed symbolically as the product of n factors $\overline{12}.\overline{13}.\overline{14}$, &c. One symbol (that which when the product is expanded will give the nth differential) must be the same for every factor. The other symbol in each factor must be different, in order that when the multiplication is performed, we may have only powers of the first differentials. Writing then $\overline{12}.\overline{13} = Q_2$, $\overline{12}.\overline{13}.\overline{14} = Q_3$, &c., we propose as an exercise to express these in terms of the standard forms. Q_2 has been already calculated (Art. 275).

Ex. 1. To calculate $Q_3 = \overline{12.13.14}$. Multiply equation B by $\overline{14}$, then the first two terms on the right-hand side become identical while the last term vanishes, and we have

$$D_2 D_3 \overline{12.13.14} = D_2{}^2 \overline{13}^2.\overline{14}, \text{ or } (n-1)Q_3 = nGU.$$

In general, every symbol can be reduced to a more compact form by substituting for every pair of simple factors having a common figure, such as $\overline{12.13}$, their values by equation B, and so expressing the given symbol in terms of others in which this pair of factors is replaced by a single square factor.

Ex. 2. To calculate $Q_4 = \overline{12.13.14.15}$. Multiply together

$$2D_2 D_3 \overline{12.13} = D_3{}^2\overline{12}^2 + D_2{}^2\overline{13}^2 - D_1{}^2\overline{23}^2,$$
$$2D_4 D_5 \overline{14.15} = D_5{}^2\overline{14}^2 + D_4{}^2\overline{15}^2 - D_1{}^2\overline{45}^2,$$

I I

242 APPLICATIONS OF SYMBOLICAL METHODS.

and assembling the identical terms, we have

$$4(n-1)^4 Q_4 = D_1{}^4\overline{23^2.45^2} - 4D_1{}^2D_2{}^2\overline{12^2.45^2} + 4D_2{}^2D_3{}^2\overline{12^2.14^2}$$
$$= -3n(n-1)(n-2)(n-3) H^2 U + 4n^2(n-1)^2 U^2\overline{12^2.13^2}.$$

It remains to bring $\overline{12^2.13^2}$ to the standard forms. Raise to the fourth power $D_1\overline{23} = D_2\overline{13} - D_3\overline{12}$, when collecting identical terms, we get

$$6D_2{}^2D_3{}^2\overline{12^2.13^2} = 8D_3{}^3D_2\overline{12^3.13} - D_2{}^4\overline{12^4}.$$

But, as in Art. 277, $8D_3{}^3D_2\overline{12^3.13} = 4D_3{}^4\overline{12^4}$. Hence

$$2(n-2)(n-3)\{\overline{12^2.13^2}\} = n(n-1) SU.$$

Otherwise thus: We know by elementary algebra that the equation $a + b + c = 0$, implies $a^4 + b^4 + c^4 - 2a^2b^2 - 2a^2c^2 - 2b^2c^2 = 0$. Hence, from equation A, we have

$$D_1{}^4\overline{23^4} + D_2{}^4\overline{31^4} + D_3{}^4\overline{12^4} - 2D_1{}^2D_2{}^2\overline{23^2.31^2} - 2D_2{}^2D_3{}^2\overline{31^2.12^2} - 2D_3{}^2D_1{}^2\overline{12^2.23^2} = 0$$

$$\dots\dots\dots\dots\dots\dots(D),$$

and, as before, $\qquad 2D_2{}^2D_3{}^2\overline{12^2.13^2} = D_3{}^4\overline{12^4}.$

We have then

$$4(n-1)^4(n-2)(n-3) Q_4 = -3n(n-1)(n-2)^2(n-3)^2 UH^2 + 2n^3(n-1)^3 U^3 S.$$

It is easily seen in general that when in a symbol $\overline{12,13,14,15}$, &c. we substitute for each pair of factors $\overline{12.13}$ by equation (B), the effect is to diminish the number of figures which enter into the symbol, and therefore U is a factor. The values of Q_5 and Q_6 are calculated (*Cambridge and Dublin Mathematical Journal*, vol. IX. p. 23).

Ex. 3. To reduce $\overline{12^4.13^2}$. Multiply equation (D) by $\overline{12^2}$, when we have

$$2D_2{}^4\overline{13^4.12^2} + D_3{}^4\overline{12^6} = 2D_1{}^2D_2{}^2\overline{12^2.23^2.31^2} + 4D_2{}^2D_3{}^2\overline{12^4.13^2};$$

or $\quad 2(n-2)(n-3)(n-4)(n-5)\{\overline{12^4.13^2}\} = n(n-1)(n-2)(n-3) AU$

$$- 2(n-4)^2(n-5)^2 T.$$

Ex. 4. To reduce $\overline{12^2.13^2.14^2}$. Multiply the three equations

$$2D_2D_3\overline{12.13} = D_3{}^2\overline{12^2} + D_2{}^2\overline{13^2} - D_1{}^2\overline{23^2},$$
$$2D_3D_4\overline{13.14} = D_4{}^2\overline{13^2} + D_3{}^2\overline{14^2} - D_1{}^2\overline{34^2},$$
$$2D_4D_2\overline{14.12} = D_2{}^2\overline{14^2} + D_4{}^2\overline{12^2} - D_1{}^2\overline{24^2},$$

when, after assembling the identical terms, and reducing, we have

$$6D_2{}^2D_3{}^2D_4{}^2\overline{12^2.13^2.14^2} = -4D_1{}^6\overline{12^2.23^2.31^2} - 3D_1{}^2D_2{}^2D_3{}^2\overline{14^4.23^2} + 6D_3{}^2D_4{}^2\overline{12^4.13^2},$$

where, as the value of $\overline{12^4.13^2}$ has been already calculated, the value of $\overline{12^2.13^2.14^2}$ is determined.

Ex. 5. To reduce $\overline{12^2.13^2.34^2}$. We may either by multiplying equation (D) by $\overline{34^2}$ obtain an expression for the derivative in question in terms of $\overline{12^4.13^2}$, and $\overline{12^2.13^2.14^2}$, which have been already calculated, or else proceed directly as follows. Multiply by $\overline{14^2}$ the product of

$$2D_2D_3\overline{12.13} = D_3{}^2\overline{12^2} + D_2{}^2\overline{13^2} - D_1{}^2\overline{23^2}; \; 2D_2D_3\overline{24.34} = D_3{}^2\overline{24^2} + D_2{}^2\overline{34^2} - D_4{}^2\overline{23^2},$$

and we have

$$4D_2{}^2D_3{}^2\overline{12.13.24.34.14^2} = 2D_3{}^4\overline{12^4.24^2.41^2} + D_1{}^2D_4{}^2\overline{23^4.14^2} - 2D_2{}^2D_4{}^2\overline{14^2.13^2.23^2}.$$

But from equation (C)

$$2.\overline{14^2}\,(\overline{12.13.24.34}) = \overline{14^2}\,(\overline{12^2.34^2} + \overline{13^2.24^2} - \overline{14^2.23^2}),$$

therefore, the left-hand side of the preceding equation becomes

$$4D_2{}^2D_3{}^2\overline{12^2}.\overline{14^2}.\overline{34^2} - 2D_2{}^2D_3{}^2\overline{14^4}.\overline{23^2},$$

and hence reducing, we have

$$6D_2{}^2D_3{}^2\overline{12^2}.\overline{14^2}.\overline{34^2} = 3D_2{}^2D_3{}^2\overline{14^4}.\overline{23^2} + 2D_3{}^4\overline{12^2}.\overline{24^2}.\overline{41^2},$$

or $\quad 6(n-2)(n-3)\overline{12^2}.\overline{14^2}.\overline{34^2} = 3(n-2)(n-3)SH + 2n(n-1)TU.$

280. We wish next to show how to form the symbol expressing any derivative of a derivative. In the first place it is obvious that if we have any function of x_1, x_2, x_3, &c., we shall get the same result whether we suppress all the suffixes and then differentiate with regard to x, or whether we take the sum of the differentials with regard to x_1, x_2, x_3, &c., and then suppress the suffixes. The differential, then, with regard to x of any derivative symbol containing the figures 1, 2, 3, &c. (such as $\overline{12^2}.\overline{13^2}$) is got by operating on this symbol with $\dfrac{d}{dx_1} + \dfrac{d}{dx_2} + \dfrac{d}{dx_3} + $&c. The manner in which other derivatives are expressed will sufficiently appear if we take a particular example, and show how to express symbolically the Hessian of the Hessian. It will be remembered that the Hessian itself is expressed by forming the product of U_1 by the same function written with different letters U_2, and then operating with $(\xi_1\eta_2 - \xi_2\eta_1)^2$ where ξ, η denote differentials. To form the Hessian of $\overline{12^2}$ we form the product $\overline{12^2}.\overline{34^2}$ and operate on it with $(\xi_1\eta_2 - \xi_2\eta_1)^2$ where $\xi_1 = \dfrac{d}{dx_1} + \dfrac{d}{dx_2}$, $\xi_2 = \dfrac{d}{dx_3} + \dfrac{d}{dx_4}$, &c. And thus it is easy to see that the Hessian of the Hessian is $(\overline{13} + \overline{14} + \overline{23} + \overline{24})^2 \overline{12^2}.\overline{34^2}$, which expanded is equivalent to

$$4(\overline{12^2}.\overline{13^2}.\overline{34^2}) + 4(\overline{12^2}.\overline{34^2}.\overline{13}.\overline{24}) + 8(\overline{13}.\overline{14}.\overline{12^2}.\overline{34^2}).$$

Or, to take a more complicated example, let us endeavour to express symbolically the result of performing the operation $\overline{12^2}.\overline{23^2}.\overline{31^2}$ on three different functions, the first being the Hessian $\overline{12^2}$, the second being T, or $\overline{12^2}.\overline{23^2}.\overline{31^2}$, and the third the function U itself. Then we must use different figures for the two derivatives, and we therefore form the product of $\overline{12^2}$ by $\overline{34^2}.\overline{45^2}.\overline{53^2}$, and operate on it with

$$(\xi_1\eta_2 - \xi_2\eta_1)^2 (\xi_2\eta_3 - \xi_3\eta_2)^2 (\xi_3\eta_1 - \xi_1\eta_3)^2,$$

244   APPLICATIONS OF SYMBOLICAL METHODS.

where we are to put for ξ_1, $\dfrac{d}{dx_1} + \dfrac{d}{dx_2}$; for ξ_2, $\dfrac{d}{dx_3} + \dfrac{d}{dx_4} + \dfrac{d}{dx_5}$;

and for ξ_3, $\dfrac{d}{dx_6}$; and the symbol required is found by expanding

$$(\overline{13}+\overline{14}+\overline{15}+\overline{23}+\overline{24}+\overline{25})^2(\overline{36}+\overline{46}+\overline{56})^2(\overline{16}+\overline{26})^2\,\overline{12}^2.\overline{34}^2.\overline{45}^2.\overline{53}^2.$$

Ex. To reduce the Hessian of the Hessian to the form $aSH + \beta TU$ (see p. 178). We have already seen, Ex. 5, p. 242, that $\overline{12}^2.\overline{13}^2.\overline{34}^2$ can be expressed in this form. It therefore only remains to prove the same thing for the other terms in the expression for the Hessian of the Hessian given in this article. Multiply by $\overline{12}^2$ the product of the equations

$$-2D_4D_1\,\overline{13}.\overline{34} = D_1{}^2\overline{34}^2 + D_4{}^2\overline{13}^2 - D_3{}^2\overline{14}^2;$$

$$2D_2D_3\,\overline{24}.\overline{34} = D_2{}^2\overline{34}^2 + D_3{}^2\overline{24}^2 - D_4{}^2\overline{23}^2;$$

when we have

$$-4D_1D_2D_3D_4\,\overline{12}^2.\overline{34}^2.\overline{13}.\overline{24} = D_1{}^2D_2{}^2\overline{12}^2.\overline{34}^4 + 2D_2{}^2D_4{}^2\overline{12}^2.\overline{13}^2.\overline{24}^2 - 2D_4{}^4\,\overline{12}^2.\overline{23}^2.\overline{31}^2,$$

which, putting in for $\overline{12}^2.\overline{13}^2.\overline{24}^2$, its value already found, gives

$$-12\,(n-3)^2\{\overline{12}^2.\overline{34}^2.\overline{13}.\overline{24}\} = 6\,(n-2)^2\,(n-3)\,SH - 4n\,(n-1)\,(n-2)\,TU.$$

Again, to calculate $\overline{12}^2.\overline{34}^2.\overline{13}.\overline{14}$, we have only to substitute for $\overline{13}.\overline{14}$ from equation (B), when we have

$$2D_3D_4\{\overline{12}^2.\overline{34}^2.\overline{13}.\overline{14}\} = 2D_4{}^2\overline{12}^2.\overline{34}^2.\overline{13}^2 - D_1{}^2\overline{12}^2.\overline{34}^4,$$

and substituting as before, we get

$$6\,(n-3)^2\{\overline{12}^2.\overline{34}^2.\overline{13}.\overline{14}\} = 2n\,(n-1)\,TU,$$

by the help of which values the Hessian of the Hessian is expressed in the desired form.

281. It would evidently be convenient if a general symbolical expression could be given for the result of elimination between two equations. When one equation is simple it is easily seen (as in Art. 279) that the result of elimination between it and an equation of the nth degree is $\overline{12}.\overline{13}.\overline{14}.\overline{15}$, &c., where the symbol (1) relates to the equation of the nth degree, and the remaining symbols to the simple equation. Let us examine whether any similar general formula can be found if one equation is a quadratic.* The result is to be of the second degree in the coefficients of the general equation, and of the nth in those of the quadratic; there must be therefore in its symbolical

---

* The general formula for the resultant of a quadratic and an equation of the nth degree was given by me in 1853 (*Cambridge and Dublin Mathematical Journal*, vol. IX. p. 32). The theorem was re-discovered by Clebsch in 1860, and extended by him to the case of a system of any number of equations, one of the second, one of the nth, and the rest of the first degree (*Crelle*, vol. LVIII.). In this article the formula is in Mr. Cayley's notation as originally given; in Art. 285 I give Clebsch's extension.

APPLICATIONS OF SYMBOLICAL METHODS. 245

expression two symbols relating to the n^{tic} and n to the quadratic; each of the former symbols being repeated n times, each of the latter twice. For distinctness we denote the latter symbols by letters, and, writing the first two symbols 1, 2, write the others $A, B, \ldots L, M, N$. Now we can make different sets of $2n$ factors, each set containing each of the one kind of symbols n times and of the other twice. Thus we may have the sets

$$(1A)(2A)(1B)(2B)\ldots(1M)(2M)(1N)(2N),$$
$$(1A)^2(2N)^2(1B)(2B)\ldots \qquad (1M)(2M),$$
$$(1A)^2(1B)^2(2M)^2(2N^2)(1C)(2C)\ldots \qquad (1L)(2L), \&c.$$

In the first, every symbol A is joined to both of the other set of symbols, $(1A)(2A)$; in the second, one symbol A is only joined with 1, $(1A)^2$; in the next, two symbols A, B are joined only with 1, $(1A)^2(1B)^2$, and so on. Now the following considerations show how to combine these sets so as to form a symbol which will be the eliminant required. Resolve the quadratic into its factors, and let a and a' refer to the differentials of the several factors, then the result of elimination is

$$(1a)(1b)(1c)\ldots(1n)(2a')(2b')(2c')\ldots(2n').$$

But plainly $A = a + a'$, $(1A) = (1a) + (1a')$, $(1B) = (1b) + (1b')$, &c. Multiplying and remembering that $(1a)^2$, $(1a)(2a)$ vanish, since if the symbol a occurs twice, it implies that a simple equation is twice differentiated, we have

$$(1A)^2 = 2(1a)(1a'), \quad (1A)(2A) = (1a)(2a') + (1a')(2a).$$

We shall write the last equation $(1A)(2A) = [1, 0] + [0, 1]$, where the first figure denotes the number of unaccented letters combined with 1, and the second denotes the number combined with 2. In this notation then

$(1A)(2A)(1B)(2B)$ &c. $\ldots(1N)(2N)$

$$= [n, 0] + n[n-1, 1] + \tfrac{1}{2}n(n-1)[n-2, 2] + \&c.$$

Similarly

$(1A)^2(2N)^2(1B)(2B)\ldots(1M)(2M) = 4\{[n-1, 1] + (n-2)[n-2, 2] + \&c.\}$

$(1A)^2(1B)^2(2M)^2(2N)^2(1C)(2C)$ &c. $= 16\{[n-2, 2] + (n-4)$ &c.$\}$.

The eliminant then which in this notation is $[n, 0]$, is

$$(1A)(2A) \&c. - n\frac{(1A)^2(2N)^2 \&c.}{4}$$
$$+ \frac{n(n-3)}{1.2} \frac{(1A)^2(1B)^2(2M)^2(2N)^2 \&c.}{16} - \&c.,$$

the coefficients being those which occur in the expansion of the sum of the inverse n^{th} powers of the roots of a quadratic, the next coefficients for instance being

$$\frac{n(n-4)(n-5)}{1.2.3}, \quad \frac{n(n-5)(n-6)(n-7)}{1.2.3.4}.$$

282. The preceding series may be transformed so as to proceed by powers of the discriminant of the quadratic. For, since $(1A)(2B) - (1B)(2A) = (12)(AB)$, we have

$$2(1A)(2A)(1B)(2B) = (1A)^2(2B)^2 + (1B)^2(2A)^2 - (12)^2(AB)^2$$
$$= 2(1A)^2(2B)^2 - (12)^2(AB)^2,$$

by the help of which substitution $(AB)^2$, which is the discriminant of the quadratic, is introduced. Thus, in the case of two quadratics, the formula given by the last article is

$$2(1A)(2A)(1B)(2B) - (1A)^2(2B)^2,$$

which, by the substitution, becomes $(1A)^2(2B)^2 - (12)^2(AB)^2$, which is, in other words, Mr. Boole's formula (Art. 191). In the case of the quadratic and cubic, the formula of the last article is

$$4(1a)(2a)(1b)(2b)(1c)(2c) - 3(1a)^2(2b)^2(1c)(2c),$$

which, in like manner, becomes

$$(1a)^2(2b)^2(1c)(2c) - (ab)^2(12)^2(1c)(2c).$$

283. In like manner, it may be investigated whether a general formula can be given for the discriminant of an equation. Formula for invariants of the same degree can easily be written down. The discriminant is to be of the degree $2(n-1)$. Take, then, two sets of $(n-1)$ symbols, 1, 3, 5, &c., 2, 4, 6, &c., and form a product of $n-1$ factors, each factor containing one of each set. Thus, when $n = 4$, we form $(12)(34)(56) = A$. Then by cyclically permuting one of the

sets of symbols we form $(n-1)$ products in all. Thus, for $n=4$, we have $(14)(36)(52) = B$, $(16)(32)(54) = C$. Then any function of the nth degree of these quantities A, B, &c., will denote an invariant of the same degree as the discriminant. The two difficulties in the general theory are, first, to ascertain which of these functions is the discriminant, and secondly, to discover the relations which exist between the different invariants which may be so formed, and to find the simplest forms to which they may be reduced. Thus, for instance, when n is even, it is easy to see that $A^n = 2^k A^{n-1} B$; $k+1$ being the number of factors in the product A.

In the case then of an equation of the fourth degree, the different invariants that can be formed are A^4, $A^2 B^2$, $A^2 BC$, omitting $A^3 B$ which we know to be only $\frac{1}{4} A^4$. The first problem is to form a combination of these invariants, which will vanish identically if the given function be of the form $u^3 v$, where u is of the first degree. To do this, suppose the function to be of this form; substitute for 1, $a + \alpha$ (a referring to v and α to u^3), for 2, $b + \beta$, &c., and examine the effect of this substitution on each of the invariants in question. It is obvious that the terms in which either a or α enter above the second degree vanish; and denoting by M the condition $(a\alpha)^2$, that u should be a factor in v, we can find without much trouble that, on this substitution, we should have

$$A^4 = 216 M^6, \quad A^2 B^2 = 66 M^6, \quad A^2 BC = 42 M^6;$$

and hence that the discriminant may be expressed by any of the three equations

$$\frac{A^4}{36} = \frac{A^2 B^2}{4} = \frac{A^2 BC}{7}.*$$

284. The methods here explained apply equally to quantics in any number of variables. The identical equations principally used in the case of ternary quantics are

$$D_4 \overline{123} = D_1 \overline{234} - D_2 \overline{134} + D_3 \overline{124} \ldots\ldots\ldots\ldots (E),$$
$$\overline{123}.\overline{145} + \overline{124}.\overline{153} + \overline{125}.\overline{134} = 0 \ldots\ldots\ldots\ldots (F),$$

---

* For the reduction of these to the standard form, and the deduction thence of the ordinary form of the discriminant of a quartic, see *Cambridge and Dublin Mathematical Journal*, vol. IX. p. 33.

248   APPLICATIONS OF SYMBOLICAL METHODS.

to which may be added the corresponding equations for contravariant symbols (Art. 156)

$$P\overline{123} = D_1\overline{a23} + D_2\overline{a31} + D_3\overline{a12} \quad \ldots\ldots\ldots\ldots(G),$$

$$\overline{a12}.\overline{a34} + \overline{a23}.\overline{a41} + \overline{a31}.\overline{a24} = 0 \quad \ldots\ldots\ldots\ldots(H),$$

where P is $\alpha x + \beta y + \gamma z$. For reasons already given (Art. 183) I touch but slightly, in this volume, on those parts of the subject which admit of geometrical explanations, and therefore only add one or two examples; referring the reader, for the present, to the *Cambridge and Dublin Mathematical Journal*, vol. IX. p. 27.

Ex. 1. It has been stated (p. 129) that a ternary cubic has an invariant of the fourth order $\overline{123}.\overline{124}.\overline{234}.\overline{314}$. It is now required to find the symbolical expression for another got by operating on the Hessian with the evectant of this invariant. The evectant (p. 129) is $123.\overline{a12}.\overline{a23}.\overline{a31}$; and the method of Art. 280 requires us now to combine with this the additional factor $\overline{456}^2$ and substitute for a, $4 + 5 + 6$. And as we are to omit any term which contains any of the symbols 4, 5, 6 more than three times the required symbol reduces to

$$\overline{123}.\overline{124}.\overline{235}.\overline{316}.\overline{456}^2.$$

Ex. 2. In the theory of double tangents to plane curves, explained, *Higher Plane Curves*, p. 81, it is necessary to calculate the result of substituting in the successive emanants, $\gamma\dfrac{dU}{dy} - \beta\dfrac{dU}{dz}$, $\alpha\dfrac{dU}{dz} - \gamma\dfrac{dU}{dx}$, $\beta\dfrac{dU}{dx} - \alpha\dfrac{dU}{dy}$, for x, y, z; and to show that each result is of the form $P_n U + Q_n(\alpha x + \beta y + \gamma z)^2$. We shall in this and in the next two examples perform the calculation of Q_2, Q_3, Q_4. The symbolical expression for the result of substitution, it is easy to see, is $\overline{a12}.\overline{a13}.\overline{a14}.\overline{a15}$, &c.

Now first to calculate $\overline{a12}.\overline{a13}$, we have only to square equation (G), when we get

$$P^2\overline{123}^2 = 3D_3{}^2\overline{a12}^2 - 6D_2 D_3\overline{a12}.\overline{a13}.$$

Or if we denote the Hessian $\overline{123}^2$ by H, and the bordered Hessian $\overline{a12}^2$ by G, we have

$$6(n-1)^2\overline{a12}.\overline{a13} = 3n(n-1)\, GU - P^2 H.$$

Ex. 3. To calculate $\overline{a12}.\overline{a13}.\overline{a14}$. From equation ($G$) we have

$$-2D_2 D_3\overline{a12}.\overline{a13} = P^2\overline{123}^2 - 2PD_1\overline{123}.\overline{a23} + D_1{}^2\overline{a23}^2 - D_2{}^2\overline{a31}^2 - D_3{}^2\overline{a12}^2.$$

Multiply by $\overline{a14}$, two of the terms vanish identically, and

$$-2(n-1)^2\overline{a12}.\overline{a13} = P^2\overline{a123}^2.\overline{a14} - 2n(n-1)\, U\overline{a12}^2.\overline{a13}.$$

To prove that $\overline{123}.\overline{a23}.\overline{a14}$ vanishes identically, we have only to multiply equation (H) by $\overline{123}$, when, since the terms in it differ only by a permutation of the figures 1, 2, 3, each must separately $= 0$. In like manner, it is proved that $\overline{123}.\overline{a23}.\overline{a14}.\overline{a15}$ is identically $= 0$.

285. In the preceding articles I have used exclusively Mr. Cayley's notation. To illustrate now the method of working with Mr. Aronhold's notation, explained (Art. 158), I give Clebsch's investigation of the problem (Art. 281) to eliminate

between a system of equations, of which one is of the nth, one of the second, the rest of the first degree. To fix the ideas, I write only a system of four homogeneous equations in four variables, but it will be understood that the method is equally applicable to any number of variables. Let the equations then be

$$\alpha = \alpha_1 x_1 + \alpha_2 x_2 + \alpha_3 x_3 + \alpha_4 x_4 = 0, \quad \beta = \beta_1 x_1 + \beta_2 x_2 + \beta_3 x_3 + \beta_4 x_4 = 0,$$

$$U = u_{11} x_1^2 + 2 u_{12} x_1 x_2 + \&c. = 0,$$

and $\phi = 0$, where ϕ is an equation of the nth order in x_1, x_2, x_3, x_4, which may be written symbolically

$$(a_1 x_1 + a_2 x_2 + a_3 x_3 + a_4 x_4)^n = 0,$$

the meaning of which notation has been explained (Art. 158). The method of elimination employed is to solve between the linear equations and the quadratic, and substituting in ϕ the two systems of values found, to multiply the results together. Now we may in an infinity of ways combine the quadratic with the linear equations multiplied by arbitrary factors, so as to obtain a result resolvable into factors: that is to say, so that

$$U + (\lambda_1 x_1 + \lambda_2 x_2 + \&c.)(\alpha_1 x_1 + \&c.) + (\mu_1 x_1 + \&c.)(\beta_1 x_1 + \&c.)$$
$$= (p_1 x_1 + \&c.)(q_1 x_1 + \&c.).$$

We shall imagine this transformation effected, but it will not be necessary to determine the actual values of λ_1, μ_1, &c., for it will be found that these quantities disappear from the result. Taking then the coefficient of any term $x_i x_k$ in the quadratic, the equation written implies that we always have

$$2 u_{ik} + (\alpha_i \lambda_k + \alpha_k \lambda_i) + (\beta_i \mu_k + \beta_k \mu_i) = p_i q_k + p_k q_i \ldots (A).$$

Instead then of solving between the quadratic and the linear equations, we get the two systems of values by combining with the linear equations successively $p_1 x_1 + \&c. = 0$, $q_1 x_1 + \&c. = 0$. And by the theory of linear equations the resulting values of x_1, x_2, &c. are the determinants of the systems

$$\begin{Vmatrix} p_1, & p_2, & p_3, & p_4 \\ \alpha_1, & \alpha_2, & \alpha_3, & \alpha_4 \\ \beta_1, & \beta_2, & \beta_3, & \beta_4 \end{Vmatrix}, \quad \begin{Vmatrix} q_1, & q_2, & q_3, & q_4 \\ \alpha_1, & \alpha_2, & \alpha_3, & \alpha_4 \\ \beta_1, & \beta_2, & \beta_3, & \beta_4 \end{Vmatrix}.$$

If then we substitute the first set of values in $a_1 x_1 + \&c.$, we get the determinant

$$\begin{vmatrix} a_1, & a_2, & a_3, & a_4 \\ p_1, & p_2, & p_3, & p_4 \\ \alpha_1, & \alpha_2, & \alpha_3, & \alpha_4 \\ \beta_1, & \beta_2, & \beta_3, & \beta_4 \end{vmatrix},$$

which we may write $(a_1 p_2 \alpha_3 \beta_4)$. The result of elimination then may be written symbolically $R = (a_1 p_2 \alpha_3 \beta_4)^n (b_1 q_2 \alpha_3 \beta_4)^n$. We use in the second factor the symbol b instead of a, for the reason explained (Art. 159), in order to obtain powers of the coefficients of ϕ; but it is understood that the b symbols have exactly the same meaning as the a, since after expansion we equally replace the products $a^i a^k a^l a^m$, $b^i b^k b^l b^m$, by the corresponding coefficient of ϕ, a_{iklm}. We may then write the result of elimination in the more symmetrical form

$$2R = (a_1 p_2 \alpha_3 \beta_4)^n (b_1 q_2 \alpha_3 \beta_4)^n + (a_1 q_2 \alpha_3 \beta_4)^n (b_1 p_2 \alpha_3 \beta_4)^n,$$

for this after expansion will be only double the former expression.

Let us now write

$$(a_1 p_2 \alpha_3 \beta_4)(a_1 q_2 \alpha_3 \beta_4) = A,$$

$$(b_1 p_2 \alpha_3 \beta_4)(b_1 q_2 \alpha_3 \beta_4) = B,$$

$$(a_1 p_2 \alpha_3 \beta_4)(b_1 q_2 \alpha_3 \beta_4) + (a_1 q_2 \alpha_3 \beta_4)(b_1 p_2 \alpha_3 \beta_4) = 2C,$$

then R may be easily expressed in terms of A, B, C. For we have

$$2R = \{C + \sqrt{(C^2 - AB)}\}^n + \{C - \sqrt{(C^2 - AB)}\}^n,$$

or $\quad R = C^n + \dfrac{n(n-1)}{1.2} C^{n-2} (C^2 - AB)$

$$+ \dfrac{n(n-1)(n-2)(n-3)}{1.2.3.4} C^{n-4} (C^2 - AB)^2 + \&c.$$

286. It remains now to examine more closely the expressions for A, B, C, and to get rid of the quantities p and q which we have introduced, so that the result may be expressed in terms of the coefficients of the given quadratic.

Now A, which is the product of two determinants, may be written as a single determinant,

$$\begin{vmatrix} p_1q_1 & , \tfrac{1}{2}(p_1q_2+p_2q_1), & \tfrac{1}{2}(p_1q_3+p_3q_1), & \tfrac{1}{2}(p_1q_4+p_4q_1), & \alpha_1, & \beta_1, & a_1 \\ \tfrac{1}{2}(p_1q_2+p_2q_1), & p_2q_2 & , \tfrac{1}{2}(p_2q_3+p_3q_2), & \tfrac{1}{2}(p_2q_4+p_4q_2), & \alpha_2, & \beta_2, & a_2 \\ \tfrac{1}{2}(p_1q_3+p_3q_1), & \tfrac{1}{2}(p_2q_3+p_3q_2), & p_3q_3 & , \tfrac{1}{2}(p_3q_4+p_4q_3), & \alpha_3, & \beta_3, & a_3 \\ \tfrac{1}{2}(p_1q_4+p_4q_1), & \tfrac{1}{2}(p_2q_4+p_4q_2), & \tfrac{1}{2}(p_3q_4+p_4q_3), & p_4q_4 & , \alpha_4, & \beta_4, & a_4 \\ \alpha_1 & , \alpha_2 & , \alpha_3 & , \alpha_4 & , & & \cdots \\ \beta_1 & , \beta_2 & , \beta_3 & , \beta_4 & , & & \cdots \\ a_1 & , a_2 & , a_3 & , a_4 & , & & \cdots \end{vmatrix}$$

multiplied however by $(-1)^{m-1}$, where m is the number of variables, that is to say, in the present case, 4.

For every constituent of this determinant must contain a constituent from each of the last three rows and columns; it is therefore of the first degree in the terms p_1q_1, $\tfrac{1}{2}(p_1q_2+p_2q_1)$, &c.; and if the coefficient of any of these terms be examined, it will be found, according to the number of variables, to be either the same, or the same with sign changed, as in the product of the two determinants. Now in this determinant we are to substitute from equation (A),

$$p_1q_1 = u_{11} + \alpha_1\lambda_1 + \beta_1\mu_1,$$

$$\tfrac{1}{2}(p_1q_2+p_2q_1) = u_{12} + \tfrac{1}{2}(\alpha_1\lambda_2+\alpha_2\lambda_1) + \tfrac{1}{2}(\beta_1\mu_2+\beta_2\mu_1),\ \&c.$$

But when this change has been made, if we subtract from each of the first four rows and columns the α row and column each multiplied by $\tfrac{1}{2}\lambda_1$, and the β row and column each multiplied by $\tfrac{1}{2}\mu_1$, the additional terms disappear, and the determinant reduces to

$$\begin{vmatrix} u_{11}, & u_{12}, & u_{13}, & u_{14}, & \alpha_1, & \beta_1, & a_1 \\ u_{21}, & u_{22}, & u_{23}, & u_{24}, & \alpha_2, & \beta_2, & a_2 \\ u_{31}, & u_{32}, & u_{33}, & u_{34}, & \alpha_3, & \beta_3, & a_3 \\ u_{41}, & u_{42}, & u_{43}, & u_{44}, & \alpha_4, & \beta_4, & a_4 \\ \alpha_1, & \alpha_2, & \alpha_3, & \alpha_4 & & & \cdots \\ \beta_1, & \beta_2, & \beta_3, & \beta_4 & & & \cdots \\ a_1, & a_2, & a_3, & a_4 & & & \cdots \end{vmatrix}.$$

Clebsch denotes the above determinant, in which the matrix

of the discriminant of a quadratic function is bordered by rows and columns, α, β, &c., by the abbreviation, (see p. 15)

$$\begin{pmatrix} \alpha, & \beta, & a \\ \alpha, & \beta, & a \end{pmatrix},$$

the upper line denoting the columns, the second the rows by which the matrix is bordered. Thus, then, we find for any number of variables

$$A = (-1)^{m-1} \begin{pmatrix} \alpha, & \beta, & \ldots a \\ \alpha, & \beta, & \ldots a \end{pmatrix}.$$

In like manner $\quad B = (-1)^{m-1} \begin{pmatrix} \alpha, & \beta, & \ldots b \\ \alpha, & \beta, & \ldots b \end{pmatrix}.$

And in the same way

$$C = (-1)^{m-1} \begin{pmatrix} \alpha, & \beta, & \ldots a \\ \alpha, & \beta, & \ldots b \end{pmatrix}.$$

Now it was proved, Ex. 2, p. 26, that $\begin{pmatrix} a \\ a \end{pmatrix}\begin{pmatrix} b \\ b \end{pmatrix} - \begin{pmatrix} a \\ b \end{pmatrix}^2 = \Delta \begin{pmatrix} a, b \\ a, b \end{pmatrix}$, where Δ is the discriminant of the quadratic function. And it is proved in the same way, in general, that

$$\begin{pmatrix} \alpha, & \beta, & \ldots a \\ \alpha, & \beta, & \ldots a \end{pmatrix} \begin{pmatrix} \alpha, & \beta, & \ldots b \\ \alpha, & \beta, & \ldots b \end{pmatrix} - \begin{pmatrix} \alpha, & \beta, & \ldots a \\ \alpha, & \beta, & \ldots b \end{pmatrix}^2 = \Delta \begin{pmatrix} \alpha, & \beta, & \ldots a, b \\ \alpha, & \beta, & \ldots a, b \end{pmatrix}.$$

If, then, we call the last written function D, we have

$$C^2 - AB = -\Delta D,$$

and the formula of Art. 285 becomes

$$R = C^n - \frac{n(n-1)}{1.2} C^{n-2} D\Delta + \frac{n(n-1)(n-2)(n-3)}{1.2.3.4} C^{n-4} D^2 \Delta^2 - \&c.*$$

---

* The reader will find some applications of this to geometrical problems in Clebsch's paper (*Crelle*, vol. LVIII.). In the case of binary quantics, to translate this result into the notation we have used, we must write for C^n, the product of n pairs of factors $1A.2A, 1B.2B$, &c., and for D, $\overline{12}^2$. I refer also to Clebsch's paper (*Crelle*, vol. LIX.), "Ueber symbolische Darstellung algebraischer Formen," for a rule for obtaining a general symbolic formula for the resultant of two binary quantics or for the discriminant of a binary quantic. The method of proceeding is to apply Cayley's form of Bezout's method of elimination, explained (Art. 83) to two quantics written symbolically $(a_1 x_1 + a_2 x_2)^n$, $(a_1 x_1 + a_2 x_2)^n$; but the resulting rule is, as may be expected, very complicated.

APPENDIX.

The following is the value of the invariant E, p. 210:

	$a^6de^3g^5$	$-$	$54a^5c^2def^2g^4$	$+$	$30a^5de^3f^6$
$-$	$3a^6de^2f^2g^4$	$+$	$33a^5c^3df^4g^3$	$+$	$72a^5e^7fg^2$
$+$	$3a^6def^4g^3$	$+$	$15a^5c^2e^3fg^4$	$-$	$36a^5e^6f^3g$
$-$	$a^6df^6g^2$	$+$	$9a^5c^2e^2f^3g^3$	$-$	$9a^5e^5f^5$
$-$	$3a^5e^4fg^4$	$-$	$18a^5c^2ef^5g^2$	$+$	$2a^4b^3e^3g^5$
$+$	$11a^6e^3f^3g^3$	$+$	$3a^5c^2f^7g$	$-$	$6a^4b^3e^2f^2g^4$
$-$	$15a^6e^2f^5g^2$	$-$	$48a^5cd^3f^2g^4$	$+$	$6a^4b^3ef^4g^3$
$+$	$9a^6ef^7g$	$+$	$60a^5cd^3e^2fg^4$	$-$	$2a^4b^3f^6g^2$
$-$	$2a^6f^9$	$+$	$168a^5cd^2ef^3g^3$	$+$	$3a^4b^2c^3dg^6$
$-$	$3a^5bce^3g^5$	$-$	$108a^5cd^2f^5g^2$	$-$	$9a^4b^2c^2efg^5$
$+$	$9a^5bce^2f^2g^4$	$-$	$69a^5cde^4g^4$	$+$	$6a^4b^2c^2f^3g^4$
$-$	$9a^5bcef^4g^3$	$-$	$84a^5cde^3f^2g^3$	$-$	$24a^4b^2cd^2fg^5$
$+$	$3a^5bcf^6g^2$	$+$	$6a^5cde^2f^7g^3$	$+$	$54a^4b^2cde^3g^5$
$-$	$12a^5bd^2e^2g^5$	$+$	$48a^5cdef^6g$	$-$	$12a^4b^2cdf^4g^3$
$+$	$24a^5bd^2ef^2g^4$	$+$	$132a^5ce^5fg^3$	$-$	$102a^4b^2ce^3fg^4$
$-$	$12a^5bd^2f^4g^3$	$-$	$201a^5ce^4f^3g^2$	$+$	$144a^4b^2ce^2f^3g^3$
$+$	$48a^5bde^3fg^4$	$+$	$141a^5ce^3f^5g$	$-$	$72a^4b^2cef^5g^2$
$-$	$120a^5bde^2f^3g^3$	$-$	$45a^5ce^2f^7$	$+$	$12a^4b^2cf^7g$
$+$	$96a^5bdef^5g^2$	$+$	$64a^5d^7f^3g^3$	$+$	$48a^4b^2d^2eg^5$
$-$	$24a^5bdf^7g$	$+$	$24a^5d^3e^2g^4$	$-$	$288a^4b^2d^2e^2fg^4$
$+$	$15a^5be^5g^4$	$-$	$192a^5d^3e^2f^2g^3$	$+$	$144a^4b^2d^2ef^3g^3$
$-$	$93a^5be^4f^2g^3$	$-$	$168a^5d^3ef^4g^2$	$+$	$24a^4b^2d^2f^5g^2$
$+$	$177a^5be^3f^4g^2$	$+$	$128a^5d^3f^6g$	$-$	$201a^4b^2de^4g^4$
$-$	$135a^5be^2f^6g$	$+$	$60a^5d^2e^4fg^3$	$+$	$1152a^4b^2de^3f^2g^3$
$+$	$36a^5bef^8$	$+$	$528a^5d^2e^3f^3g^2$	$-$	$1188a^4b^2de^2f^4g^2$
$-$	$a^5c^3dg^6$	$-$	$336a^5d^2e^2f^5g$	$+$	$336a^4b^2def^6g$
$+$	$3a^5c^3efg^5$	$-$	$24a^5de^6g^3$	$+$	$153a^4b^2efg^3$
$-$	$2a^5c^3f^3g^4$	$-$	$300a^5de^5f^2g^2$	$-$	$132a^4b^2e^4f^3g^2$
$+$	$12a^5c^2d^2fg^5$	$+$	$159a^5def^4g$	$+$	$822a^4b^2e^3f^5g$

254 APPENDIX.

−	$270a^4b^2e^2f^7$	−	$3000a^4bd^2e^3f^2g^2$	−	$2082a^4c^2e^6fg^2$
+	$3a^4bc^4g^8$	−	$480a^4bd^2e^4f^3g$	+	$93a^4c^2e^5f^3g$
−	$48a^4bc^3dfg^5$	+	$1656a^4bd^2e^2f^6$	+	$225a^4c^2e^4f^5$
−	$15a^4bc^3e^2g^5$	+	$1416a^4bde^6fg^2$	−	$1152a^5cd^5f^3g^3$
+	$102a^4bc^2ef^2g^4$	+	$1284a^4bde^5f^3g$	+	$1440a^4cd^4e^7fg^5$
−	$51a^4bc^2f^4g^3$	−	$1674a^4bde^4f^5$	−	$1920a^4cd^4ef^3g^2$
−	$60a^4bc^2d^2eg^5$	−	$360a^4be^8g^2$	−	$3456a^4cd^4f^5g$
+	$288a^4bc^3d^2f^2g^4$	−	$324a^4be^7f^2g$	−	$1200a^4cd^3e^4g^3$
−	$444a^4bc^2def^3g^3$	+	$441a^4be^6f^4$	+	$6960a^4cd^3e^3f^2g^2$
+	$126a^4bc^2df^5g^2$	−	$15a^4c^5fg^5$	+	$12240a^4cd^3e^2f^4g$
+	$132a^4bc^2e^4g^4$	+	$69a^4c^4deg^5$	−	$1920a^4cd^3ef^6$
−	$243a^4bc^2e^3f^2g^3$	+	$201a^4c^4df^2g^4$	−	$3972a^4cd^2e^5fg^2$
+	$270a^4bc^2e^2f^4g^2$	−	$132a^4c^4e^2fg^4$	−	$17760a^4cd^2e^4f^3g$
−	$33a^4bc^2ef^6g$	−	$195a^4c^4ef^3g^3$	+	$3720a^4cd^2e^3f^5$
−	$18a^4bc^2f^8$	+	$129a^4c^4f^5g^2$	+	$1128a^4cde^7g^2$
−	$288a^4bcd^2f^3g^3$	−	$24a^4c^3d^3g^5$	+	$8808a^4cde^5f^2g$
+	$588a^4bcd^2e^6g^4$	−	$588a^4c^3d^2efg^4$	−	$1575a^4cde^5f^4$
−	$1224a^4bcd^2e^5f^2g^3$	−	$1160a^4c^3d^2f^3g^3$	−	$1584a^4ce^8fg$
+	$1140a^4bcd^2ef^4g^2$	+	$2820a^4c^3de^2f^2g^3$	+	$180a^4ce^7f^3$
−	$48a^4bcd^2f^6g$	+	$255a^4c^3def^4g^2$	+	$1536a^4d^6f^3g^2$
−	$504a^4bcde^4fg^3$	−	$534a^4c^3df^6g$	+	$192a^4d^5e^3g^3$
+	$1440a^4bcde^3f^3g^2$	−	$660a^4c^3e^4fg^5$	+	$3456a^4d^5e^2f^2g^2$
−	$1188a^4bcde^2f^5g$	−	$655a^4c^3e^3f^3g^2$	−	$1536a^4d^5ef^4g$
+	$72a^4bcdef^7$	+	$363a^4c^3e^2f^5g$	+	$2816a^4d^5f^6$
−	$588a^4bce^6g^3$	+	$45a^4c^3ef^7$	+	$1920a^4d^4e^4fg^2$
+	$1539a^4bce^5f^2g^2$	+	$288a^4c^3d^4fg^4$	+	$2880a^4d^4e^3f^3g$
−	$1590a^4bce^4f^4g$	+	$1800a^4c^3d^3ef^2g^3$	−	$8640a^4d^4e^2f^5$
+	$639a^4bce^3f^6$	+	$2940a^4c^3d^3f^4g^2$	−	$464a^4d^3e^6g^2$
−	$288a^4bd^4e^2g^4$	−	$2940a^4c^3d^2e^3fg^3$	−	$792a^4d^3e^5f^2g$
+	$576a^4bd^3ef^2g^3$	−	$8760a^4c^3d^2e^2f^2g^2$	+	$11100a^4d^3e^4f^4$
−	$480a^4bd^4f^4g^2$	−	$36a^4c^3d^2ef^5g$	−	$336a^4d^2e^7fg$
+	$384a^4bd^3e^3fg^3$	+	$960a^4c^3d^2f^7$	−	$7800a^4d^2e^6f^3$
+	$480a^4bd^3e^2f^3g^2$	+	$1674a^4c^3de^9g^3$	+	$144a^4de^9g$
+	$864a^4bd^3ef^5g$	+	$5670a^4c^3de^4f^3g^2$	+	$2880a^4de^8f^2$
−	$768a^4bd^3f^7$	+	$1965a^4c^3de^3f^4g$	−	$432a^4e^{10}f$
+	$276a^4bd^2e^8g^2$	−	$1665a^4c^2de^2f^6$	−	$3a^3b^4cdg^6$

APPENDIX. 255

+	$9a^3b^4cefg^5$	−	$1152a^3b^2c^3df^2g^4$	−	$504a^3b^3d^2e^5fg^2$
−	$6a^3b^4cf^3g^4$	+	$243a^3b^2c^3e^3fg^4$	+	$27720a^3b^3d^2e^4f^2g$
+	$12a^3b^4d^2fg^5$	+	$732a^3b^2c^6ef^2g^3$	−	$13728a^3b^3d^2e^3f^5$
−	$33a^3b^4de^2g^5$	−	$246a^3b^2c^3f^5g^2$	+	$1092a^3b^3de^7g^2$
+	$12a^3b^4def^2g^4$	+	$192a^3b^2c^3d^3g^5$	−	$22398a^3b^3def^2g$
+	$51a^3b^4e^3fg^4$	+	$1224a^3b^2c^2d^2efg^4$	+	$13356a^3b^3def^4$
−	$66a^3b^4e^2f^3g^3$	+	$2208a^3b^2c^2d^2f^2g^3$	+	$5364a^3b^2e^8fg$
+	$24a^3b^4ef^5g^2$	−	$2820a^3b^2c^2de^3g^4$	−	$3618a^3b^3e^5f^3$
−	$11a^3b^3c^3g^5$	−	$2844a^3b^2c^2def^4g^2$	−	$132a^3bc^5eg^5$
+	$120a^3b^3c^4dfg^5$	+	$1056a^3b^2c^2df^5g$	−	$153a^3bc^5f^2g^4$
−	$9a^3b^3c^4e^3g^5$	+	$2367a^3b^2c^2e^4fg^3$	−	$60a^3bc^4d^2g^5$
−	$144a^3b^3c^4ef^2g^4$	−	$708a^3b^2c^3e^2f^3g^2$	+	$504a^3bc^4defg^4$
+	$66a^3b^3c^3f^4g^5$	−	$432a^3b^2c^3e^3f^5g$	+	$2916a^3bc^4df^6g^3$
−	$168a^3b^3cd^2eg^5$	+	$216a^3b^2c^3ef^7$	+	$660a^3bc^4e^3g^4$
−	$144a^3b^3cd^2f^2g^4$	−	$576a^3b^2cd^3fg^4$	−	$2367a^3bc^4e^3f^2g^3$
+	$444a^3b^3cde^2fg^4$	−	$1800a^3b^3cd^3e^2g^4$	−	$480a^3bc^4ef^4g^3$
−	$48a^3b^3cdf^2g^2$	−	$2688a^3b^3cd^2f^4g^2$	+	$54a^3bc^4f^6g$
+	$195a^3b^3ce^4g^4$	+	$7248a^3b^2cd^2e^3fg^3$	−	$384a^3bc^3d^3fg^4$
−	$732a^3b^3ce^3f^2g^3$	+	$6336a^3b^2cd^2e^2f^3g^2$	+	$2940a^3bc^3d^2e^2g^4$
+	$564a^3b^3ce^2f^4g^2$	−	$5184a^3b^2cd^2ef^5g$	−	$7248a^3bc^3d^2ef^2g^3$
−	$144a^3b^3cef^6g$	+	$1536a^3b^2cd^2f^7$	−	$6600a^3bc^3d^2f^4g^2$
−	$64a^3b^3d^4g^5$	+	$4140a^3b^2cde^5g^3$	+	$4560a^3bc^3de^2f^6g^2$
+	$288a^3b^3d^3efg^4$	−	$23994a^3b^2cde^4f^2g^2$	+	$10272a^3bc^3def^5g$
+	$1160a^3b^3d^3e^3g^4$	+	$16440a^3b^2cde^3f^4g$	−	$2736a^3bc^6df^7$
−	$2208a^3b^3d^2ef^2g^3$	−	$3168a^3b^2cde^2f^6$	−	$1722a^3bc^3e^5g^3$
+	$1344a^3b^3d^2ef^4g^2$	−	$363a^3b^2ce^6fg^2$	+	$4695a^3bc^3e^4f^2g^2$
−	$512a^3b^3d^2f^6g$	+	$4554a^3b^2ce^5f^3g$	−	$7110a^3bc^3e^3f^4g$
−	$2916a^3b^3de^4fg^3$	−	$2754a^3b^2ce^4f^5$	+	$1242a^3bc^3e^2f^6$
+	$3952a^3b^3de^3f^3g^2$	+	$1152a^3b^2d^5eg^4$	−	$1440a^3bc^2d^4eg^4$
−	$1104a^3b^3de^2f^5g$	−	$3168a^3b^2d^4e^2fg^3$	+	$3168a^3bc^2d^4f^2g^3$
+	$137a^3b^3e^6g^3$	−	$384a^3b^2d^4ef^3g^2$	+	$1920a^3bc^2d^3ef^3g^2$
+	$1716a^3b^3e^5f^2g^2$	+	$4608a^3b^2d^4f^5g$	+	$6336a^3bc^2d^3f^5g$
−	$2862a^3b^3e^4f^4g$	−	$2520a^3b^2d^3e^4g^3$	−	$7980a^3bc^2d^4e^4g^3$
+	$1080a^3b^3e^3f^6$	+	$9984a^3b^2d^3e^3f^2g^2$	+	$30960a^3bc^2d^2e^3f^2g^2$
+	$93a^3b^2c^4fg^5$	−	$23040a^3b^2d^3e^2f^4g$	−	$27720a^3bc^2d^2e^2f^4g$
+	$84a^3b^2c^3deg^5$	+	$6144a^3b^2d^3ef^6$	−	$2784a^3bc^2d^2ef^6$

APPENDIX.

−	$13032a^3bc^2de^5fg^2$	+	$24a^3c^6dg^5$	+	$2304a^3c^2d^6fg^3$
−	$2640a^3bc^2de^4f^3g$	+	$588a^3c^6efg^4$	+	$33984a^3c^2d^5ef^2g^2$
+	$12240a^3bc^2de^3f^5$	−	$137a^3c^6f^2g^3$	+	$19968a^3c^2d^5f^4g$
+	$7026a^3bc^2e^7g^2$	−	$276a^3c^5d^4fg^4$	−	$45600a^3c^2d^4e^3fg^2$
−	$2835a^3bc^2e^6f^2g$	−	$1674a^3c^5de^2g^4$	+	$5760a^3c^2d^4e^3f^3g$
−	$1458a^3bc^2e^5f^4$	−	$4140a^3c^5def^2g^3$	+	$18912a^3c^2d^3e^5g^2$
+	$768a^3bcd^5f^3g^2$	−	$753a^3c^5df^4g^2$	−	$81360a^3c^2d^3e^4f^2g$
+	$9120a^3bcd^4e^3g^3$	+	$1722a^3c^5e^3fg^3$	−	$22800a^3c^2d^3e^3f^4$
−	$28800a^3bcd^4e^2f^2g^2$	+	$4641a^3c^5e^2f^3g^2$	+	$61764a^3c^2d^2e^6fg$
+	$34560a^3bcd^4ef^4g$	−	$2781a^3c^5ef^5g$	+	$49560a^3c^2d^2e^5f^3$
−	$13824a^3bcd^4f^6$	+	$756a^3c^5f^7$	−	$14280a^3c^3de^8g$
+	$8400a^3bcd^3e^4fg^2$	+	$1200a^3c^4d^3eg^4$	−	$29340a^3c^2de^7f^2$
−	$78720a^3bcd^3e^3f^3g$	+	$2520a^3c^4d^3f^2g^3$	+	$5400a^3c^2e^8f$
+	$43200a^3bcd^3e^2f^5$	+	$7980a^3c^4d^2e^2fg^3$	−	$9216a^3cd^7f^2g^2$
−	$7548a^3bcd^2e^6g^2$	+	$21480a^3c^4d^2ef^3g^2$	+	$11520a^3cd^6e^3fg^2$
+	$109800a^3bcd^2e^5f^2g$	+	$1020a^3c^4d^2f^5g$	−	$49152a^3cd^6ef^3g$
−	$54840a^3bcd^2e^4f^4$	−	$41160a^3c^4de^3f^2g^2$	−	$6384a^3cd^5e^4g^2$
−	$48912a^3bcde^7fg$	+	$615a^3c^4de^2f^4g$	+	$103296a^3cd^5e^3f^2g$
+	$20412a^3bcde^6f^3$	−	$720a^3c^4def^6$	+	$7680a^3cd^5e^2f^4$
+	$7488a^3bce^9g$	+	$8610a^3c^4e^5fg^2$	−	$63216a^3cd^4e^5fg$
−	$1836a^3bce^8f^2$	+	$4635a^3c^4e^4f^3g$	−	$4800a^3cd^4e^4f^3$
−	$2304a^3bd^8e^2g^3$	−	$675a^3c^4e^3f^5$	+	$12960a^3cd^3e^7g$
+	$4608a^3bd^6ef^2g^2$	−	$192a^3c^3d^5g^4$	−	$15720a^3cd^3e^6f^2$
−	$12288a^3bd^6f^4g$	−	$9120a^3c^3d^4efg^3$	+	$15120a^3cd^2e^8f$
+	$384a^3bd^5e^3fg^2$	−	$12640a^3c^3d^4f^3g^2$	−	$3600a^3cde^{10}$
+	$37632a^3bd^5e^2f^3g$	−	$9120a^3c^3d^3e^2f^2g^2$	+	$12288a^3d^8f^3g$
−	$6144a^3bd^5ef^5$	−	$41280a^3c^3d^3ef^4g$	+	$512a^3d^7e^3g^2$
+	$1536a^3bd^4e^5g^2$	+	$5760a^3c^3d^3f^6$	−	$24576a^3d^7e^2f^2g$
−	$50400a^3bd^4e^4f^2g$	+	$39900a^3c^3d^2e^4fg^2$	+	$14784a^3d^6e^4fg$
+	$21120a^3bd^4e^3f^4$	+	$88880a^3c^3d^2e^3f^3g$	−	$5120a^3d^6e^3f^3$
+	$22896a^3bd^3e^6fg$	−	$6600a^3c^3d^2e^2f^5$	−	$2960a^3d^5e^6g$
−	$30048a^3bd^3e^5f^3$	−	$16186a^3c^3de^6g^2$	+	$11136a^3d^5e^5f^2$
−	$3504a^3bd^2e^8g$	−	$58884a^3c^3de^5f^2g$	−	$7440a^3d^4e^7f$
+	$26856a^3bd^2e^7f^2$	−	$1350a^3c^3de^4f^4$	+	$1600a^3d^3e^9$
−	$12384a^3bde^9f$	+	$13428a^3c^3e^7fg$		$a^3b^6dg^6$
+	$2160a^3be^{11}$	+	$945a^3c^3e^6f^3$	−	$3a^3b^6efg^5$

APPENDIX.

+	$2a^2b^6f^3g^4$	−	$4725a^3b^4e^6fg^2$	−	$1536a^3b^3d^5g^4$
+	$15a^4b^5c^2g^6$	+	$6885a^3b^4e^5f^2g$	−	$768a^2b^3d^5efg^3$
−	$96a^4b^5cdfg^5$	−	$2430a^3b^4e^4f^5$	+	$12640a^2b^3d^4e^2g^3$
+	$18a^2b^5ce^2g^5$	+	$201a^3b^5c^4eg^5$	+	$192a^2b^3d^4e^2f^2g$
+	$72a^2b^5cef^2g^4$	+	$732a^3b^3c^4f^2g^4$	+	$1536a^2b^3d^4ef^4g$
−	$24a^2b^5cf^4g^3$	−	$528a^3b^3c^3d^2g^5$	−	$8192a^2b^3d^4f^6$
+	$108a^2b^5d^2eg^5$	−	$1440a^3b^3c^3defg^4$	−	$20400a^2b^3d^3e^4fg^2$
−	$24a^2b^5d^2f^2g^4$	−	$3952a^2b^3c^3df^2g^3$	+	$18304a^2b^3d^3e^3f^3g$
−	$126a^2b^5de^3fg^4$	+	$655a^2b^3c^3e^3g^4$	+	$10752a^2b^3d^3e^2f^5$
+	$48a^2b^5def^3g^3$	+	$708a^2b^3c^3e^2f^2g^3$	−	$4320a^2b^3d^2e^6g^2$
−	$129a^2b^5e^4g^4$	+	$1782a^3b^3c^2ef^4g^2$	−	$30600a^2b^3d^2e^4f^2g$
−	$246a^2b^5e^3f^2g^3$	−	$216a^2b^3c^2f^6g$	+	$6480a^2b^3d^2e^3f^4$
−	$108a^3b^4e^2f^4g^2$	−	$480a^2b^3c^2d^3fg^4$	+	$45684a^2b^3de^5fg$
−	$177a^2b^4c^3fg^5$	+	$8760a^2b^3c^2d^2e^2g^4$	−	$24084a^2b^3de^5f^3$
	$6a^2b^4c^3deg^5$	−	$6336a^2b^3c^2d^2ef^2g^3$	−	$13932a^2b^3e^7g$
+	$1188a^3b^4c^3df^4g^2$	+	$11808a^3b^3c^2d^3f^4g^2$	+	$8586a^2b^3e^5f^2$
−	$270a^2b^4c^3e^2fg^4$	+	$4560a^2b^3c^2de^3fg^5$	+	$300a^3b^3c^5dg^5$
−	$564a^2b^4c^2ef^2g^3$	−	$8208a^3b^3c^2def^5g$	−	$1539a^2b^3c^5efg^4$
+	$108a^2b^4c^2f^5g^2$	−	$4641a^2b^3c^2e^2g^3$	−	$1716a^2b^3c^5f^2g^3$
+	$168a^3b^4cd^3g^5$	+	$3942a^3b^3c^2e^4f^2g^2$	+	$3000a^3b^3c^4d^3fg^4$
−	$1140a^2b^4cd^3efg^4$	+	$4212a^3b^3c^2e^3f^4g$	−	$5670a^2b^3c^4de^2g^4$
−	$1344a^2b^4cd^2f^3g^3$	−	$972a^2b^3c^2e^2f^6$	+	$23994a^2b^3c^4def^2g^3$
−	$255a^2b^4cde^2g^4$	+	$1920a^2b^3cd^4eg^4$	−	$1719a^2b^3c^4df^4g^2$
+	$2844a^2b^4cde^2f^2g^3$	+	$384a^2b^3cd^4f^2g^2$	−	$4695a^2b^3c^4e^3fg^3$
+	$480a^2b^3ce^4fg^5$	−	$1920a^2b^3cd^3e^2fg^3$	−	$3942a^2b^3c^4e^2f^3g^2$
−	$1782a^2b^4ce^3f^3g^2$	−	$9216a^2b^3cd^3f^5g$	+	$567a^2b^3c^4ef^5g$
+	$648a^2b^4ce^2f^5g$	−	$21480a^2b^3cd^2e^4g^3$	+	$162a^2b^3c^4f^7$
+	$480a^2b^3d^4fg^4$	−	$18432a^2b^3cd^2e^3f^2g^2$	−	$6960a^3b^3c^3d^3eg^4$
−	$2940a^2b^4d^3e^2g^4$	+	$58176a^2b^3cd^2e^2f^4g$	−	$9984a^2b^3c^3d^3f^2g^3$
+	$2688a^2b^4d^3ef^2g^3$	−	$13824a^2b^3cd^2ef^6$	−	$30960a^2b^3c^3d^2e^3fg^2$
+	$6600a^2b^4d^2e^3fg^3$	+	$76428a^2b^3cde^3fg^2$	+	$18432a^2b^3c^3d^2ef^3g^2$
−	$11808a^2b^4d^2e^2f^3g^2$	−	$95904a^2b^3cde^4f^3g$	−	$13536a^2b^3c^3d^2f^5g$
+	$4608a^2b^4d^2ef^5g$	+	$24624a^2b^3cde^3f^5$	+	$41160a^2b^3c^3de^5g^2$
+	$753a^2b^5de^6g^3$	−	$9855a^2b^3ce^7g^2$	−	$32400a^2b^3c^3de^3f^4g$
+	$1719a^2b^4de^4f^3g^2$	+	$4860a^2b^3ce^6f^2g$	+	$23328a^2b^2c^3def^6$
−	$1728a^2b^4de^3f^4g$	+	$2430a^2b^3ce^5f^4$	−	$22869a^2b^2c^3e^5fg^2$

L L

APPENDIX.

$+\ 28350a^2b^2c^3e^4f^2g$	$-\ 170496a^2b^2d^5e^2f^4$	$+\ 17280a^2bc^3d^7f^2g$
$-\ 12960a^2b^2c^3e^8f^5$	$+\ 31248a^2b^2d^5e^3fg$	$-\ 200640a^2bc^3d^6e^3f^3g$
$+\ 3456a^2b^2c^4d^3g^4$	$+\ 245280a^2b^2d^4e^4f^3$	$+\ 127680a^2bc^3d^6ef^5$
$+\ 28800a^2b^2c^4d^4efg^3$	$-\ 14136a^2b^2d^4e^7g$	$+\ 18756a^2bc^3d^5e^5g^2$
$-\ 192a^2b^2c^3d^5f^3g^2$	$-\ 203592a^2b^2d^3e^6f^2$	$+\ 97920a^2bc^3d^5e^4f^2g$
$+\ 9120a^2b^2c^3d^2e^2g^3$	$+\ 84024a^2b^2d^2e^5f$	$-\ 141480a^2bc^3d^3e^3f^4$
$-\ 63360a^2b^2c^3d^2ef^4g$	$-\ 12960a^2b^2de^{10}$	$+\ 36504a^2bc^5de^5fg$
$+\ 52224a^2b^2c^2d^7f^6$	$-\ 72a^2bc^7g^5$	$+\ 60264a^2bc^5de^3f^3$
$-\ 110160a^2b^2c^2d^5e^2fg^2$	$-\ 1416a^2bc^6dfg^4$	$-\ 24300a^2bc^5e^6g$
$+\ 200160a^2b^2c^2d^2e^3f^2g$	$+\ 2082a^2bc^6e^2g^4$	$-\ 6723a^2bc^5e^3f^2$
$-\ 123120a^2b^2c^2d^2e^2f^5$	$+\ 363a^2bc^5ef^2g^3$	$-\ 11520a^2bc^4d^5eg^3$
$+\ 5418a^2b^2c^4de^9g^2$	$+\ 4725a^2bc^5f^4g^2$	$+\ 16128a^2bc^4d^5f^2g^2$
$-\ 20736a^2b^2c^4de^5f^2g$	$+\ 3972a^2bc^5d^5eg^4$	$+\ 31488a^2bc^4d^5ef^3g$
$+\ 38070a^2b^2c^4de^3f^4$	$+\ 504a^2bc^5d^7f^2g^3$	$-\ 119808a^2bc^4d^5f^5$
$+\ 5103a^2b^2c^2e^7fg$	$+\ 13032a^2bc^5de^7fg^3$	$-\ 54480a^2bc^4d^4e^5g^2$
$-\ 10368a^2b^2c^2e^6f^3$	$-\ 76428a^2bc^5def^3g^2$	$+\ 213120a^2bc^4d^4e^3f^3g$
$-\ 4608a^2b^2cd^8fg^3$	$+\ 14040a^2bc^5df^5g$	$+\ 123840a^2bc^4d^4e^2f^4$
$-\ 33984a^2b^2cd^5e^2g^3$	$-\ 8610a^2bc^5e^3g^3$	$-\ 137088a^2bc^4d^3e^5fg$
$-\ 3072a^2b^2cd^5f^4g$	$+\ 22869a^2bc^5e^5f^2g^2$	$+\ 10560a^2bc^4d^3e^4f^3$
$+\ 83040a^2b^2cd^4e^3fg^2$	$+\ 11988a^2bc^5e^2f^4g$	$+\ 31500a^2bc^4d^2e^7g$
$+\ 5760a^2b^2cd^4e^2f^3g$	$-\ 7614a^2bc^5ef^6$	$-\ 150984a^2bc^4d^2e^6f^2$
$-\ 23040a^2b^2cd^4ef^5$	$-\ 1920a^2bc^4d^7g^4$	$+\ 93960a^2bc^5de^8f$
$+\ 22536a^2b^2cd^3e^5g^2$	$-\ 8400a^2bc^4d^7efg^3$	$-\ 16200a^2bc^2e^{10}$
$-\ 92880a^2b^2cd^2e^4f^2g$	$+\ 20400a^2bc^4d^3f^3g^2$	$+\ 12288a^2bcd^7f^3g$
$+\ 53760a^2b^2cd^2e^2f^4$	$-\ 39900a^2bc^4d^4e^3g^3$	$+\ 33792a^2bcd^6e^3g^3$
$-\ 45504a^2b^2cd^2e^6fg$	$+\ 110160a^2bc^4d^4e^2f^3g^2$	$-\ 124416a^2bcd^6e^2f^3g$
$+\ 10152a^2b^2cd^2e^5f^3$	$+\ 83340a^2bc^4d^4ef^4g$	$+\ 233472a^2bcd^6ef^4$
$+\ 26568a^2b^2cde^8g$	$-\ 55080a^2bc^4d^3f^6$	$+\ 41856a^2bcd^5e^4fg$
$+\ 1134a^2b^2cde^7f^2$	$-\ 125820a^2bc^4de^7f^2g$	$-\ 604416a^2bcd^5e^3f^3$
$-\ 5508a^2b^2ce^9f$	$+\ 65610a^2bc^4de^5f^5$	$-\ 2400a^2bcd^4e^9g$
$+\ 9216a^2b^2d^7eg^3$	$+\ 5508a^2bc^4e^6g^2$	$+\ 661536a^2bcd^4e^5f^2$
$-\ 16128a^2b^2d^7e^2fg^2$	$+\ 3807a^2bc^4e^5f^2g$	$-\ 315120a^2bcd^3e^7f$
$-\ 6144a^2b^2d^6ef^3g$	$-\ 6885a^2bc^4e^4f^4$	$+\ 54000a^2bcd^2e^9$
$+\ 49152a^2b^2d^6f^5$	$-\ 384a^2bc^3d^6fg^3$	$-\ 6144a^2bd^9e^2g^2$
$-\ 9936a^2b^2d^5e^6g^2$	$+\ 45600a^2bc^3d^4e^2g^3$	$+\ 12288a^2bd^8ef^2g$
$+\ 15360a^2b^2d^5e^3f^2g$	$-\ 83040a^2bc^3d^3ef^2g^2$	$-\ 73728a^2bd^7f^4$

APPENDIX.   259

$+\ 3072a^2bd^7e^3fg$	$+\ 166320a^2c^4d^4ef^3g$	$+\ 32768a^2d^{10}f^3$
$+\ 208896a^2bd\,e^2f^3$	$+\ 76800a^2c^4d^7f^5$	$-\ 61440a^2d^9e^2f^2$
$-\ 2880a^2bd^6e^5g$	$-\ 139320a^2c^4d^3e^3f^2g$	$+\ 38400a^2d^9e^4f$
$-\ 234624a^2ba^4e^4f^2$	$+\ 136500a^2c^4d^3e^7f^4$	$-\ 8000a^2d^7e^8$
$+\ 114720a^2bd^5e^6f$	$-\ 93780a^2c^4d^6efg$	$-\ 9ab^7cg^6$
$-\ 20400a^2bd^4e^8$	$-\ 248400a^2c^4d^5e^4f^3$	$-\ 96ab^7dfg^5$
$+\ 360a^2c^8fg^4$	$+\ 47925a^2c^4de^7g$	$-\ 3ab^7e^3g^5$
$-\ 1128a^2c^7deg^4$	$+\ 130005a^2c^4de^6f^2$	$-\ 12ab^7ef^3g^4$
$-\ 1092a^2c^7df^2g^3$	$-\ 22275a^2c^4e^8f$	$+\ 135ab^6c^3fg^5$
$-\ 7026a^2c^7e^2fg^3$	$-\ 512a^2c^3d^7g^3$	$-\ 48ab^6cdeg^5$
$+\ 9855a^2c^7ef^3g^2$	$-\ 33792a^2c^3d^6efg^2$	$-\ 336ab^6cdf^2g^4$
$-\ 8667a^2c^7f^5g$	$-\ 57344a^2c^3d^6f^3g$	$+\ 33ab^6ce^2fg^4$
$+\ 464a^2c^6d^3g^4$	$-\ 131712a^2c^3d^5e^2f^2g$	$+\ 144ab^6cef^3g^3$
$+\ 7548a^2c^6d^2efg^3$	$-\ 314880a^2c^3d^5ef^4$	$-\ 128ab^6d^3g^5$
$+\ 4320a^2c^6d^2f^3g^2$	$+\ 272400a^2c^3d^4e^4fg$	$+\ 48ab^6d^2efg^4$
$+\ 16186a^2c^6de^2g^3$	$+\ 320000a^2c^3d^4e^3f^3$	$+\ 512ab^6d^2f^3g^3$
$-\ 5418a^2c^6de^2f^2g^2$	$-\ 91600a^2c^3d^3e^6g$	$+\ 534ab^6de^3g^4$
$+\ 25488a^2c^6def^4g$	$+\ 8160a^2c^3d^3e^5f^2$	$-\ 1056ab^6de^2f^2g^3$
$+\ 20250a^2c^6df^6$	$-\ 99900a^2c^3d^2e^7f$	$-\ 54ab^6e^4fg^3$
$-\ 5508a^2c^6e^4fg^2$	$+\ 27000a^2c^3de^9$	$+\ 216ab^6e^3f^3g^2$
$-\ 18009a^2c^6e^3f^3g$	$+\ 6144a^2c^4d^5fg^2$	$-\ 141ab^5c^3eg^5$
$-\ 5670a^2c^6e^2f^5$	$+\ 129024a^2c^4d^4ef^2g$	$-\ 822ab^5c^3f^2g^4$
$-\ 1536a^2c^5d^4fg^3$	$+\ 92160a^2c^2d^7f^4$	$+\ 336ab^5c^3d^2g^5$
$-\ 18912a^2c^5d^3e^2g^3$	$-\ 168960a^2c^2d^6e^3fg$	$+\ 1188ab^5c^4defg^4$
$-\ 22536a^2c^5d^3ef^2g^2$	$+\ 168960a^2c^2d^6e^2f^3$	$+\ 1104ab^5c^3df^3g^3$
$-\ 40152a^2c^5d^3f^4g$	$+\ 52800a^2c^2d^5e^5g$	$-\ 363ab^5c^2e^3g^4$
$-\ 18756a^2c^5d^2e^3fg^2$	$-\ 482400a^2c^2d^5e^4f^2$	$+\ 432ab^5c^2e^2f^2g^3$
$-\ 35136a^2c^5d^2e^2f^3g$	$+\ 319200a^2c^2d^4e^6f$	$-\ 648ab^5c^2ef^4g^2$
$-\ 105300a^2c^5d^2ef^5$	$-\ 66000a^2c^2d^3e^8$	$-\ 864ab^5cd^3fg^4$
$+\ 122364a^2c^5de^4f^2g$	$-\ 24576a^2cd^7f^2g$	$+\ 36ab^5cd^2e^2g^4$
$+\ 73305a^2c^5de^3f^4$	$+\ 30720a^2cd^7efg$	$+\ 5184ab^5cd^2ef^2g^3$
$-\ 27540a^2c^5e^6fg$	$-\ 184320a^2cd^6ef^3$	$-\ 4608ab^5cd^2f^4g^2$
$-\ 18549a^2c^5e^5f^3$	$-\ 9600a^2cd^7e^4g$	$-\ 10272ab^5cde^2fg^3$
$+\ 6384a^2c^4d^5eg^3$	$+\ 337920a^2cd^7e^3f^2$	$+\ 8208ab^5cde^2f^3g^2$
$+\ 9936a^2c^4d^5f^2g^2$	$-\ 206400a^2cd^5e^5f$	$+\ 2781ab^5ce^5g^3$
$+\ 54480a^2c^4d^4e^2fg^2$	$+\ 42000a^2cd^8e^7$	$-\ 567ab^5ce^4f^2g^2$

−	$1296ab^5ce^2f^4g$	+	$41280ab^4cd^2e^2g^3$	+	$3483ab^3c^4e^2f^4g$
+	$3456ab^5d^4eg^4$	+	$63360ab^4cd^3e^2f^2g^2$	−	$972ab^3c^4ef^6$
−	$4608ab^5d^4f^2g^3$	−	$83340ab^4cd^2e^4fg^2$	−	$2880ab^3c^4d^2g^4$
−	$6336ab^5d^3e^2fg^3$	−	$93312ab^4cd^2e^3f^3g$	+	$78720ab^3c^3d^3efg^3$
+	$9216ab^5d^2ef^2g^4$	+	$41472ab^4cd^2e^2f^5$	−	$18304ab^3c^3d^3f^3g^2$
−	$1020ab^5d^2e^4g^3$	−	$25488ab^4cde^5g^2$	−	$88880ab^3c^3d^2e^2g^3$
+	$13536ab^5d^2e^3f^2g^2$	+	$167832ab^4cde^3f^2g$	−	$200160ab^3c^3d^2e^2f^2g^2$
−	$13824ab^5d^2e^2f^4g$	−	$69984ab^4cde^2f^4$	+	$93312ab^3c^3d^2ef^4g$
−	$14040ab^5de^5fg^2$	−	$21951ab^4ce^7fg$	−	$13824ab^3c^3d^2f^6$
+	$13608ab^5de^4f^2g$	+	$8991ab^4ce^5f^3$	+	$125820ab^3c^3de^5fg^2$
+	$8667ab^5e^7g^2$	+	$12288ab^4d^5fg^3$	−	$45360ab^3c^3de^4f^5$
−	$11259ab^5e^5f^2g$	−	$19968ab^4d^5e^2g^2$	+	$18009ab^3c^3e^8g^2$
+	$2916ab^5e^3f^4$	+	$3072ab^4d^5ef^2g^2$	−	$39690ab^3c^3e^6f^2g$
−	$159ab^4c^4dg^5$	−	$17280ab^4d^4e^3fg^2$	+	$31590ab^3c^3e^4f^4$
+	$1590ab^4c^4efg^4$	−	$66048ab^4d^4e^2f^3g$	−	$37632ab^3c^3d^5fg^3$
+	$2862ab^4c^3f^2g^3$	+	$49152ab^4d^4ef^5$	−	$5760ab^3c^3d^4e^2g^3$
+	$480ab^4c^3d^2fg^4$	+	$40152ab^4d^3e^5g^2$	−	$5760ab^3c^3d^4ef^2g^2$
−	$1965ab^4c^3de^2g^4$	+	$192000ab^4d^3ef^2g$	+	$66048ab^3c^3d^4f^4g$
−	$16440ab^4c^3def^2g^3$	−	$147456ab^4d^3e^3f^4$	+	$200640ab^3c^3d^3e^2fg^2$
+	$1728ab^4c^3df^4g^2$	−	$207576ab^4d^2e^6fg$	−	$230400ab^3c^3d^3ef^5$
+	$7110ab^4c^3e^3fg^3$	+	$152928ab^4d^2e^2f^3$	+	$35136ab^3c^3d^2e^5g^2$
−	$4212ab^4c^3e^2f^2g^2$	+	$54999ab^4de^3g$	−	$339120ab^3c^3d^2e^3f^2g$
+	$1296ab^4c^3ef^5g$	−	$47466ab^4de^7f^2$	+	$544320ab^3c^3d^2ef^4$
−	$12240ab^4c^2d^3eg^4$	+	$2187ab^4e^9f$	+	$17820ab^3c^3de^5fg$
+	$23040ab^4c^2d^2f^2g^3$	+	$36ab^3c^6g^5$	−	$270216ab^3c^3de^4f^3$
+	$27720ab^4c^2d^2e^2fg^2$	−	$1284ab^3c^5dfg^4$	+	$13527ab^3c^5e^6g$
−	$58176ab^4c^2d^2ef^3g^2$	−	$93ab^3c^6e^2g^4$	+	$53946ab^3c^5e^7f^2$
+	$13824ab^4c^2d^2f^5g$	−	$4554ab^3c^5ef^2g^3$	+	$49152ab^3cd^6eg^3$
−	$615ab^4c^2de^4g^3$	−	$6885ab^3c^5f^4g^2$	+	$6144ab^3cd^5f^2g^2$
+	$32400ab^4c^2de^3f^2g^2$	+	$17760ab^3c^4d^2eg^4$	−	$31488ab^3cd^5e^2fg^2$
−	$11988ab^4c^2e^5fg^2$	−	$27720ab^3c^4d^2f^2g^3$	−	$98304ab^3cd^5f^5$
−	$3483ab^4c^2e^4f^3g$	+	$2640ab^3c^4de^2fg^3$	−	$166320ab^3cd^4e^2g^3$
+	$1944ab^4c^2e^3f^5$	+	$95904ab^3c^4def^2g^2$	−	$226560ab^3cd^4e^2f^2g$
+	$1536ab^4cd^5g^4$	−	$13608ab^3c^4df^4g$	+	$652800ab^3cd^4e^2f^4$
−	$34560ab^4cd^4efg^3$	−	$4635ab^3c^4e^4g^3$	+	$641280ab^3cd^3e^5fg$
−	$1536ab^4cd^4f^3g^2$	−	$28350ab^3c^4e^3f^2g^2$	−	$1235520ab^3cd^3e^3f^3$

APPENDIX. 261

$- 207144 ab^3 cd^2 e^7 g$	$+ 339120 ab^3 c^4 d^2 e^2 f^3 g$	$- 387600 ab^2 cd^4 e^8 f$
$+ 841752 ab^3 cd^2 e^5 f^2$	$+ 265680 ab^3 c^4 d^2 ef^5$	$+ 70200 ab^2 cd^3 e^8$
$- 284148 ab^3 cde^8 f$	$- 122364 ab^3 c^4 de^5 g^2$	$+ 24576 ab^2 d^9 eg^2$
$+ 43740 ab^3 ce^{10}$	$- 329670 ab^3 c^4 de^3 f^4$	$- 73728 ab^2 d^8 e^2 fg$
$- 12288 ab^3 d^6 g^3$	$+ 30861 ab^3 c^4 e^5 fg$	$+ 49152 ab^2 d^8 ef^3$
$- 12288 ab^3 d^7 efg^2$	$+ 59778 ab^3 c^4 e^5 f^3$	$+ 30720 ab^2 d^7 e^4 g$
$+ 57344 ab^3 d^6 e^3 g^2$	$- 103296 ab^3 c^3 d^5 eg^3$	$- 36864 ab^2 d^7 e^3 f^2$
$+ 116736 ab^3 d^6 e^2 f^2 g$	$- 15360 ab^3 c^3 d^5 f^2 g^2$	$+ 11520 ab^2 d^6 e^5 f$
$- 98304 ab^3 d^6 ef^4$	$- 213120 ab^3 c^3 d^4 e^2 fg^2$	$- 2400 ab^2 d^5 e^7$
$- 247680 ab^3 d^5 e^4 fg$	$+ 226560 ab^3 c^3 d^4 ef^3 g$	$+ 1584 abc^8 eg^4$
$+ 200704 ab^3 d^5 e^3 f^3$	$+ 284160 ab^3 c^3 d^4 f^5$	$- 5364 abc^7 f^2 g^3$
$+ 81456 ab^3 d^4 e^6 g$	$+ 139320 ab^3 c^3 d^3 e^4 g^2$	$+ 336 abc^7 d^2 g^4$
$- 101184 ab^3 d^4 e^5 f^2$	$- 1434240 ab^3 c^3 d^3 e^3 f^4$	$+ 48912 abc^7 defg^3$
$+ 20736 ab^3 d^3 e^7 f$	$- 48168 ab^3 c^3 d^2 e^5 fg$	$- 45684 abc^7 df^2 g^2$
$- 3240 ab^3 d^2 e^9$	$+ 1464480 ab^2 c^3 d^2 e^2 f^3$	$- 13428 abc^7 e^3 g^3$
$+ 324 ab^2 c^7 fg^4$	$+ 11340 ab^2 c^3 de^7 g$	$- 5103 abc^7 e^2 f^2 g^2$
$- 8808 ab^2 c^6 deg^4$	$- 534438 ab^2 c^3 de^5 f^2$	$+ 21951 abc^7 ef^4 g$
$+ 22398 ab^2 c^6 df^2 g^3$	$+ 54675 ab^2 c^8 e^3 f$	$- 8748 abc^7 f^6$
$+ 2835 ab^2 c^6 e^2 fg^3$	$+ 24576 ab^2 c^2 d^7 g^3$	$- 22896 abc^6 d^3 fg^3$
$- 4860 ab^2 c^6 ef^3 g^2$	$+ 124416 ab^2 c^2 d^6 efg^2$	$- 61764 abc^6 d^2 e^2 g^3$
$+ 11259 ab^2 c^6 f^5 g$	$- 116736 ab^2 c^2 d^5 f^3 g$	$+ 45504 abc^6 d^2 ef^2 g^2$
$+ 792 ab^2 c^5 d^3 g^4$	$+ 131712 ab^2 c^2 d^5 e^2 g^2$	$+ 207576 abc^6 d^2 f^4 g$
$- 109800 ab^2 c^5 d^2 efg^3$	$- 248832 ab^2 c^2 d^5 ef^4$	$- 36504 abc^6 de^3 fg^2$
$+ 30600 ab^2 c^5 d^2 f^3 g^2$	$- 498960 ab^2 c^2 d^4 e^3 fg$	$- 17820 abc^6 de^2 f^3 g$
$+ 58884 ab^2 c^5 de^6 g^3$	$+ 1610880 ab^2 c^2 d^4 e^2 f^3$	$- 83268 abc^6 def^5$
$+ 20736 ab^2 c^5 de^2 f^2 g^2$	$+ 190920 ab^2 c^2 d^3 e^9 g$	$+ 27540 abc^6 e^5 g^2$
$- 167832 ab^2 c^5 def^4 g$	$- 1849824 ab^2 c^2 d^3 e^5 f^2$	$- 30861 abc^6 e^4 f^2 g$
$+ 16848 ab^2 c^5 df^6$	$+ 874800 ab^2 c^2 d^2 e^7 f$	$+ 29889 abc^6 e^3 f^4$
$- 3807 ab^2 c^5 e^4 fg^2$	$- 153900 ab^2 c^5 de^9$	$+ 63216 abc^5 d^4 eg^3$
$+ 39690 ab^2 c^5 e^3 f^3 g$	$- 12288 ab^2 cd^8 fg^2$	$- 31248 abc^5 d^4 f^2 g^2$
$+ 13851 ab^2 c^5 e^2 f^5$	$- 129024 ab^2 cd^7 e^2 g^2$	$+ 137088 abc^5 d^3 e^2 fg^2$
$+ 50400 ab^2 c^4 d^4 fg^3$	$+ 196608 ab^2 cd^7 f^4$	$- 641280 abc^5 d^3 ef^3 g$
$+ 81360 ab^2 c^4 d^3 e^2 g^3$	$+ 390144 ab^2 cd^6 e^3 fg$	$- 259488 abc^5 d^3 f^5$
$+ 92880 ab^2 c^4 d^3 ef^2 g^2$	$- 866304 ab^2 cd^6 e^2 f^3$	$+ 93780 abc^5 d^2 e^4 g^2$
$- 192000 ab^2 c^4 d^2 f^4 g$	$- 155520 ab^2 cd^5 e^5 g$	$+ 48168 abc^5 d^2 e^3 f^2 g$
$- 97920 ab^2 c^4 d^2 e^3 fg^2$	$+ 868608 ab^2 cd^5 e^4 f^2$	$+ 735156 abc^5 d^2 e^2 f^4$

APPENDIX.

$-\ 338904abc^5de^4f^8$ $\quad+\ 13932ac^9f^8g^2$ $\quad-\ 50625ac^5de^8$

$-\ 6075abc^5e^7g$ $\quad+\ 3504ac^9d^2fg^3$ $\quad+\ 9600ac^4d^7eg^2$

$+\ 53460abc^5e^6f^2$ $\quad+\ 14280ac^8de^2g^5$ $\quad-\ 30720ac^4d^7f^2g$

$-\ 14784abc^4d^9g^3$ $\quad-\ 26568ac^8def^2g^4$ $\quad+\ 9600ac^4d^6e^2fg$

$-\ 41856abc^4d^5efg^2$ $\quad-\ 54999ac^8df^4g$ $\quad-\ 184320ac^4d^6ef^3$

$+\ 247680abc^4d^5f^3g$ $\quad+\ 24300ac^8e^3fg^2$ $\quad+\ 357600ac^4d^5e^3f^2$

$-\ 272400abc^4d^4e^2g^4$ $\quad-\ 13527ac^8e^2f^3g$ $\quad-\ 222000ac^4d^4e^5f$

$+\ 498960abc^4d^4e^2f^2g$ $\quad+\ 21870ac^8ef^5$ $\quad+\ 45000ac^4d^3e^7$

$+\ 713280abc^4d^4ef^4$ $\quad-\ 12960ac^7d^3eg^3$ $\quad+\ 40960ac^3d^8f^3$

$-\ 1759680abc^4d^3e^3f^3$ $\quad+\ 14136ac^7d^3f^2g^2$ $\quad-\ 76800ac^3d^7e^2f^2$

$-\ 48600abc^4d^2e^5g$ $\quad-\ 31500ac^7d^2e^2fg^2$ $\quad+\ 48000ac^3d^6e^4f$

$+\ 1095120abc^4d^2e^3f^2$ $\quad+\ 207144ac^7d^2ef^3g$ $\quad-\ 10000ac^3d^5e^6$

$-\ 275400abc^4de^7f$ $\quad+\ 61560ac^7d^2f^5$ $\quad+\ 2b^9g^6$

$+\ 30375abc^4e^9$ $\quad-\ 47925ac^7de^4g^2$ $\quad-\ 36b^8cfg^5$

$-\ 3072abc^3d^7fg^2$ $\quad-\ 11340ac^7de^3f^2g$ $\quad+\ 18b^8e^2fg^4$

$+\ 168960abc^3d^6e^2g^2$ $\quad-\ 163215ac^7de^2f^4$ $\quad+\ 45b^7c^2eg^5$

$-\ 390144abc^3d^6ef^2g$ $\quad+\ 6075ac^7e^5fg$ $\quad+\ 270b^7c^2f^2g^4$

$-\ 331776abc^3d^6f^4$ $\quad+\ 42525ac^7e^4f^3$ $\quad-\ 72b^7cdefg^4$

$+\ 206592abc^3d^5e^2f^3$ $\quad+\ 2960ac^6d^5g^3$ $\quad-\ 45b^7ce^3g^4$

$+\ 44400abc^3d^4e^5g$ $\quad+\ 2400ac^6d^4efg^2$ $\quad-\ 216b^7ce^2f^2g^3$

$+\ 638400abc^3d^4e^3f^2$ $\quad-\ 81456ac^6d^4f^3g$ $\quad+\ 768b^7d^3fg^4$

$-\ 565200abc^3d^3e^5f$ $\quad+\ 91600ac^6d^3e^3g^2$ $\quad-\ 960b^7d^2e^3g^4$

$+\ 121500abc^3d^2e^6$ $\quad-\ 190920ac^6d^3e^2f^2g$ $\quad-\ 1536b^7d^2ef^2g^3$

$-\ 30720abc^2d^8eg^2$ $\quad-\ 213130ac^6d^3ef^4$ $\quad+\ 2736b^7de^3fg^3$

$+\ 73728abc^2d^8f^2g$ $\quad+\ 48600ac^6d^2e^4fg$ $\quad-\ 756b^7e^5g^3$

$+\ 454656abc^2d^7ef^3$ $\quad+\ 468720ac^6d^2e^3f^3$ $\quad-\ 162b^7e^4f^2g^2$

$-\ 9600abc^2d^6e^4g$ $\quad-\ 230850ac^6de^5f^2$ $\quad-\ 30b^6c^3dg^5$

$-\ 898560abc^2d^6e^2f^2$ $\quad+\ 30375ac^6e^7f$ $\quad-\ 639b^6c^3efg^4$

$+\ 561600abc^2d^5e^5f$ $\quad+\ 2880ac^5d^9fg^2$ $\quad-\ 1080b^6c^3f^3g^3$

$-\ 114000abc^2d^4e^7$ $\quad-\ 52800ac^5d^5e^3g^2$ $\quad-\ 1656b^6c^2d^2fg^4$

$-\ 98304abcd^9f^3$ $\quad+\ 155520ac^5d^5ef^2g$ $\quad+\ 1665b^6c^2de^2g^4$

$+\ 184320abcd^8e^2f^2$ $\quad+\ 101376ac^5d^5f^4$ $\quad+\ 3168b^6c^2def^2g^3$

$-\ 115200abcd^7e^4f$ $\quad-\ 44400ac^5d^4e^3fg$ $\quad-\ 1242b^6c^2e^3fg^3$

$+\ 24000abcd^6e^6$ $\quad+\ 6960ac^5d^4e^2f^3$ $\quad+\ 972b^6c^2e^2f^3g^2$

$-\ 144ac^9dg^4$ $\quad-\ 316800ac^5d^3e^4f^2$ $\quad+\ 1920b^6cd^3eg^4$

$-\ 7488ac^9efg^3$ $\quad+\ 243000ac^5d^4e^5f$ $\quad-\ 6144b^6cd^3f^2g^3$

APPENDIX. 263

$+\ 2784b^6cd^2e^2fg^3$ $+\ 45360b^5c^3de^2f^3g$ $-\ 6480b^4c^4d^2f^3g^u$
$+\ 13824b^6cd^2ef^3g^2$ $+\ 5670b^5c^3e^6g^u$ $+\ 1350b^4c^4de^3g^3$
$+\ 720b^6cde^4g^3$ $-\ 13851b^5c^3e^3f^u g$ $-\ 38070b^4c^4de^2f^u g^u$
$-\ 23328b^6cde^3f^2g^2$ $-\ 1458b^5c^3e^3f^4$ $+\ 69984b^4c^4def^4g$
$+\ 7614b^6ce^5fg^2$ $+\ 6144b^5cd^5fg^3$ $+\ 6885b^4c^4e^5fg^u$
$+\ 972b^6ce^4f^3g$ $+\ 23040b^5cd^4ef^2g^2$ $-\ 31590b^4c^4e^3f^3g$
$-\ 2816b^6d^5g^4$ $-\ 49152b^5cd^4f^5g$ $+\ 1458b^4c^4e^3f^5$
$+\ 13824b^6d^4efg^3$ $-\ 127680b^5cd^3e^3fg^2$ $-\ 21120b^4c^3d^4fg^3$
$+\ 8192b^6d^4f^3g^u$ $+\ 230400b^5cd^3e^2f^3g$ $+\ 22800b^4c^3d^3e^4g^3$
$-\ 5760b^6d^3e^3g^3$ $+\ 105300b^5cd^4e^5g^2$ $-\ 53760b^4c^3d^3ef^2g^2$
$-\ 52224b^6d^3e^2f^4g^2$ $-\ 265680b^5cd^4e^4f^4g$ $+\ 147456b^4c^3d^3f^4g$
$+\ 55080b^6d^2e^5fg^2$ $-\ 41472b^5cd^4ef^4$ $+\ 141480b^4c^3d^4e^3fg^2$
$+\ 13824b^6d^3e^3f^3g$ $+\ 83268b^5cde^5fg$ $-\ 544320b^4c^3d^3e^2f^3g$
$-\ 20250b^5de^6g^2$ $+\ 68040b^5cde^3f^3$ $+\ 41472b^4c^3d^4ef^5$
$-\ 16848b^5de^5f^3g$ $-\ 21870b^5ce^8g$ $-\ 73305b^4c^3de^5g^2$
$+\ 8748b^6e^7fg$ $-\ 15309b^5ce^7f^2$ $+\ 329670b^4c^3de^4f^2g$
$-\ 1458b^6e^8f^3$ $-\ 49152b^5d^2f^4g^2$ $-\ 29889b^4c^3e^6fg$
$+\ 9b^5c^5g^5$ $+\ 119808b^5d^2e^3fg^2$ $-\ 15309b^4c^3e^6f^3$
$+\ 1674b^5c^4dfg^4$ $+\ 98304b^5d^3ef^3g$ $-\ 7680b^4c^2d^5eg^u$
$-\ 225b^5c^4e^3g^4$ $-\ 76800b^5d^4e^3g^u$ $+\ 170496b^4c^2d^3f^3g^2$
$+\ 2754b^5c^4ef^4g^3$ $-\ 284160b^5d^4e^3f^4g$ $-\ 123840b^4c^2d^4e^3fg^u$
$+\ 2430b^5c^4f^4g^2$ $-\ 737280b^5d^4e^2f^4$ $-\ 65280b^4c^2d^4ef^3g$
$-\ 3720b^5c^3d^2eg^4$ $+\ 259488b^5d^3e^3fg$ $+\ 73728b^4c^2d^4f^5$
$+\ 13728b^5c^3d^u f^2g^3$ $+\ 241920b^5d^3e^3f^3$ $-\ 136500b^4c^2d^3e^4g^2$
$-\ 12240b^5c^3de^2fg^3$ $-\ 61560b^5d^2e^7g$ $+\ 143424b^4c^2d^3e^3f^u g$
$-\ 24624b^5c^3def^3g^2$ $-\ 280584b^5d^2e^6f^2$ $-\ 73515b^4c^2d^3e^5fg$
$+\ 675b^5c^3e^4g^3$ $+\ 126846b^5def$ $-\ 278640b^4c^2d^2e^4f^3$
$+\ 12960b^5c^3e^3f^2g^u$ $-\ 19683b^5e^{10}$ $+\ 163215b^4c^2de^7g$
$-\ 1944b^5c^3e^2f^4g$ $-\ 441b^5c^2fg^4$ $+\ 147987b^4c^2de^6f^2$
$+\ 8640b^5c^2d^4g^4$ $+\ 1575b^5c^5deg^4$ $-\ 24057b^4c^2e^8f$
$-\ 43200b^5c^2d^3efg^3$ $-\ 13356b^5c^2df^u g^3$ $-\ 233472b^4cd^6efg^u$
$-\ 10752b^5c^2d^3f^3g^2$ $+\ 1458b^5c^5e^2fg^3$ $+\ 98304b^4cd^5f^3g$
$+\ 6600b^5c^2d^u e^2g^3$ $-\ 2430b^5c^5ef^3g^2$ $+\ 314880b^4cd^5e^3g^2$
$+\ 123120b^5c^2d^u e^2f^u g^u$ $-\ 2916b^5c^5f^5g$ $+\ 248832b^4cd^5e^2f^2g$
$-\ 41472b^5c^2d^u ef^4g$ $-\ 11100b^5c^4d^3g^4$ $-\ 71328b^4cd^4e^5fg$
$-\ 65610b^5c^3de^2fg^2$ $+\ 54840b^5c^4d^u efg^u$ $-\ 345600b^4cd^4e^3f^3$

APPENDIX.

$+\ 213120b^4cd^3e^6g$
$+\ 774144b^4cd^3e^5f^2$
$-\ 441936b^4cd^2e^5f$
$+\ 76545b^4cde^9$
$+\ 73728b^4d^8fg^2$
$-\ 92160b^4d^7e^3g^2$
$-\ 196608b^4d^7ef^2g$
$+\ 331776b^4d^6e^3fg$
$+\ 147456b^4d^6e^2f^3$
$-\ 101376b^4d^5e^5g$
$-\ 309504b^4d^5e^4f^2$
$+\ 179424b^4d^4e^6f$
$-\ 31860b^4d^3e^8$
$-\ 180b^3c^7eg^4$
$+\ 3618b^3c^7f^2g^3$
$+\ 7800b^3c^6d^3g^4$
$-\ 20412b^3c^6defg^2$
$+\ 24084b^3c^6df^3g^2$
$-\ 945b^3c^6e^3g^3$
$+\ 10368b^3c^6e^2f^2g^2$
$-\ 8991b^3c^6ef^4g$
$+\ 30048b^3c^5d^3fg^3$
$-\ 49560b^3c^5d^2e^2g^3$
$-\ 10152b^3c^5d^2ef^2g^2$
$-\ 152928b^3c^5d^2f^4g$
$-\ 60264b^3c^5de^3fg^2$
$+\ 270216b^3c^5de^2f^3g$
$-\ 68040b^3c^5def^5$
$+\ 18549b^3c^5e^5g^2$
$-\ 59778b^3c^5e^4f^2g$
$+\ 15309b^3c^5e^3f^4$
$+\ 4800b^3c^4d^4eg^3$
$-\ 245280b^3c^4d^4f^2g^2$
$-\ 10560b^3c^4d^3e^2fg^2$
$+\ 1235520b^3c^4d^3ef^3g$
$-\ 241920b^3c^4d^3f^5$

$+\ 248400b^3c^4d^2e^4g^2$
$-\ 1464480b^3c^4d^2e^3f^2g$
$+\ 278640b^3c^4d^2e^2f^4$
$+\ 338904b^3c^4de^5fg$
$-\ 42525b^3c^4e^7g$
$-\ 4374b^3c^4e^6f^2$
$+\ 5120b^3c^3d^6g^3$
$+\ 604416b^3c^3d^5efg^2$
$-\ 200704b^3c^3d^5f^3g$
$-\ 320000b^3c^3d^4e^3g^2$
$-\ 1610880b^3c^3d^4e^2f^2g$
$+\ 345600b^3c^3d^4ef^4$
$+\ 1759680b^3c^3d^3e^4fg$
$-\ 468720b^3c^3d^2e^6g$
$-\ 3214408b^3c^3d^2e^5f^2$
$+\ 160380b^3c^3de^7f$
$-\ 18225b^3c^3e^9$
$-\ 208896b^3c^2d^7fg^2$
$-\ 168960b^3c^2d^6e^2g^2$
$+\ 866304b^3c^2d^6ef^2g$
$-\ 147456b^3c^2d^6f^4$
$-\ 206592b^3c^2d^5e^3fg$
$-\ 6960b^3c^2d^4e^5g$
$-\ 344160b^3c^2d^4e^4f^2$
$+\ 300240b^3c^2d^3e^6f$
$-\ 64800b^3c^2d^2e^8$
$+\ 184320b^3cd^8eg^2$
$-\ 49152b^3cd^8f^2g$
$-\ 454656b^3cd^7e^2fg$
$+\ 184320b^3cd^7e^4g$
$+\ 374784b^3cd^6e^3f^2$
$-\ 301440b^3cd^5e^5f$
$+\ 61200b^3cd^4e^7$
$-\ 32768b^3d^{10}g^2$
$-\ 983204b^3d^9efg$
$-\ 40960b^3d^9e^3g$

$-\ 73728b^3d^8e^2f^2$
$+\ 61440b^2d^7e^5f$
$-\ 2880b^2c^8dg^4$
$+\ 1836b^2c^8efg^3$
$-\ 8586b^2c^8f^3g^2$
$-\ 26856b^2c^7d^2fg^3$
$+\ 29340b^2c^7de^2g^3$
$-\ 1134b^2c^7def^2g^2$
$+\ 47466b^2c^7df^4g$
$+\ 6723b^2c^7e^3fg^2$
$-\ 53946b^2c^7e^2f^3g$
$+\ 15309b^2c^7ef^5$
$+\ 15720b^2c^6d^5eg^3$
$+\ 203592b^2c^6d^3f^2g^2$
$+\ 150984b^2c^6d^2e^2fg^2$
$-\ 841752b^2c^6d^2ef^3g$
$+\ 280584b^2c^6d^2f^5$
$-\ 130005b^2c^6de^4g^2$
$+\ 534438b^2c^6de^2f^2g$
$-\ 147987b^2c^6de^2f^4$
$-\ 53460b^2c^6e^6fg$
$+\ 4374b^2c^6e^4f^3$
$-\ 11136b^2c^5d^5g^3$
$-\ 661536b^2c^5d^4efg^2$
$+\ 101184b^2c^5d^4f^3g$
$-\ 8160b^2c^5d^3e^3g^2$
$+\ 1849824b^2c^5d^3e^2f^2g$
$-\ 774144b^2c^5d^3ef^4$
$-\ 1095120b^2c^5d^2e^4fg$
$+\ 321408b^2c^5d^2e^3f^3$
$+\ 230850b^2c^5de^8g$
$-\ 18225b^2c^5e^7f$
$+\ 234624b^2c^4d^6fg^2$
$+\ 482400b^2c^4d^5e^2g^2$
$-\ 868608b^2c^4d^5ef^2g$
$+\ 309504b^2c^4d^5f^4$

APPENDIX. 265

$- 638400b^2c^4d^4e^2fg$ $+ 7440bc^7d^4g^3$ $+ 12960c^{10}df^2g^2$
$+ 344160b^2c^4d^4e^2f^3$ $+ 315120bc^7d^3efg^2$ $+ 16200c^{10}e^2fg^2$
$+ 316800b^2c^4d^3e^5g$ $- 20736bc^7d^3f^3g$ $- 43740c^{10}ef^3g$
$- 145800b^2c^4d^3e^3f$ $+ 99900bc^7d^2e^3g^2$ $+ 19683c^{10}f^5$
$+ 30375b^2c^4de^6$ $- 874800bc^7d^2e^2f^2g$ $- 1600c^9d^3g^3$
$- 337920b^2c^3d^7eg^2$ $+ 441936bc^7d^2ef^4$ $- 54000c^9d^2efg^2$
$+ 36864b^2c^3d^7f^2g$ $+ 275400bc^7de^4fg$ $+ 3240c^9d^2f^3g$
$+ 898560b^2c^3d^6e^2fg$ $- 160380bc^7de^3f^3$ $- 27000c^9de^2g^2$
$- 374784b^2c^3d^6ef^3$ $- 30375bc^7e^6g$ $+ 153900c^9def^2g$
$- 357600b^2c^3d^5e^4g$ $+ 18225bc^7e^5f^2$ $- 76545c^9def^4$
$+ 133200b^2c^3d^4e^3f$ $- 114720bc^6d^8fg^2$ $- 30375c^8e^4fg$
$- 27000b^2c^3d^3e^7$ $- 319200bc^6d^4e^2g^2$ $+ 18225c^8e^3f^3$
$+ 61440b^2c^3d^9g^2$ $+ 387600bc^6d^4ef^2g$ $- 360c^8d^4fg^2$
$- 184320b^2c^3d^8efg$ $- 179424bc^6d^4f^4$ $+ 66000c^8d^3e^3g^2$
$+ 73728b^2c^3d^8f^3$ $+ 565200bc^6d^3e^3fg$ $- 70200c^8d^3ef^2g$
$+ 76800b^2c^3d^7e^2g$ $- 300240bc^6d^3e^2f^3$ $+ 31860c^8d^3f^4$
$- 28800b^2c^3d^4e^4f$ $- 243000bc^6d^2e^5g$ $- 121500c^8d^2e^3fg$
$+ 6000b^2c^3d^5e^6$ $+ 145800bc^6d^2e^4f^2$ $+ 64800c^8d^2e^2f^3$
$+ 432bc^{10}g^4$ $+ 20640bc^5d^6eg^2$ $+ 50625c^8de^5g$
$+ 12384bc^9dfg^3$ $- 11520bc^5d^5f^2g$ $- 30375c^8de^4f^2$
$- 5400bc^9e^2g^3$ $- 561600bc^5d^5e^2fg$ $- 42000c^7d^5eg^2$
$+ 5508bc^9ef^2g^2$ $+ 301440bc^5d^5ef^3$ $+ 2400c^7d^5f^2g$
$- 2187bc^9f^4g$ $+ 222000bc^5d^4e^3g$ $+ 114000c^7d^4e^2fg$
$- 15120bc^8d^2eg^3$ $- 133200bc^5d^4e^2f^2$ $- 61200c^7d^4ef^3$
$- 84024bc^8d^2f^2g^2$ $- 38400bc^4d^8g^2$ $- 45000c^7d^3e^4g$
$- 93960bc^8defg^2$ $+ 115200bc^4d^7efg$ $+ 27000c^7d^3e^3f^2$
$+ 284148bc^8def^2g$ $- 61440bc^4d^7f^3$ $+ 8000c^6d^7g^2$
$- 126846bc^8df^5$ $- 48000bc^4d^6e^2g$ $- 24000c^6d^6efg$
$+ 22275bc^8e^3g^2$ $+ 28800bc^4d^6ef^2$ $+ 10000c^6d^6e^3g$
$- 54675bc^8e^2f^2g$ $- 2160c^{11}fg^3$ $- 6000c^6d^5e^2f^2$
$+ 24057bc^8e^2f^4$ $+ 3600c^{10}deg^3$

M M

(266)

NOTES.

HISTORY OF DETERMINANTS. (Page 1).

THE following historical notices are taken from Baltzer's *Theory of Determinants*; and from the sketch prefixed to Spottiswoode's *Elementary Theorems relating to Determinants*. The first idea of determinants is due to Leibnitz, as Dirichlet has pointed out. In Leibnitz's letter to L'Hôpital, 28 April, 1693, (Leibnitz's *Mathematical Works*, published by Gerhardt, vol. II., p. 239), is to be found the first example of the formation of these functions, and of their application to the solution of linear equations; the double suffix notation (p. 7) is employed, and he expresses his conviction of the fertility of his idea. But nowhere else in his writings is there to be found any proof that he sought to draw any new fruits from his discovery; and the method was lost until re-discovered by Cramer in 1750. Cramer in his *Introduction à l'Analyse des lignes Courbes* (Appendix), has exhibited the determinants arising from linear equations in the case of two and three variables, and has indicated the law according to which they would be formed in the case of a greater number. The rule of signs by the method of displacements (Note, p. 5) is given by Cramer. The equivalence of the other method by permutation of suffixes was afterwards proved by Bezout and Laplace. In the *Histoire de l'Académie Royale des Sciences*, Année 1764, (published in 1767), Bezout has investigated the degree of the equation resulting from the elimination of unknown quantities from a given system of equations, and has at the same time noticed several cases of determinants, without however entering upon the general law of formation, or the properties of these functions. The *Histoire de l'Académie*, An. 1772, part II. (published in 1776), contains papers by Laplace and Vandermonde relating to determinants of the second, third, fourth, &c. orders. The former, in discussing a system of simultaneous differential equations, has given the law of formation, and shown that when two rows or columns are interchanged, the sign of the determinant is changed, and that when two are identical, the determinant vanishes. The latter employs a notation in substance identical with that which, after Mr. Sylvester, we have called the umbral notation, and explained p. 7. In his Memoir on Pyramids (*Memoires de l'Académie de Berlin*, 1773), Lagrange made an extensive use of determinants of the third order, and demonstrated that the square of such a determinant can itself be expressed as a determinant. The next impulse to the study was given by Gauss, *Disquisitiones Arithmeticæ*, 1801, who showed, in the case of the second and third orders, that the product of two determinants is a determinant, and very completely discussed the case of determinants of the second order arising from quadratic functions, *i.e.* of the form $b^2 - ac$. In 1812 Binet published a memoir on this subject (*Journal de l'Ecole Polytechnique*, tome IX., cahier 16), in which he establishes the principal theorems for determinants of the second, third, and fourth orders, and applies them to geometrical problems. The next volume of the same series contains a paper, written at the same time, by Cauchy, on functions which only change sign when the variables which they contain are transposed. The second part of this paper refers immediately to determinants, and contains a large number of very general theorems. Cauchy introduced the name "determinants," already applied by Gauss to the functions considered by him, and called by him "determinants of quadratic forms." In 1826 Jacobi took possession of the new calculus, and the volumes of Crelle's *Journal* contain brilliant proofs of the power of the instrument in the hand of such a master. By his memoirs in 1841, *De formatione et proprietatibus determinantium*, and *De determinantibus functionalibus* (*Crelle*, vol. XXII.), determinants first became easily accessible to all mathematicians. Of later papers on this subject, perhaps the most important are Cayley's papers on Skew Determinants (*Crelle*, vols. XXXII. and XXXVIII.). Of elementary treatises on this subject, I have to mention Spottiswoode's *Elementary Theorems relating to Determinants*, London (1851); Brioschi, *La teorica dei determinanti*, Pavia, 1854; and Baltzer, *Theorie und Anwendung der Determinanten*, Leipzig, 1857; second edition, 1864. French translations both of Brioschi's and Baltzer's works have been published.

COMMUTANTS. (Page 7).

In connection with the umbral notation may be explained what is meant by *commutants*, which are but an extension of the same idea. If we write for brevity ξ, η, for $\dfrac{d}{dx}$, $\dfrac{d}{dy}$, it is easy to see what, according to the rule of the umbral notation, is meant by

$$\xi, \eta, \qquad \xi^2, \xi\eta, \eta^2,$$
$$\xi, \eta, \qquad \xi^2, \xi\eta, \eta^2.$$

We compound the partial constituents in each column in order to find the factors in the product we want to form, and we take the sum with proper signs of all possible products obtained by permuting the terms in the lower row. Thus the first example denotes $\xi^2.\eta^2 - \xi\eta.\xi\eta$, which is the Hessian; and the second denotes

$$\xi^4.\xi^2\eta^2.\eta^4 - \xi^4.\xi\eta^3.\xi\eta^3, \&c.,$$

which is the ordinary cubinvariant of a quartic.

Again, since multiplication is performed by addition of indices, it will be readily understood that we can equally form commutants where the partial constituents are combined by addition instead of by multiplication. Thus, considering the quantics

$$(a_2, a_1, a_0 \!\! \left.\right) \!\!\! \left(\right. x, y)^2, \quad (a_4, a_3, a_2, a_1, a_0 \!\! \left.\right) \!\!\! \left(\right. x, y)^4,$$

the invariants in the last two examples may be written

$$1, 0, \qquad 2, 1, 0,$$
$$1, 0, \qquad 2, 1, 0,$$

which expanded are $a_2 a_0 - a_1 a_1$; $a_4 a_2 a_0 - a_4 a_1 a_1 + \&c.$

All these commutants with only two rows may be written as determinants, but it is a natural extension of the above notation to form commutants with more than two rows, such as

$$\xi, \eta, \qquad 1, 0, \qquad \xi^2, \xi\eta, \eta^2.$$
$$\xi, \eta, \qquad 1, 0, \qquad \xi^2, \xi\eta, \eta^2.$$
$$\xi, \eta, \qquad 1, 0, \qquad \xi^2, \xi\eta, \eta^2.$$
$$\xi, \eta, \qquad 1, 0, \qquad \xi^2, \xi\eta, \eta^2.$$

These all denote the sum of a number of products, each product consisting of as many factors as there are columns in the commutant, and each factor being formed by compounding the constituents of the same column; and where we permute in every possible way the constituents in each row after the first. Thus the first and second examples denote the same thing, namely, the quadrinvariant of a quartic expressed in either of the forms $\xi^4 . \eta^4 - 4\xi^3\eta . \xi\eta^3 + 3\xi^2\eta^2 . \xi^2\eta^2$ or $a_4 a_0 - 4a_3 a_1 + 3a_2 a_2$, while the third example $\xi^3 . \xi^4\eta^4 . \eta^3 -$ &c. denotes the cubinvariant of an octavic given at length, p. 124.

We have seen that the two invariants of a binary quartic can be expressed as commutants, but it will be found impossible to express in the same way the discriminant of a cubic. Thus, the leading term in it being $a_1^2 a_0^2$ or $\xi_2 \xi_3 \eta_2 \eta_3$, we are naturally led to expect that it might be the commutant

$$\xi, \eta, \xi, \eta,$$
$$\xi, \eta, \xi, \eta,$$
$$\xi, \eta, \xi, \eta,$$

but this commutant, instead of giving the discriminant, will be found to vanish identically. It may, however, be made to yield the discriminant by placing certain restrictions on the permutations which are allowable. For further details I refer to the papers of Messrs. Cayley and Sylvester in the *Cambridge and Dublin Mathematical Journal*, 1852.

HESSIANS. (Page 15).

The name was given by Sylvester after Professor Otto Hesse of Heidelberg, who has made much use of the functions in question, which he called functional determinants. They are a particular case of those studied under the same name by Jacobi, (*Crelle*, vol. XXII.), the constituents of which are the differentials of a series of n homogeneous functions in n variables. It is so convenient to have short distinctive names for the functions of which we have repeatedly occasion to speak, that I have followed Sylvester in calling the former Hessians, the latter Jacobians, see p. 69.

SYMMETRIC FUNCTIONS. (Page 45).

The rules for the weight and order of symmetric functions I believe to be Mr. Cayley's, though I cannot give the reference. The formula, Art. 55, I have taken from Serret's *Lessons on Higher Algebra*. The differential equation, Art. 56, is an application of the differential equation for invariants, of which I speak afterwards. Brioschi's expression, Art. 61, I know from the use made of it by Mr. M. Roberts, *Quarterly Journal*, vol. IV., p. 168.

ELIMINATION. (Page 53).

The name 'eliminant' was introduced I think by Professor De Morgan: I believe I have done wrong in using a second appellation when a name to which there was no objection was already in use. The older name 'resultant' was employed by Bezout, *Histoire de l'Académie de Paris*, 1764. The method of elimination by symmetric functions is due to Euler (*Berlin Memoirs*, 1748). The reduction of the resultant to a linear system was made simultaneously by Euler (*Berlin Memoirs*, 1764) and Bezout (*Paris Memoirs*, 1764). The theorem as to the degree of the resultant is Bezout's.

NOTES. 269

The method used in Art. 70 of forming symmetric functions of the common values of a system of two or more equations is Poisson's (see *Journal de l'Ecole Polytechnique*, Cahier XI.). Sylvester's mode of elimination was given by him in the *Philosophical Magazine* for 1840, and called by him 'dialytical,' because the process as it were *dissolves* the relations which connect the different combinations of powers of the variables and treats them as simple independent quantities. Cayley's statement of Bezout's method is to be found, *Crelle*, vol. LIII., p. 366. Sylvester's results in Art. 87 are to be found in the *Cambridge and Dublin Mathematical Journal* for 1852, vol. VII., p. 68; and Cayley's general theory (Art. 88, &c.) in the same Journal, vol. III., p. 116. It was noticed by Lagrange that when two equations have two sets of common roots, the differential of the resultant with respect to the last term vanishes (see *Berlin Memoirs*, 1770). Mr. Sylvester showed, in January, 1853, that the same was true of all the differentials, *Cambridge and Dublin Mathematical Journal*, vol. VIII., p. 64. He showed at the same time, that the common roots were given by the ratios of the differentials. The proof in Art. 95 is, I believe, my own. The theorem, Art. 95, is Jacobi's, *Crelle*, vol. XV., p. 105. In this part I have made some use of the *Treatise on Elimination* by Faà de Bruno. The theorem of Art. 98 is Mr. Cayley's.

DISCRIMINANTS. (Page 83).

The word 'discriminant' was introduced by Mr. Sylvester in 1852, *Cambridge and Dublin Mathematical Journal*, vol. VI., p. 52. The word 'determinant' had been previously used, and had come to have a perplexing variety of significations. The theorem referred to, Note, p. 87, was the basis of my investigations (*Cambridge and Dublin Mathematical Journal*, 1847 and 1849) on the nature of cones circumscribing surfaces having multiple lines. If the equation of a surface be $b_0 + b_1 x + b_2 x^2 + $ &c., and if xy be a double line, b_0 must contain y in the second, and b_1 in the first degree. The discriminant with respect to x is a tangent cone which has y^2 for a factor. Instead of the note, p. 90, which was added without sufficient consideration, I substitute a reference to the extension of Sturm's theorem contained in Mr. Sylvester's memoir in the *Philosophical Transactions* for 1853.

LINEAR TRANSFORMATIONS. (Page 92).

The germ of the principle of invariance may be traced to Lagrange, who, in the *Berlin Memoirs*, 1773, p. 265, established the invariance of the discriminant of the quadratic form $ax^2 + 2bxy + cy^2$, when for x is substituted $x + \lambda y$. Gauss, in his *Disquisitiones Arithmeticæ* (1801), investigated very completely the theory of the general linear transformation as applied to binary and ternary quadratic forms, and, in particular, established the invariance of their discriminants. This property of invariance was shown to belong to discriminants generally by the late Professor Boole, who, in a remarkable paper, *Cambridge Mathematical Journal*, 1841, vol. III., pp. 1, 106, applied it to the theory of orthogonal substitutions. He there showed how to form simultaneous invariants of a system of two functions of the same degree by performing on the discriminant of one of them the operation $a' \dfrac{d}{da} + b' \dfrac{d}{db} + $ &c. Boole's paper led to Mr. Cayley's proposing to himself the problem to determine *à priori* what functions of the coefficients of an equation possess this property of invariance. He found that it was not peculiar to discriminants, and he discovered other functions of the coefficients of an equation, at first called by him 'hyper-determinants,' possessing the same property. Mr. Cayley's first results were published in 1845 (*Cambridge Mathematical Journal*, vol. IV., p. 193). From this discovery of Cayley's, the modern algebra which forms the subject of the bulk of this volume may be said to take its rise. Among the first invariants distinct from discriminants, which were thus brought to light, were the quadrinvariants of binary quantics, and in particular the invariant S of a quartic. Mr. Boole next discovered the other invariant T of a quartic, and the expression of the discriminant in terms of S and T (*Cambridge Mathematical Journal*, vol. IV., p. 208). It is worthy of notice that both the functions S and T had been used by Eisenstein (*Crelle*, 1844, XXVII., p. 81) in his expression for the general solution of a quartic, but their property of invariance was unknown to him, as well as the expression for the discriminant in terms of them. Mr. Cayley next (1846) published the symbolical method of finding invariants, explained in Lesson XIV. (*Cambridge and Dublin Mathematical Journal*, vol. I., p. 104, *Crelle*, vol. XXX.). The next important paper was by Aronhold, 1849, (*Crelle*, vol. XXXIX., p. 140), in which the existence of the invariants S and T of a ternary cubic was demonstrated. Early in 1851 Mr. Boole reproduced, with additions, his paper on Linear Transformations (*Cambridge and Dublin Mathematical Journal*, vol. VI., p. 87), and Mr. Sylvester began his series of papers in the same Journal on the Calculus of Forms, after which discoveries followed in rapid succession. I can scarcely pretend to be able to assign to their proper authors the merits of the several steps; and, as between Messrs. Cayley and Sylvester, perhaps these gentlemen themselves, who were in constant communication with each other at the time, would now find it hard to say how much properly belongs to each. To Mr. Boole is, I believe, due the principle that in a binary quantic the operative symbols $\frac{d}{dy}$, $-\frac{d}{dx}$ may be substituted for x and y (*Cambridge and Dublin Mathematical Journal*, vol. VI., p. 95, January, 1851). The principle was extended to quantics in general by Mr. Sylvester, to whom is to be ascribed the general statement of the theory of contravariants, *Cambridge and Dublin Mathematical Journal*, (1857), vol. VI., p. 291; although particular applications of contravariants had previously been made in Geometry in the theory of Polar Reciprocals, and in the theory of ternary quadratic forms by Gauss (*Disquisitiones Arithmeticæ*, Art. 267), who gives the reciprocal under the name of the adjunctive form, and establishes its invariance under what he calls the "transformed substitution." Mr. Sylvester also remarked that we might not only replace contravariant by operative symbols, but also by the actual differentials $\frac{du}{dx}$, $\frac{du}{dy}$, &c. To Boole I would ascribe the principle (Art. 121) that invariants of emanants are covariants of the quantic 1842, *Cambridge Mathematical Journal*, vol. III., p. 110, though Boole's methods were generalized by Mr. Sylvester, *Cambridge and Dublin Mathematical Journal*, vol. VI., p. 190. Some of the first steps in the general theory of covariants may thus be ascribed to Boole, though a remarkable use of such a function had been made by Hesse in determining the points of inflexion of plane curves. I had myself been led to study the same functions both for curves and surfaces, in ignorance of what Hesse had done (*Cambridge and Dublin Mathematical Journal*, vol. II., p. 74). The discovery of evectants (Art. 130) is Hermite's, *Cambridge and Dublin Mathematical Journal*, vol. VI., p. 292. In Mr. Cayley's first paper he gave a system of partial differential equations satisfied by invariants of functions linear in any number of sets of variables. The partial differential equations (p. 113) satisfied by the invariants and covariants of binary quantics were, as far as I know, first given in print by Mr. Sylvester (*Cambridge and Dublin Mathematical Journal*, vol. VII., p. 211). Mr. Sylvester there acknowledges himself to have been indebted to an idea communicated to him in conversation by Mr. Cayley; and he also speaks of having heard it said that Aronhold was also in possession of a system of differential equations. These are not made use of in Aronhold's paper (*Crelle*, vol. XXXIX.) already referred to, but he refers, *Crelle*, vol. LXII., to a communication made by him in 1851 to the Philosophical Faculty at Königsberg, which, if it ever appeared in print, I have not seen. Very probably there may be other parts of the theory to which Aronhold may justly lay claim. After the publication in *Crelle*, vol. XXX., of Mr. Cayley's paper, in which the symbolical method of forming invariants was fully explained, Aronhold worked at the theory in Germany simultaneously with the labours of Cayley and Sylvester in England; and the mastery of the subject exhibited by his papers leads me to suppose that of some of the principles he must be able to claim independent if not prior discovery. The method in which the subject is introduced (Art. 117) is taken from his paper (*Crelle*, vol. LXII). I refer in a subsequent note to the valuable paper by Hermite (*Cambridge and Dublin Mathematical Journal*, vol. IX., p. 172) in which the theorem of reciprocity was established, which had at first suggested itself to Sylvester, but was hastily rejected by him; and in which the whole theory of quintics received important additions. Mixed concomitants are Mr. Sylvester's (*Cambridge and Dublin Mathematical Journal*, vol. VII., p. 80). The theorem, Art. 131, is Cayley's and Sylvester's. The application of symmetric functions to the invariants of binary quartics was, I believe, first made in the Appendix to my *Higher Plane Curves* (1851). The method (Art. 134) of thence finding conditions for systems of equalities between the roots is Mr. Cayley's (*Philosophical Transactions*, 1857, p. 703). With regard to the subject generally, reference must be made to the important series of papers by Mr. Sylvester, beginning in the sixth volume of the *Cambridge and Dublin Mathematical Journal;* to a series of papers on Quantics published by Mr. Cayley in the *Philosophical Transactions;* and to Aronhold's Memoir on Invariants (*Crelle*, vol. LXII). The name 'invariant,' as well as much of the rest of the nomenclature, is Mr. Sylvester's.

ON THE NUMBER OF INVARIANTS OR COVARIANTS OF A BINARY QUANTIC. (Page 116).

The following is an abridged sketch of the method pursued in Mr. Cayley's investigation (*Philosophical Transactions*, 1855, p. 101). The following illustration will shew the kind of formulæ obtained and the interpretation to be put on them. A cubic we have seen has three distinct covariants, viz. U, H, J, whose orders in the coefficients are 1, 2, 3, and degrees in the variables 3, 2, 3. To these we may add the discriminant Δ which is of the order 4 and of the degree 0 in the variables. These covariants are not independent; but H^3, J^2, and ΔU^2 are connected by a linear relation (Art. 193). Assuming then that these are the only distinct covariants, any other covariant must be of the form either $U^p H^q \Delta^r$ or $J U^p H^q \Delta^r$. The number of the covariants of the order θ of the first form is equal to the number of ways in which θ can be expressed in the form $p + 2q + 4r$; that is to say, it is the coefficient of x^θ in the expansion of $\dfrac{1}{(1-x)(1-x^2)(1-x^4)}$. In like manner, the number of covariants of the second form is the coefficient of x^θ in $x^3 \div (1-x)(1-x^2)(1-x^4)$. The total number then is the coefficient of x^θ in the expansion of $1 + x^3 \div (1-x)(1-x^2)(1-x^4)$; or, what comes to the same thing, in

$$\dfrac{1-x^6}{(1-x)(1-x^2)(1-x^3)(1-x^4)}.$$

272   NOTES.

And conversely, if this expression for the number of distinct covariants were established independently, it would indicate on inspection that there were four irreducible covariants of the orders 1, 2, 3, 4 respectively, and connected by an equation of the order 6.

Now it was proved (Art. 144) that the number of covariants of the order θ and weight q is equal to the difference of the number of ways in which q and $q-1$ can be expressed as the sum of θ numbers from 0 to n inclusive. Now the number of ways in which q may be so expressed may easily be seen to be the coefficient of $x^q z^\theta$ in the development of

$$\frac{1}{(1-z)(1-xz)(1-x^2z)\ldots(1-x^nz)},$$

where the expansion is to be effected in ascending powers of z. This will be equal to

$$1 + \frac{1-x^{n+1}}{1-x}z + \frac{(1-x^{n+1})(1-x^{n+2})}{(1-x)(1-x^2)}z^2 + \&c.,$$

the general term being $\dfrac{(1-x^{n+1})(1-x^{n+2})\ldots(1+x^{n+\theta})}{(1-x)(1-x^2)\ldots 1-x^\theta}z^\theta,$

or, what is the same thing,

$$\frac{(1-x^{\theta+1})(1-x^{\theta+2})\ldots(1-x^{n+\theta})}{(1-x)\ldots(1-x^n)}z^\theta.$$

It remains then to find the coefficient of x^q in the part multiplying z^θ. To transform this expression, the equation is used

$$(1+xz)(1+x^2z)\ldots(1+x^nz) = 1 + \frac{x(1-x^n)}{1-x}z + \frac{x^3(1-x^n)(1-x^{n-1})}{(1-x)(1-x^2)}z^2 + \&c.,$$

the general term being

$$x^{\frac{1}{2}s(s+1)}\frac{(1-x^n)(1-x^{n-1})\ldots(1-x^{n-s+1})}{(1-x)(1-x^2)\ldots(1-x^s)}z^s,$$

and the series is a finite one, the last term being that corresponding to $s = n$; viz. $x^{\frac{1}{2}n(n+1)}z^n$. Writing $-x^\theta$ for z, and substituting the resulting value of

$$(1-x^{\theta+1})(1-x^{\theta+2})\ldots(1-x^{\theta+n})$$

in the preceding formula, the number we are investigating is found to be

$$\Sigma_s (-)^s \text{ coefficient of } x^q \text{ in } \frac{x^{sn+\frac{1}{2}s(s+1)}}{(1-x)(1-x^2)\ldots(1-x^s)(1-x)(1-x^2)\ldots(1-x^{n-s})},$$

where the sum extends from $s = 0$ to $s = n$, but it is of course unnecessary to include any value of s which makes the index of s in the numerator greater than q. If we write $q = \frac{1}{2}(n\theta - a)$, the formula last written may be transformed into

$$\Sigma_s (-)^s \text{ coefficient of } x^{\frac{1}{2}(n-s)\theta} \text{ in } \frac{x^{\frac{1}{2}sa+\frac{1}{2}s(s+1)}}{(1-x)\ldots(1-x^s)(1-x)\ldots(1-x^{n-s})},$$

where the sum extends when n is even from $s = 0$ to $s = \frac{1}{2}n - 1$, and when n is odd from $s = 0$ to $s = \frac{1}{2}(n-1)$.

Thus, suppose it were required to find the number of terms in an invariant of a cubic of the order θ, we have to calculate the number of ways in which the weight $\frac{3}{2}\theta$ can be made up as the sum of θ numbers from 0 to 3 inclusive. We have then $a = 0$, and the sum consists of two terms, viz.

the coeff. of $x^{\frac{3}{2}\theta}$ in $\dfrac{1}{(1-x)(1-x^2)(1-x^3)}$ — the coeff. of $x^{\frac{1}{2}\theta}$ in $\dfrac{x}{(1-x)^2(1-x^2)}$.

The following will illustrate the process by which these are transformed. Writing x^2 for x, the first of these numbers is the coefficient of $x^{3\theta}$ in $\dfrac{1}{(1-x^2)(1-x^4)1-x^6)}$.

NOTES. 273

In order that the indices in the denominator may be divisible by 3, we multiply numerator and denominator by $\dfrac{(1-x^6)(1-x^{12})}{(1-x^2)(1-x^4)}$; and we find the number required is the coefficient of $x^{3\theta}$ in $\dfrac{1+x^2+2x^4+x^6+2x^8+x^{10}+x^{12}}{(1-x^6)^2(1-x^{12})}$.

Now we may reject all the terms in the numerator whose exponents are not divisible by 3, and then writing θ for 3θ, the first term in the sum is the coefficient of x^θ in

$$\frac{1+x^2+x^4}{(1-x^2)^2(1-x^4)}.$$

And subtracting the second term, which is the coefficient of x^θ in x^2 divided by the same denominator, the number sought is the coefficient in

$$\frac{1+x^4}{(1-x^2)^2(1-x^4)}.$$

This specimen will illustrate the process by which he has proved the number of invariants of the first six degrees to be those assigned in the preceding pages. Thus, for a sextic, the total number of distinct invariants is found to be the coefficient of x^6 in

$$\frac{(1-x)(1+x-x^3-x^4-x^5+x^7+x^8)}{(1-x)^2(1-x^3)(1-x^4)(1-x^5)}.$$

Now the second factor of the numerator is the irreducible factor of $1-x^{30}$, i.e. it is equal to $(1-x^{30})(1-x^5)(1-x^3)(1-x^2) \div (1-x^{15})(1-x^{10})(1-x^6)(1-x)$, and, substituting this value, the number becomes the coefficient of x^θ in

$$\frac{1-x^{30}}{(1-x^2)(1-x^4)(1-x^6)(1-x^{10})(1-x^{15})},$$

from which is inferred the existence of five invariants of the orders 2, 4, 6, 10, 15 respectively, but connected by a relation of the order 30; that is, the square of the last invariant is an integral function of the others. But when this process is applied to the seventh and higher degrees, the numerator can be no longer expressed as a quotient whose denominator is the product of a finite number of factors of the form $1 - x^a$. Mr. Cayley concludes, therefore, that the number of invariants for quantics of the seventh and higher degrees is infinite, and, in like manner, that the number of covariants for the fifth and higher degrees is infinite.

CANONICAL FORMS. (Page 130).

The name is Hermite's: the theory explained in this Lesson is Mr. Sylvester's, see a paper (*Philosophical Magazine*, November, 1851) published separately, with a supplement, in the same year, with the title *An Essay on Canonical Forms*.

COMBINANTS. (Page 144).

The theory of combinants is Sylvester's, *Cambridge and Dublin Mathematical Journal* (1853), vol. VIII., p. 63. In the case of the resultant of two equations it had, I think, been previously shown by Jacobi, that the resultant of $\lambda u + \mu v$, $\lambda' u + \mu' v$

was the resultant of u, v multiplied by a power of $(\lambda\mu' - \lambda'\mu)$. Mr. Sylvester's results, Arts. 182, 186, 189, are given in the *Comptes Rendus*, vol. LVIII., p. 1071. The reference to Lesson XVIII., p. 158, was made before I had decided on omitting the Lesson on the Applications of the Theory to Ternary Quantics.

APPLICATIONS TO BINARY QUANTICS. (Page 158).

The discussion in this Lesson of the quadratic, cubic and quartic, is mainly Mr. Cayley's. See his Memoirs on Quantics in the *Philosophical Transactions*, 1854. The second form of the resultant of two quadratics, p. 160, is as elsewhere stated, Dr. Boole's. The discussion of the systems of quadratic and cubic, two cubics, and two quartics, is I believe for the most part new. The form for the resultant of two cubics, p. 165, has been published by Clebsch (*Crelle*, vol. LXIV., p. 95), and was obtained by him by a different method, but had been previously in my possession by the method here given. The proof (Art. 202) that every invariant of a quartic is a rational function of S and T is slightly modified from Mr. Sylvester's (*Philosophical Magazine*, April, 1853). The theorem, p. 176, that the quartic may be reduced to its canonical form by real substitutions, is Legendre's (*Traité des Fonctions Elliptiques*, chap. II.). The canonical form of the quintic $ax^5 + by^5 + cz^5$, which so much facilitates its discussion, was given by Mr. Sylvester in his Essay on Canonical Forms, 1851. The invariants J and K were calculated by Mr. Cayley. The value of the discriminant and its resolution into the sum of products (p. 186), was given by me in 1850 (*Cambridge and Dublin Mathematical Journal*, vol. V. p. 154). Some most important steps in the theory of the quintic were made in Hermite's paper in the *Cambridge and Dublin Mathematical Journal*, 1854, vol. IX. p. 172, where the number of independent invariants was established, the invariant I [in which it may be stated the highest power of a, a^7, has for multiplier $f(df - e^2)^3$] was discovered; attention was called to the linear covariants, and the possibility demonstrated of expressing by invariants the conditions of the reality of the roots of all equations of odd degrees. The theory of the quintic was further advanced by Mr. Sylvester's "Trilogy," (*Philosophical Transactions*, 1864, p. 579); and in Hermite's series of papers in the *Comptes Rendus* for the present year (1866) already referred to. The values of the invariants A, B, C of the sextic were given by Mr. Cayley in his papers on Quantics, and the existence of the invariant E pointed out. The rest of what is stated in the text about the sextic is new.

THE QUINTIC. (Page 198).

With respect to the special form $x(x^2 - a^2)(x^2 - b^2)$ used, pp. 198, 201, I have noticed since that its characteristic is that Hermite's invariant I vanishes. This form may therefore be safely used in calculating any invariant functions whose order is divisible by 4 and is below 36, since such forms cannot contain I. For the calculation therefore at p. 201, it was sufficient to use this special form. More generally, if the alternate terms be wanting in any equation, every skew invariant vanishes. For the weight of a skew invariant is an odd number; and if the degree of the equation be odd, the order of every invariant is even. Now an odd number can neither be made up as the sum of an even number of odd numbers, nor of any number of even numbers. In the special form just referred to, x and y are the linear covariants.

What has been just stated leads to a simple proof for the expression of I in terms of the roots. It has in fact appeared that when I vanishes, one of the roots is one of the foci of the involution determined by the other two pair. I is therefore the product of the fifteen determinants of the form $\begin{vmatrix} 2\alpha - \beta - \gamma, & \alpha(\beta+\gamma) - 2\beta\gamma \\ 2\alpha - \delta - \epsilon, & \alpha(\delta+\epsilon) - 2\delta\epsilon \end{vmatrix}$, since if any of them vanish, the equation is reducible to the special form in question. And the vanishing of any of these determinants may be expressed as at p. 202,

$$(\alpha - \beta)(\alpha - \delta)(\gamma - \epsilon) + (\alpha - \gamma)(\alpha - \epsilon)(\beta - \delta) = 0.$$

Mr. Sylvester had also communicated a simple proof of the same thing depending on the fact (p. 194) that a quintic for which I vanishes, is linearly transformable into a recurring equation. In like manner, what is stated (p. 210) gives at once the expression for the skew invariant of the sextic in terms of the roots: viz. that it is the product of the fifteen determinants of the form

$$\begin{vmatrix} 1, & 1, & 1 \\ \alpha+\beta, & \gamma+\delta, & \epsilon+\phi \\ \alpha\beta, & \gamma\delta, & \epsilon\phi \end{vmatrix}.$$

Until my attention was called to it by Mr. Sylvester, I had omitted to notice (Art. 224) the use made by M. Hermite of the fact, that the quintic as well as every equation of odd degree is reducible to a *forme-type*, in which the x and y are the linear covariants and the coefficients are invariants. It follows immediately, that by applying Sturm's theorem to the *forme-type*, the conditions for reality of roots may be expressed by invariants. Hermite extends his theorem to equations of even degree above the fourth, by the method explained at the end of the next note. I think it therefore worth while now to give the coefficients of the *forme-type* of the quintic. They were given by Hermite (*Cambridge and Dublin Mathematical Journal*, vol. IX. p. 193), and re-calculated by me before I found out the key for the translation of Hermite's notation into Mr. Sylvester's, which is $\Delta = J$, $J_2 = -K$, $J_3 = JK + 9L$. I write now $J^2 - 3K = M$, $JK + 9L = N$; and Q a numerical multiple of Hermite's I, such that

$$Q^2 = JK^2M^2 - 2MNK(J^2 + 12K) + JN^2(J^2 + 72K) - 48N^3,$$

then the coefficients of the *forme-type* are

$A = QM,$

$B = JKM^2 - MN(J^2 + 18K) + 30JN^2,$

$C = Q(JM - 12N),$

$D = J^2KM^2 - JMN(J^2 + 30K) + N^2(42J^2 + 144K),$

$E = Q(J^2M - 24JN),$

$F = J^3KM^2 - J^2MN(J^2 + 42K) + N^2J(54J^2 + 288K) - 1152N^3.$

I thus find the first Sturmian constant $B^2 - AC$ to be

$$36N^2\{(MK - 5JN)^2 - 16MN^2\}.$$

The Sturmian constants being essentially unsymmetrical, there seems no reason to expect that the discussion of these forms would lead to any results of practical interest. The coefficients of the *forme-type*, as M. Hermite remarked, satisfy the relations

$$AJ^2 - 2CJ + E = 0, \quad BJ^2 - 2DJ + F = -1152N^3,$$

$$AE - 4BD + 3C^2 = -12^4N^3, \quad AF - 3BE + 2CD = 0, \quad BF - 4CE + 3D^2 = 12^4JN^3.$$

Thus then the quadratic covariant is $N^5 (x^2 - Jy^2)$; and operating with this on the quintic, we get the canonizant in the form

$$N^3 (AJ - C, BJ - D, CJ - E, DJ - F\rangle x, y)^3;$$

the coefficients inside the parentheses being all further divisible by N. Hence, we have

$$ACE + 2BCD - AD^2 - EB^2 - C^3 = -4.12^4 N^6 Q,$$

and the second Sturmian constant is got immediately by substituting the values just found for $B^2 - AC$, $AE - 4BD + 3C^2$, $ACE + 2BCD -$ &c., in the formula of Art. 225. I have not thought it worth while to calculate the third constant.

THE TSCHIRNHAUSEN TRANSFORMATION. (Page 200).

The Tschirnhausen transformation consists in taking a new variable

$$y = a + \beta x + \gamma x^2 + \ldots + \lambda x^{n-1};$$

then there are n values of y corresponding to the n values of x, and the coefficients of the new equation in y are readily found in terms of those of the given equation by the method of symmetric functions, the first for example being $as_0 + \beta s_1 + \gamma s_2 +$ &c. The coefficient of y^{n-1} is evidently a linear homogeneous function of a, β, &c., that of y^{n-2} a quadratic, of y^{n-3} a cubic function, and so on. In the case of the quintic, the transformation is $y = a + \beta x + \gamma x^2 + \delta x^3$, and we have four constants a, β, γ, δ at our disposal. Mr. Jerrard pointed out that the coefficient of y^3 being a quadratic function of a, β, γ, δ was (Art. 162) capable of being written as the algebraic sum of four squares, say $t^2 - u^2 + v^2 - w^2$. It can therefore be made to vanish, by assuming two linear relations between a, β, γ, δ; $t - u = 0$, $v - w = 0$. If we combine with these two that linear relation which makes the coefficient of y^4 vanish, we have three relations enabling us to express three of the constants a, β, γ, δ linearly in terms of the fourth. We can then by solving a cubic only make the coefficient of y^2 also vanish, or else by solving a biquadratic make the coefficient of y vanish. In this way Mr. Jerrard showed, that by the solution of equations of inferior orders, a quintic may be reduced to either of the trinomial forms $y^5 + by = c$, or $y^5 + by^2 = c$. The actual performance of the transformations would be a work of great labour, but M. Hermite showed how by somewhat altering the form of substitution, we can avail ourselves of the help of the calculus of invariants.

If we have to transform the equation $ax^n + bx^{n-1} + cx^{n-2} +$ &c., Hermite's form is to take

$$y = a\lambda + (ax + b) a + (ax^2 + bx + c) \beta + (ax^3 + bx^2 + cx + d) \gamma + \&c.,$$

then in the first place the transformed equation will be divisible by a; and secondly, if the given equation be linearly transformed, and if the corresponding substitution for the transformed equation be

$$Y = A\lambda' + (AX + B) a' + (AX^2 + BX + C) \beta' + \&c.;$$

then he has shewn that the expressions for a', β', &c. in terms of a, β, &c. involve only the coefficients of linear transformation, and not those of the given equation. It is not so with respect to the first coefficient λ, which we have therefore designated by a special letter. But the theory of linear substitutions will be directly applicable to all functions of the coefficients of the transformed equation which do not contain λ.

NOTES. 277

Such, for example, will be all symmetric functions of the differences of the roots of the new equation, since, on subtracting

$$y_1 = a\lambda + (ax_1 + b) a + \&c., \quad y_2 = a\lambda + (ax_2 + b) a + \&c.,$$

λ disappears. Or, what comes to the same thing, if we take λ such that the coefficient of y^{n-1} in the new equation shall vanish, then the theory of linear substitutions is applicable to all the coefficients of the transformed. I give Cayley's proof of Hermite's theorem, and, after his example, take, to fix the ideas, the quartic

$$(a, b, c, d, e \!)\!(x, 1)^4.$$

Then, as we have used binomial coefficients, the equation of transformation is

$$y = a\lambda + (ax + 4b) a + (ax^2 + 4bx + 6c) \beta + (ax^3 + 4bx^2 + 6cx + 4d) \gamma.$$

Adding the 4 values of y, and observing Newton's formulæ for the sums of powers of the roots, we see that the coefficient of y^{n-1} in the transformed equation will vanish if

$$a\lambda + 3ba + 3c\beta + d\gamma = 0.$$

This reduces the value of y to

$$(ax + b) a + (ax^2 + 4bx + 3c) \beta + (ax^3 + 4bx^2 + 6cx + 3d) \gamma.$$

[In general it will be observed that in this substitution all the terms have the binomial coefficients corresponding to the order of the given equation, except the terms not involving x which have the binomial coefficients answering to the order one lower.] Now what is asserted is that all the coefficients of the transformed equation will be invariants of the system

$$(a, b, c, d, e \!)\!(x, y)^4, \quad (a, \beta, \gamma \!)\!(y, -x)^3,$$

and of course if we regard y as constant, the whole transformed function will be such an invariant.

This will be proved by showing that it is made to vanish by either of the operations

$$a \frac{d}{db} + 2b \frac{d}{dc} + 3c \frac{d}{dd} + 4d \frac{d}{de} - \left(\gamma \frac{d}{d\beta} + 2\beta \frac{d}{da} \right),$$

$$4b \frac{d}{da} + 3c \frac{d}{db} + 2d \frac{d}{dc} + e \frac{d}{dd} - \left(2\beta \frac{d}{d\gamma} + a \frac{d}{d\beta} \right).$$

Let the general substitution be $y = V$, and let $V_1, V_2, \&c.$ be what V becomes when we substitute for x each of the roots of the given equation, the transformed in y is the product of the factors $y - V_1, y - V_2, \&c.$, and it is sufficient to prove that each of these factors is reduced to zero by this differentiation. We may, as in Art. 60, write the first part of the first operation $\frac{d}{d\zeta}$, and in order to calculate $\frac{dV}{d\zeta}$, we must first find $\frac{dx}{d\zeta}$. Operating on the given equation, we get

$$(a, b, c, d \!)\!(x, 1)^3 \frac{dx}{d\zeta} + (a, b, c, d \!)\!(x, 1)^3 = 0, \text{ or } \frac{dx}{d\zeta} = -1.$$

The part then of the differential of V which depends on the variation of x is

$$- \{aa + (2ax + 4b) \beta + (3ax^2 + 8bx + 6c) \gamma\},$$

and the part got by directly operating on the $a, b, \&c.$ which explicitly appear in V is

$$aa + (4ax + 6b) \beta + (4ax^2 + 12bx + 9c) \gamma.$$

Adding, we have

$$\frac{dV}{d\zeta} = 2 (ax + b) \beta + (ax^2 + 4bx + 3c) \gamma = 2\beta \frac{dV}{da} + \gamma \frac{dV}{d\beta},$$

which proves that the effect of the first operation on V is zero.

In like manner, for the second operation, we have, by operating on the original equation,

$$(a, b, c, d \!\!\not\!\!\!(x, 1)^3 \frac{dx}{d\eta} + x (b, c, d, e \!\!\not\!\!\!(x, 1)^3 = 0.$$

But the original equation may be written

$$x (a, b, c, d \!\!\not\!\!\!(x, 1)^3 + (b, c, d, e \!\!\not\!\!\!(x, 1)^3 = 0.$$

Hence $\frac{dx}{d\eta} = x^2$. The part of $\frac{dV}{d\eta}$ due to the variation of x is therefore

$$ax^2\alpha + (2ax^3 + 4bx^2) \beta + (3ax^4 + 8bx^3 + 6cx^2) \gamma.$$

The remaining part is

$$(4bx + 3c) \alpha + (4bx^2 + 12cx + 6d) \beta + (4bx^3 + 12cx^2 + 12dx + 3e) \gamma.$$

Adding, the coefficient of γ vanishes in virtue of the original equation, and the remaining part is found to be

$$\alpha \frac{dV}{d\beta} + 2\beta \frac{dV}{d\gamma},$$

which completes the proof of the theorem.

When this transformation is applied to a cubic, if we consider α, β as variables, the coefficients of the transformed equation in y will be covariants of the given equation. The transformed in fact has been calculated by Mr. Cayley, and found to be $y^3 + 3Hy + J$, where H is the Hessian $(ac - b^2) a^2 +$ &c., and J is the covariant (Art. 138), $(a^2d - 2abc + 2b^3) a^3 +$ &c.

Mr. Cayley has also calculated the result of transformation as applied to a quartic. Take the two quantics

$$(a, b, c, d, e \!\!\not\!\!\!(x, y)^4, \quad (\alpha, \beta, \gamma \!\!\not\!\!\!(y, -x)^2.$$

Let A denote the invariant got by squaring the second equation, introducing differential symbols and operating on the first, viz.

$$a\alpha^2 + 4b\alpha\beta + c (2\alpha\gamma + 4\beta^2) + 4d\beta\gamma + e\gamma^2;$$

and let B denote the invariant got by operating similarly on the Hessian of the first, viz.

$$(ac - b^2) \alpha^2 + 2 (ad - bc) \alpha\beta + (ae - 2bd + c^2) \alpha\gamma + 4 (bd - c^2) \beta^2 + 2 (be - cd) \beta\gamma + (ce - d^2) \gamma^2;$$

let C denote the result of operating with the cube of the quadratic on the covariant J (p. 173) of the quartic, viz.

$$(a^2d - 3abc + 2b^3) \alpha^3 + (a^2e + 2abd - 9ac^2 + 6b^2c) \alpha^2\beta + (abe - 3acd + 2b^2d) \alpha^2\gamma$$
$$+ (4abe - 12acd + 8b^2d) \alpha\beta^2 - 6 (ad^2 - b^2e) \alpha\beta\gamma - 4 (ad^2 - b^2e) \beta^3$$
$$- (ade - 3bce + 2bd^2) \alpha\gamma^2 - (4ade - 12bce + 8bd^2) \beta^2\gamma$$
$$- (ae^2 + 2bde - 9c^2e + 6cd^2) \beta\gamma^2 - (be^2 - 3cde + 2d^3) \gamma^3;$$

let S and T denote the two invariants of the quartic, and Δ the discriminant $\alpha\gamma - \beta^2$ of the quadratic, then the transformed equation in y is

$$y^4 + (6B - 2S\Delta) y^2 + 4Cy + SA^2 - 3B^2 + S^2\Delta^2 + 12TA\Delta + 2SB\Delta.$$

Mr. Cayley has also calculated the S and T of the transformed equation. In making the calculation, it is useful to observe that since the square of J, from which C was derived, can be expressed in terms of the other invariants, so also may the square of C; the actual expression found by him being

$$- C^2 = TA^3 - SA^2B + 4B^3 + (S^2A^2 - 12TAB - 4SB^2) \Delta + 8STA\Delta^2 + 16T^2\Delta^3.$$

The result then is that the new S is

$$SA^2 + \tfrac{4}{3}S^2\Delta^2 + 12TA\Delta,$$

and the new T is $\quad TA^3 + \tfrac{2}{3}S^2A^2\Delta + 4STA\Delta^2 + \Delta^3(16T^2 - \tfrac{8}{27}S^3).$

Finally, he has observed that these are the S and T of $AU + 4\Delta H$, as may be verified by the formulæ of Art. 211. It follows, then, that the effect of the Tschirnhausen transformation is always to change a quartic into an equation having the same invariants as one of the form $U + \lambda H$, and therefore reducible by linear transformation to the latter form. Mr. Cayley has not, as yet at least, made the corresponding calculations for the quintic. The following is the form in which M. Hermite has applied his methods to the quintic.

Let u be a quantic $(x, y)^n$; u_1, u_2 its differentials with regard to x and y; let ϕ be a covariant, which we take of the degree $n - 2$ in order that the equation we are about to use may be homogeneous in x and y; then the coefficients of the transformed equation, obtained by putting $z = \dfrac{y\phi}{u_1}$, are all invariants of u. The equation in z is got by eliminating x and y between $zu_1 - y\phi = 0$, and $u = 0$, or, what comes to the same thing, $zu_2 + x\phi = 0$, which follows from the other two. If we linearly transform x and y, the new equation in z is got, in like manner, by eliminating between $zU_1 - Y\Phi = 0$, $zU_2 + X\Phi = 0$. But, if $x = \lambda X + \mu Y$, $y = \lambda'X + \mu'Y$, $\Delta = \lambda\mu' - \lambda'\mu$, we have $\Delta X = \mu'x - \mu y$, $\Delta Y = \lambda y - \lambda'x$, and, Art. 126, $U_1 = \lambda u_1 + \lambda'u_2$, $U_2 = \mu u_1 + \mu'u_2$, and, since ϕ is a covariant, we have $\Phi = \Delta^i \phi$. Making these substitutions, the equation in z, corresponding to the transformed equation, is got by eliminating between

$$z(\lambda u_1 + \lambda' u_2) - \Delta^{i-1}\phi(\lambda y - \lambda'x) = 0, \quad z(\mu u_1 + \mu' u_2) + \Delta^{i-1}\phi(\mu'x - \mu y) = 0.$$

Multiply the first by μ', the second by λ', and subtract, and we have $\Delta z u_1 - \Delta^i y \phi = 0$. In like manner, multiplying the first by μ, the second by λ, and subtracting, we get $\Delta z u_2 + \Delta^i x \phi = 0$. In other words, we have the two original equations, except that z is replaced by $\dfrac{z}{\Delta^{i-1}}$. Consequently, the equations in z corresponding to the original equation, and to the same linearly transformed, only differ in having the powers of z multiplied by different powers of the modulus of transformation Δ, and therefore the several coefficients of the powers of z are invariants.

The actual form of the equation in z will be

$$z^n + \frac{A}{D}z^{n-2} + \frac{B}{D}z^{n-3} + \&c. = 0.$$

It is easy to see that the discriminant will appear in the denominator; and the coefficient of z^{n-1} will vanish, since, if ϕ be any function of the order $n - 2$, the sum of the results of substituting all the roots of U in $\dfrac{\phi}{U_1}$ vanishes. In fact, when the terms of this sum are brought to a common denominator, the numerator is the sum of ϕa multiplied by the differences of all the roots except a, and this is a function of the order $n - 2$ in a, which vanishes for $n - 1$ values of a, $a = \beta$, $a = \gamma$, &c., and must therefore be identically nothing.

In applying this method to the quintic $(x, 1)^5$, Hermite substitutes

$$zU_1 = a\phi_1 + \beta\phi_2 + \gamma\phi_3 + \delta\phi_4,$$

where $\phi_1, \phi_2, \phi_3, \phi_4$ are four covariant cubics of the orders 3, 5, 7, 9 respectively in the coefficients; ϕ_1 is the canonizant; ϕ_2 is the covariant cubic of the fifth order, noticed p. 191; and for the general equation, its leading term or *source* (p. 117), whence all the other terms can be derived, is

$a^2cef - 3a^2d^2f + 2a^2de^2 - ab^2ef + 14abcdf - 11abce^2 - abd^2e - 9ac^2f + 14ac^2de$
$\quad - 6acd^3 - 8b^3df + 9b^3e^2 + 6b^2c^2f - 16b^2cde + 8b^2d^3 + 3bc^3e - 2bc^2d^2.$

On inspecting this, we see that it vanishes if both a and b vanish; consequently, if the given quintic has two equal roots, their common value satisfies this covariant. We can form a covariant cubic of the seventh order from ϕ_2 in the same way that ϕ_2

was formed from ϕ_1, and by adding ϕ_1, multiplied by J and a numerical coefficient, can obtain ϕ_2, such that its source vanishes when a and b vanish; and, in like manner, ϕ_4 can be made to possess the same property. When this substitution is made, the coefficient of z^3 is a quadratic function of a, β, γ, δ. Hermite finds for its actual value (a result which may be verified by working with the special form, p. 200)

$$\{Fa^2 + 6KDa\gamma - D(F + 10JK)\gamma^2\} + D\{K\beta^2 + 2F\beta\delta - (9KD + 10AF)\delta^2\},$$

where $F = 9(16L - JK)$, and vanishes when the quintic has two distinct pairs of equal roots. By breaking up into factors each of the parts into which this coefficient has been divided, the two linear relations between $a, \gamma; \beta, \delta$, which will make it to vanish, can readily be obtained; as also by another process which I shall not delay to explain. The discussion of this coefficient is also the basis of Hermite's later investigations as to the criteria for reality of the roots. He avails himself of a principle of Jacobi's (*Crelle*, vol. L.), that if a, β, γ, &c. be the roots of a given equation, and if the quadratic function

$$(t + au + a^2v + \ldots a^{n-1}w)^2 + (t + \beta u + \beta^2 v + \&c.)^2 + \&c.,$$

be brought by real substitution to a sum of squares, the number of negative squares will be equal to the number of pairs of imaginary roots in the equation. Hermite shows, by an easy extension of this principle, that the number of pairs of imaginary roots of the quintic is found by ascertaining the number of negative squares, when the coefficient of z^3 just written is resolved into a sum of squares. And since the same process is applicable to every equation whose degree is above the fourth, he concludes that the conditions for reality of roots in every equation above the fourth can be expressed by invariants.

NOTE ON THE ORDER OF SYSTEMS OF EQUATIONS. (Page 212).

Mr. Cayley, in the *Cambridge and Dublin Mathematical Journal*, vol. IV., p. 134, determined the order of a matrix with k rows and $k + 1$ columns, in the particular case where each constituent is of the first degree. My own investigations were published, *Quarterly Journal*, vol. I., p. 246, and in the Appendix to my *Geometry of Three Dimensions*. These are, as far as I know, the only papers published on the subject of this Lesson. Since the Lesson was printed, Mr. Samuel Roberts has communicated to me some extensions of the theory there developed. His method is to suppose each quantic resolved into factors, and to deal with the combinations of the factors into which the quantics have been broken up. The method is directly applicable to binary quantics which can always be resolved into factors, and in the case of ternary and higher quantics, it would seem that the question whether or not they can be so resolved does not affect the problems here discussed, and that the orders determined in the case of quantics which are the products of factors must be generally true. Thus, to determine the order of the resultant of two binary quantics of the degrees m, n; if the order of the terms in the first be $\lambda, \lambda + a, \lambda + 2a$, &c., it may be resolved into the product of m factors $ax + by$, the orders of a and b being $\frac{\lambda}{m}, \frac{\lambda}{m} + a$ respectively; similarly, for the second quantic; and the resultant is the product of mn factors, the order of each being $\frac{\lambda}{m} + \frac{\lambda'}{n} + a$; and therefore mn times this number will be the order of the resultant. Now he argues that we may deal in the same manner

with the problem in Art. 254; that knowing, by Art. 249, the order of the matrix
$$\begin{Vmatrix} a, & b \\ a', & b' \\ a'', & b'' \end{Vmatrix}$$
to be $a^2 + (\lambda + \mu + \nu) a + \lambda\mu + \mu\nu + \nu\lambda$, the orders of the rows being supposed to be λ, $\lambda + a$; μ, $\mu + a$; ν, $\nu + a$; then we may conclude that the order of the system of conditions for the simultaneous existence of three equations of orders l, m, n is

$$lmn \left\{ a^2 + a \left(\frac{\lambda}{l} + \frac{\mu}{m} + \frac{\nu}{n} \right) + \frac{\lambda\mu}{lm} + \frac{\mu\nu}{mn} + \frac{\nu\lambda}{nl} \right\}.$$

And in like manner, that the order of conditions for the co-existence of a system of $k + 1$ binary quantics is the product of their degrees multiplied by

$$a^k + P_1 a^{k-1} + P_2 a^{k-2} + \ldots P_k,$$

where P_1, P_2, &c. are the sum, sum of products in pairs, &c. of the numbers $\frac{\lambda}{l}$, $\frac{\mu}{m}$, &c. And so more generally, the order of the conditions for the co-existence of any number of quantics in any number of variables is derived from the order determined by Art. 249 for the co-existence of a system of linear equations. He finds thus that the order of conditions for the co-existence of $k + s - 1$ homogeneous quantics in s variables, in which the order of the coefficients of x^l, $x^{l-1}y$, $x^{l-1}z$, &c. is λ, $\lambda + a$, $\lambda + \beta$, &c., is the product of their degrees multiplied by

$$p_k + p_{k-1} P + p_{k-2} Q + \&c.,$$

where p_k has the same meaning as at p. 229, and P, Q, &c. are the sum, sum of products in pairs, &c. of the numbers $\frac{\lambda}{l}$, $\frac{\mu}{m}$, &c. Thus, for instance, this formula applied to the case of ternary quantics gives the order of the conditions that a curve should have a cusp. We determine by the formula the order for the co-existence of U_1, U_2, U_3, $U_{11}U_{22} - U_{12}^2$, which system belongs either to cusps or double points on the line z, and we subtract the order for the co-existence of U_1, U_2, U_3, z which belongs to the latter. His result is

$$12 (n - 1) (n - 2) \lambda^2 + 8n (n - 1) (n - 2) (a + \beta) \lambda + 2n (n - 1) (n - 2) (n + 1) a\beta$$
$$+ 2n (n - 1)^2 (n - 2) (a^2 + \beta^2).$$

The problem of finding the order of conditions that two binary quantics should have two common roots is discussed as follows: Consider first the simpler system, formed by taking two factors from each equation, $(ax + by)(a'x + b'y)$, $(a''x + b''y)(a'''x + b'''y)$; and we have the pair of conditions $(ab'')(a'b''') = 0$, $(ab''')(a'b'') = 0$, whose order combined is $4(\lambda + \mu + a)^2$; but from this we must subtract the irrelevant systems $(ab'')(ab''')$; $(a'b'')(a'b''')$, which reduces the order to $2(\lambda + \mu + a)^2$. But if we take two factors from the first equation and one from the second, the system $(ab'') = 0$, $(a'b'') = 0$ is satisfied by $a'' = 0$, $b'' = 0$, whose order is $\mu(\mu + a)$. Now since the number of ways in which two factors of the first equation may be combined with two of the second is $\frac{1}{2}l(l-1) \times \frac{1}{2}m(m-1)$; and the number of ways in which one of the second may be combined with two of the first is $\frac{1}{2}l(l-1)m$; the resulting order in general is

$$\tfrac{1}{4}lm(l-1)(m-1)\left(\tfrac{\lambda}{l} + \tfrac{\mu}{m} + a\right)^2 + \tfrac{1}{2}lm(m-1)\tfrac{\lambda}{l}\left(\tfrac{\lambda}{l} + a\right) + \tfrac{1}{2}ml(l-1)\tfrac{\mu}{m}\left(\tfrac{\mu}{m} + a\right).$$

By the same process of reasoning he arrives at the order of the conditions (Art. 274) that three ternary quantics should have two points common, in the form

$$\{\tfrac{1}{2}lmn(l-1)(m-1)(n-1) + \Sigma \tfrac{1}{2}lmn(m-1)(n-1)\}\left\{\tfrac{\lambda}{l} + \tfrac{\mu}{m} + \tfrac{\nu}{n} + a + a'\right\}^2$$
$$+ \Sigma \tfrac{1}{2}lmn(l-1)\left\{\tfrac{\lambda^2}{l^2} + \tfrac{\mu^2}{m^2} + \tfrac{\lambda\mu}{lm} + (a + a')\left(\tfrac{\lambda}{l} + \tfrac{\mu}{m}\right) + aa'\right\}.$$

In this way the order of conditions that a curve should have two double points is found to be

$$\tfrac{1}{4}(n-1)(n-2)^2(n+1)\{3\lambda + n(a+a')\}^2$$
$$-\tfrac{1}{4}(n-1)(n-2)\{15\lambda^2 + 10n(a+a')\lambda + n(n+6)aa' + 2n(2n-3)(a^2+a'^2)\}.$$

Mr. Roberts investigates other problems by the same method; as, for instance, the order of conditions that four curves may have two points common; or that a surface may have a bi-planar double point. For these I must refer to his paper which I hope will be soon published. I only give the following result: The order of conditions that three binary quantics should have two roots common is

$$\tfrac{1}{4}lmn(l-1)(m-1)(n-1)\left\{\frac{\lambda\mu}{lm} + \frac{\mu\nu}{mn} + \frac{\nu\lambda}{nl} + a\left(\frac{\lambda}{l} + \frac{\mu}{m} + \frac{\nu}{n}\right) + a^2\right\}^2$$
$$+ \Sigma\tfrac{1}{4}lmn(m-1)(n-1)\left\{\frac{\lambda}{l}\left(\frac{\lambda}{l} + a\right)\left(\frac{\mu}{m} + \frac{\nu}{n} + a\right)^2\right\}$$
$$+ \Sigma\tfrac{1}{4}lmn(n-1)\frac{\lambda\mu}{lm}\left(\frac{\lambda}{l} + a\right)\left(\frac{\mu}{m} + a\right).$$

With regard to the other subject discussed in this Lesson, I find (see Faà de Bruno *On Elimination*, p. 94) that Bezout gave a formula for the degree of the resultant of two equations from which some terms are wanting, viz. that if m, n be the degrees of two ternary quantics, and if the highest powers of the variables x and y, which occur in the quantics, be only α, β; α', β' respectively; then the number of their common values is reduced from mn to $mn - (m-\alpha)(n-\alpha') - (m-\beta)(n-\beta')$. And the same case of quantics from which certain terms are wanting has been investigated by Minding (*Crelle*, vol. XX.) But I noticed these references too late to be able to study the papers in question, and to compare them with the theory I have given of the cases where the order of resultants falls below the ordinary number. The theory of elimination cannot be said to be perfect until rules have been given for determining in every case the exact order of the resultant.

BEZOUTIANTS. (Page 90).

It has been shown (Art. 81) that the resultant of two equations of the nth degree is expressed by Bezout's method as a symmetrical determinant. This may be considered (Art. 114) as the discriminant of a quadratic function which Mr. Sylvester has called the Bezoutiant of the system. When the quantics are the two differentials of the same quantic, then if we resolve the Bezoutiant into a sum of squares (Art. 162), the number of negative squares in this sum will indicate the number of pairs of imaginary roots in the quantic. The number of negative squares is found by adding (as in Art. 44) λ to each of the terms in the leading diagonal of the matrix of the Bezoutiant, and then determining by Des Cartes' rule the number of negative roots in the equation for λ. The result of this method is to substitute for the leading terms in Sturm's functions, terms which are symmetrical with respect to both ends of the quantic; that is to say, which do not alter when for x we substitute $\dfrac{1}{x}$ (see Mr. Sylvester's Memoir, *Philosophical Transactions*, 1853, p. 513).

(283)

TABLES.

In the preceding pages, the equations are usually written with binomial coefficients, but as in practice it is often necessary to apply formulæ to equations not so written, we give for convenience some of the principal results of elimination as applied to equations written without binomial coefficients.

(1) The resultant of the two quadratics
$(A, B, C\!\!\not\!\!(x, y)^2$, $(a, b, c\!\!\not\!\!(x, y)^2$ is $(Ac - Ca)^2 - (Ab - Ba)(Bc - Cb)$,
or $\qquad a^2C^2 - abBC + ac(B^2 - 2AC) + b^2AC - bcAB + c^2A^2$.

2. The resultant of the quadratic $(A, B, C\!\!\not\!\!(x, y)^2$ and the cubic $(a, b, c, d\!\!\not\!\!(x, y)^3$, is

$a^2C^3 - abBC^2 + acC(B^2 - 2AC) - ad(B^3 - 3ABC)$
$\qquad + b^2AC^2 - bcABC + bdA(B^2 - 2AC) + c^2A^2C - cdBA^2 + d^2A^3$.

3. The resultant of quadratic and quartic is
$a^2C^4 - abBC^3 + acC^2(B^2 - 2AC) - adC(B^3 - 3ABC)$
$+ ae(B^4 - 4B^2AC + 2A^2C^2) + b^2AC^3 - bcABC^2 + bdAC(B^2 - 2AC)$
$- beA(B^3 - 3ABC) + c^2A^2C^2 - cdA^2BC + ceA^2(B^2 - 2AC)$
$+ d^2A^3C - deA^3B + e^2A^4$.

4. The resultant of quadratic and quintic is
$a^2C^5 - abBC^4 + acC^3(B^2 - 2AC) - adC^2(B^3 - 3ABC)$
$+ aeC(B^4 - 4AB^2C + 2A^2C^2) - af(B^5 - 5B^3AC + 5A^2BC^2)$
$+ b^2AC^4 - bcABC^3 + bdAC^2(B^2 - 2AC) - beAC(B^3 - 3ABC)$
$+ bfA(B^4 - 4AB^2C + 2A^2C^2) + c^2A^2C^3 - cdA^2BC^2 + ceA^2C(B^2 - 2AC)$
$- cfA^2(B^3 - 3ABC) + d^2A^3C^2 - deA^3BC + dfA^3(B^2 - 2AC)$
$+ e^2A^4C - efBA^4 + f^2A^5$.

5. Discriminant of cubic is
$\qquad 27A^2D^2 + 4AC^3 + 4DB^3 - B^2C^2 - 18ABCD$.

6. Resultant of two cubics $(A, B, C, D\!\!\not\!\!(x, y)^3$, $(a, b, c, d\!\!\not\!\!(x, y)^3$.
The value expressed in terms of the determinants of the form $Ab - Ba$, is given in p. 63 and p. 165. Expanded it is

$a^3D^3 - a^2bCD^2 + a^2cD(C^2 - 2BD) - a^2d(C^3 - 3BCD + 3AD^2)$
$+ ab^2BD^2 - abcD(BC - 3AD) + abd(BC^2 - 2B^2D - ACD)$
$+ ac^2D(B^2 - 2AC) + acd(2AC^2 + ABD - B^2C) + ad^2(B^3 - 3ABC + 3A^2D)$
$- b^3AD^2 + b^2cACD - b^2dA(C^2 - 2BD) - bc^2ABD + bcdA(BC - 3AD)$
$- bd^2A(B^2 - 2AC) + c^3A^2D - c^2dA^2C + cd^2A^2B - d^3A^3$.

The other invariants of a system of two cubics are given (p. 165).

284   TABLES.

7. The resultant of cubic

$(A, B, C, D\)(x, y)^3$ and quartic $(a, b, c, d, e\)(x, y)^4$, is

$a^3D^4 - a^2bCD^3 + a^2cD^2(C^2 - 2BD) - a^2dD(C^3 - 3BCD + 3AD^2)$
$+ a^2e(C^4 - 4BC^2D + 2B^2D^2 + 4ACD^2) + ab^2BD^3 - abcD^2(BC - 3AD)$
$+ abdD(BC^2 - 2B^2D - ACD) - abe(BC^3 - 3B^2CD - AC^2D + 5ABD^2)$
$+ ac^2D^2(B^2 - 2AC) + acdD(2AC^2 + ABD - B^2C)$
$+ ace(B^2C^2 - 2AC^3 - 2DB^3 + 4ABCD - 3A^2D^2)$
$+ ad^2D(B^3 - 3ABC + 3A^2D) - ade(B^2C - 3ABC^2 - AB^2D + 5A^2CD)$
$+ ae^2(B^4 - 4AB^2C + 2A^2C^2 + 4A^2BD) - b^3AD^2 + b^2cACD^2$
$- b^2dAD(C^2 - 2BD) + b^2eA(C^3 - 3BCD + 3AD^2) - bc^2ABD^2$
$+ bcdAD(BC - 3AD) + bceA(2B^2D + ACD - BC^2) - bd^2AD(B^2 - 2AC)$
$+ bdeA(B^2C - 2AC^2 - ABD) - be^2A(B^3 - 3ABC + 3A^2D)$
$+ c^3A^2D^2 - c^2dA^2CD + c^2eA^2(C^2 - 2BD) + cd^2A^2BD - cdeA^2(BC - 3AD)$
$+ ce^2A^2(B^2 - 2AC) - d^3A^3D + d^2eA^3C - de^2A^3B + e^3A^4$.

8. The discriminant of a quartic written with binomial coefficients, expanded is

$a^2e^3 - 12a^2bde^2 - 18a^2c^2e^2 + 54a^2cd^2e - 27a^2d^4 + 54ab^2ce^2 - 6ab^2d^2e - 180abc^2de$
$+ 108abcd^3 + 81ac^4e - 54ac^3d^2 - 27b^4e^2 + 108b^3cde - 64b^3d^3 - 54b^2c^2e + 36b^2c^2d^2$.

9. The discriminant of a quartic written without binomial coefficients, is

$4(12ae - 3bd + c^2)^3 - (72ace + 9bcd - 27ad^2 - 27eb^2 - 2c^3)^2$,

or expanding and dividing by 27,

$256a^3e^3 - 192a^2bde^2 - 128a^2c^2e^2 + 144a^2cd^2e - 27a^2d^4 + 144ab^2ce - 6ab^2d^2e$
$- 80abc^2de + 18abcd^3 + 16ac^4e - 4ac^3d^2 - 27b^4e^2 + 18b^3cde$
$- 4b^3d^3 - 4b^2c^3e + b^2c^2d^2$.

10. The resultant of the two quartics

$(A, B, C, D, E\)(x, y)^4$, $(a, b, c, d, e\)(x, y)^4$,

expanded is (see also p. 180),

$a^4E^4 - a^3bDE^3 + a^3cE^2(D^2 - 2CE) - a^3dE(D^3 - 3CDE + 3BE^2)$
$+ a^3e(D^4 - 4CD^2E + 2C^2E^2 + 4BDE^2 - 4AE^3) + a^2b^2CE^3$
$- a^2bcE^2(CD - 3BE) + a^2bdE(CD^2 - 2C^2E - BDE + 4AE^2)$
$- a^2be(CD^3 - 3C^2DE - BD^2E + 5BCE^2 + ADE^2) + a^2c^2E^2(C^2 - 2BD)$
$- a^2cdE(C^2D - 2BD^2 - BCE + 5ADE)$
$+ a^2ce(C^2D^2 - 2BD^3 - 2C^3E + 4BCDE + 2AD^2E - 3B^2E^2 + 2ACE^2)$
$+ a^2d^2E(C^3 - 3BCD + 3AD^2 + 3B^2E - 3ACE)$
$- a^2de(C^3D - 3BCD^2 + 3AD^3 - BC^2E + 5B^2DE - 2ACDE - 5ABE^2)$

$+ a^2e^2 (C^4 - 4BC^2D + 2B^2D^2 + 4ACD^2 + 4B^2CE - 2AC^2E - 9ABDE + 4A^2E^2)$
$- ab^3BE^2 + ab^2cE^2 (BD - 4AE) - ab^2dE (BD^2 - 2BCE - ADE)$
$+ ab^2e (BD^2 - 3BCDE - AD^2E + 3B^2E^2 + 2ACE^2) - abc^2E^2 (BC - 3AD)$
$+ abcdE (BCD - 3AD^2 - 3B^2E + 4ACE)$
$- abce (BCD^2 - 3AD^3 - 2BC^2E - B^2DE + 8ACDE - 2ABE^2)$
$- abd^2E (BC^2 - 2B^2D - ACD + 5ABE)$
$+ abde (BC^2D - 2B^2D^2 - ACD^2 - B^2CE + 10ABDE - 8A^2E^2)$
$- abe^2 (BC^3 - 3B^2CD - AC^2D + 5ABD^2 + 3B^2E - 2ABCE - 5A^2DE)$
$+ ac^2E^2 (B^2 - 2AC) - ac^2dE (B^2D - 2ACD - ABE)$
$+ ac^2e (B^2D^2 - 2ACD^2 - 2B^2CE + 4AC^2E - 4A^2E^2)$
$+ acd^2E (B^2C - 2AC^2 - ABD + 4A^2E)$
$- acde (B^2CD - 2AC^2D - ABD^2 - 3B^2E + 8ABCE - 2A^2DE)$
$+ ace^2 (B^2C^2 - 2AC^3 - 2B^2D + 4ABCD - 3A^2D^2 + 2AB^2E + 2A^2CE)$
$- ad^2E (B^3 - 3ABC + 3A^2D) + ad^2e (B^3D - 3ABCD + 3A^2D^2 - AB^2E + 2A^2CE)$
$- ade^2 (B^3C - 3ABC^2 - AB^2D + 5A^2CD + A^2BE)$
$+ ae^3 (B^4 - 4AB^2C + 2A^2C^2 + 4A^2BD - 4A^2E) + b^4AE^2 - b^2cADE^2$
$+ b^2dAE (D^2 - 2CE) - b^2eA (D^3 - 3CDE + 3BE^2)$
$+ b^2c^2ACE^2 - b^2cdAE (CD - 3BE) + b^2ceA (CD^2 - 2C^2E - BDE + 4AE^2)$
$+ b^2d^2AE (C^2 - 2BD + 2AE) - b^2deA (C^2D - 2BD^2 - BCE + 5ADE)$
$+ b^2e^2A (C^3 - 3BCD + 3AD^2 + 3B^2E - 3ACE)$
$- bc^3ABE^2 + bc^2dAE (BD - 4AE) - bc^2eA (BD^2 - 2BCE - ADE)$
$- bcd^2AE (BC - 3AD) + bcdeA (BCD - 3AD^2 - 3B^2E + 4ACE)$
$- bce^2A (BC^2 - 2B^2D - ACD + 5ABE) + bd^3AE (B^2 - 2AC)$
$- bd^2eA (B^2D - 2ACD - ABE) + bde^2A (B^2C - 2AC^2 - ABD + 4A^2E)$
$- be^3A (B^3 - 3ABC + 3A^2D) + c^4A^2E^2 - c^3dA^2DE + c^3eA^2 (D^2 - 2CE)$
$+ c^2d^2A^2CE - c^2deA^2 (CD - 3BE) + c^2e^2A^2 (C^2 - 2BD)$
$- cd^3A^2BE + cd^2eA^2 (BD - 4AE) - cde^2A^2 (BC - 3AD) + ce^3A^2 (B^2 - 2AC)$
$+ d^4A^3E - d^3eA^3D + d^2e^2A^3C - de^3A^3B + e^4A^4.$

I add the following very useful tables of symmetric functions as calculated by Meyer Hirsch and verified by Mr. Cayley. The equation is supposed to be $x^n + bx^{n-1} + cx^{n-2} + $ &c. $= 0$.

I. $\Sigma a \quad = -b.$

II. $\Sigma a^2 \quad = b^2 - 2c; \quad \Sigma a\beta = c.$

III. $\Sigma a^3 \quad = -b^3 + 3bc - 3d; \quad \Sigma a^2\beta = -bc + 3d; \quad \Sigma a\beta\gamma = -d.$

IV. $\Sigma a^4 \quad = b^4 - 4b^2c + 2c^2 + 4bd - 4e; \quad \Sigma a^3\beta = b^2c - 2c^2 - bd + 4e.$
$\Sigma a^2\beta^2 = c^2 - 2bd + 2e; \quad \Sigma a^2\beta\gamma = bd - 4e; \quad \Sigma a\beta\gamma\delta = e.$

286 TABLES.

V. $\quad \Sigma a^5 \quad = - b^5 + 5b^3c - 5bc^2 - 5b^2d + 5cd + 5be - 5f.$

$\Sigma a^4\beta \quad = - b^3c + 3bc^2 + b^2d - 5cd - be + 5f.$

$\Sigma a^3\beta^2 \quad = - bc^2 + 2b^2d + cd - 5be + 5f.$

$\Sigma a^3\beta\gamma \quad = - b^2d + 2cd + be - 5f.$

$\Sigma a^2\beta^2\gamma = - cd + 3be - 5f.$

$\Sigma a^2\beta\gamma\delta = - be + 5f; \quad \Sigma a\beta\gamma\delta\epsilon = -f.$

VI. $\quad \Sigma a^6 \quad = b^6 - 6b^4c + 9b^2c^2 - 2c^3 + 6b^3d - 12bcd + 3d^2 - 6b^2e + 6ce + 6bf - 6g.$

$\Sigma a^5\beta \quad = b^4c - 4b^2c^2 + 2c^3 - b^3d + 7bcd - 3d^2 + b^2e - 6ce - bf + 6g.$

$\Sigma a^4\beta^2 \quad = b^3c^2 - 2c^3 - 2b^2d + 4bcd - 3d^2 + 2b^2e + 2ce - 6bf + 6g.$

$\Sigma a^3\beta^3 \quad = c^3 - 3bcd + 3d^2 + 3b^2e - 3ce - 3bf + 3g.$

$\Sigma a^4\beta\gamma \quad = b^2d - 3bcd + 3d^2 - b^2e + 2ce + bf - 6g.$

$\Sigma a^3\beta^2\gamma \quad = bcd - 3d^2 - 3b^2e + 4ce + 7bf - 12g.$

$\Sigma a^2\beta^2\gamma^2 = d^2 - 2ce + 2bf - 2g.$

$\Sigma a^3\beta\gamma\delta = b^2e - 2ce - bf + 6g.$

$\Sigma a^2\beta^2\gamma\delta = ce - 4bf + 9g.$

$\Sigma a^2\beta\gamma\delta\epsilon = bf - 6g.$

$\Sigma a\beta\gamma\delta\epsilon\zeta = g.$

VII. $\quad \Sigma a^7 \quad = - b^7 + 7b^5c - 14b^3c^2 + 7bc^3 - 7b^4d + 21b^2cd - 7c^2d - 7bd^2$
$\qquad + 7b^3e - 14bce + 7de - 7b^2f + 7cf + 7bg - 7h.$

$\Sigma a^6\beta \quad = - b^5c + 5b^3c^2 - 5bc^3 + b^3d - 9b^2cd + 7c^2d + 4bd^2 - b^2e + 8bce$
$\qquad - 7de + b^2f - 7cf - bg + 7h.$

$\Sigma a^5\beta^2 \quad = - b^3c^2 + 3bc^3 + 2b^2d - 6b^2cd - 3c^2d + 7bd^2 - 2b^2e + 4bce$
$\qquad - 7de + 2b^2f + 3cf - 7bg + 7h.$

$\Sigma a^4\beta^3 \quad = - bc^3 + 3b^2cd + c^2d - 5bd^2 - 3b^2e + 2bce + 5de + 7b^2f - 7cf$
$\qquad - 7bg + 7h.$

$\Sigma a^5\beta\gamma \quad = - b^4d + 4b^2cd - 2c^2d - 4bd^2 + b^2e - 3bce + 7de - b^2f + 2cf + bg - 7h.$

$\Sigma a^4\beta^2\gamma = - b^2cd + 2c^2d + bd^2 + 3b^2e - 8bce + 2de - 3b^2f + 4cf + 8bg - 14h.$

$\Sigma a^3\beta^3\gamma = - c^2d + 2bd^2 + bce - 5de - 4b^2f + 7cf + 4bg - 7h.$

$\Sigma a^3\beta^2\gamma^2 = - bd^2 + 2bce + de - 2b^2f - 3cf + 7bg - 7h.$

$\Sigma a^4\beta\gamma\delta = - b^3e + 3bce - 3de + b^2f - 2cf - bg + 7h.$

$\Sigma a^3\beta^2\gamma\delta = - bce + 3de + 4b^2f - 6cf - 9bg + 21h.$

$\Sigma a^2\beta^2\gamma^2\delta = - de + 3cf - 5bg + 7h.$

$\Sigma a^3\beta\gamma\delta\epsilon = - b^2f + 2cf + bg - 7h.$

$\Sigma a^2\beta^2\gamma\delta\epsilon = - cf + 5bg - 14h.$

$\Sigma a^2\beta\gamma\delta\epsilon\zeta = - bg + 7h.$

$\Sigma a\beta\gamma\delta\epsilon\zeta\eta = - h.$

TABLES. 287

VIII. Σa^9 $= b^9 - 8b^7c + 20b^6c^2 - 16b^5c^3 + 2c^4 + 8b^5d - 32b^3cd + 24bc^2d$
$+ 12b^4d^2 - 8cd^2 - 8b^4e + 24b^2ce - 8c^2e - 16bde + 4e^2 + 8b^3f$
$- 16bcf + 8df - 8b^2g + 8cg + 8bh - 8i.$

$\Sigma a^7\beta$ $= b^8c - 6b^6c^2 + 9b^4c^3 - 2c^4 - b^6d + 11b^4cd - 17bc^2d - 5b^3d^2$
$+ 8cd^2 + b^4e - 10b^2ce + 8c^2e + 9bde - 4e^2 - b^3f + 9bcf$
$- 8df + b^2g - 8cg - bh + 8i.$

$\Sigma a^6\beta^2$ $= b^6c^2 - 4b^4c^3 + 2c^4 - 2b^5d + 8b^3cd - 9b^2d^2 + 2cd^2 + 2b^4e$
$- 6b^2ce - 4c^2e + 16bde - 4e^2 - 2b^3f + 4bcf - 8df + 2b^2g$
$+ 4cg - 8bh + 8i.$

$\Sigma a^5\beta^3$ $= b^5c^3 - 2c^4 - 3b^5cd + 6bc^2d + 3b^3d^2 - 7cd^2 + 3b^4e - 9b^2ce + 8c^2e$
$+ bde - 4e^2 - 3b^3f + bcf + 7df + 8b^2g - 8cg - 8bh + 8i.$

$\Sigma a^4\beta^4$ $= c^4 - 4bc^2d + 2b^3d^2 + 4cd^2 + 4b^2ce - 4c^2e - 8bde + 6e^2 - 4b^3f$
$+ 8bcf - 4df + 4b^2g - 4cg - 4bh + 4i.$

$\Sigma a^6\beta\gamma$ $= b^5d - 5b^3cd + 5bc^2d + 5b^2d^2 - 5cd^2 - b^4e + 4b^2ce - 2c^2e - 9bde$
$+ 4e^2 + b^3f - 3bcf + 8df - b^2g + 2cg + bh - 8i.$

$\Sigma a^5\beta^2\gamma$ $= b^4cd - 3bc^2d - b^2d^2 + 5cd^2 - 3b^3e + 11b^2ce - 4c^2e - 10bde$
$+ 8e^2 + 3b^3f - 8bcf + df - 3b^2g + 4cg + 9bh - 16i.$

$\Sigma a^4\beta^3\gamma$ $= bc^2d - 2b^2d^2 - cd^2 - b^2ce + 10bde - 8e^2 + 4b^3f - 10bcf$
$+ df - 9b^2g + 16cg + 9bh - 16i.$

$\Sigma a^4\beta^2\gamma^2$ $= b^2d^2 - 2cd^2 - 2b^2ce + 4c^2e - 4e^2 + 2b^3f - 4bcf + 8df$
$- 2b^2g - 4cg + 8bh - 8i.$

$\Sigma a^3\beta^3\gamma^2$ $= cd^2 - 2c^2e - bde + 4e^2 + 5bcf - 7df - 5b^2g + 2cg + 8bh - 8i.$

$\Sigma a^5\beta\gamma\delta$ $= b^4e - 4b^2ce + 2c^2e + 4bde - 4e^2 - b^3f + 3bcf - 3df + b^2g$
$- 2cg - bh + 8i.$

$\Sigma a^4\beta^2\gamma\delta$ $= b^2ce - 2c^2e - bde + 4e^2 - 4b^3f + 11bcf - 9df + 4b^2g - 6cg$
$- 10bh + 24i.$

$\Sigma a^3\beta^3\gamma\delta$ $= c^2e - 2bde + 2e^2 - bcf + 3df + 5b^2g - 9cg - 5bh + 12i.$

$\Sigma a^3\beta^2\gamma^2\delta$ $= bde - 4e^2 - 3bcf + 6df + 5b^2g - 17bh + 24i.$

$\Sigma a^2\beta^2\gamma^2\delta^2$ $= e^2 - 2df + 2cg - 2bh + 2i.$

$\Sigma a^4\beta\gamma\delta\varepsilon$ $= b^3f - 3bcf + 3df - b^2g + 2cg + bh - 8i.$

$\Sigma a^3\beta^2\gamma\delta\varepsilon$ $= bcf - 3df - 5b^2g + 8cg + 11bh - 32i.$

$\Sigma a^2\beta^2\gamma^2\delta\varepsilon$ $= df - 4cg + 9bh - 16i.$

$\Sigma a^3\beta\gamma\delta\varepsilon\zeta$ $= b^2g - 2cg - bh + 8i.$

$\Sigma a^2\beta^2\gamma\delta\varepsilon\zeta$ $= cg - 6bh + 20i.$

$\Sigma a^2\beta\gamma\delta\varepsilon\zeta\eta$ $= bh - 8i.$

$\Sigma a\beta\gamma\delta\varepsilon\zeta\eta\theta = i.$

288 TABLES.

IX. Σa^9 $= -b^9 + 9b^7c - 27b^5c^2 + 30b^3c^3 - 9bc^4 - 9b^4d + 45b^2cd - 54b^2c^2d$
$+ 9c^3d - 18b^3d^2 + 27bcd^2 - 3d^3 + 9b^4e - 36b^2ce + 27bc^2e$
$+ 27b^2de - 18cde - 9be^2 - 9b^4f + 27b^2cf - 9c^2f - 18bdf$
$+ 9ef + 9b^3g - 18bcg + 9dg - 9b^2h + 9ch + 9bi - 9j.$

$\Sigma a^8\beta$ $= -b^7c + 7b^5c^2 - 14b^3c^3 + 7bc^4 + b^6d - 13b^4cd + 30b^2c^2d$
$- 9c^3d + 6b^3d^2 - 19bcd^2 + 3d^3 - b^5e + 12b^3ce - 19bc^2e$
$- 11b^2de + 18cde + 5be^2 + b^4f - 11b^2cf + 9c^2f + 10bdf$
$- 9ef - b^3g + 10bcg - 9dg + b^2h - 9ch - bi + 9j.$

$\Sigma a^7\beta^2$ $= -b^5c^3 + 5b^3c^3 - 5bc^4 + 2b^4d - 10b^4cd + 5b^2c^2d + 5c^3d$
$+ 11b^3d^2 - 13bcd^2 + 3d^3 - 2b^3e + 8b^3ce + bc^2e - 20b^2de$
$+ 4cde + 9be^2 + 2b^4f - 6b^2cf - 5c^2f + 18bdf - 9ef$
$- 2b^3g + 4bcg - 9dg + 2b^2h + 5ch - 9bi + 9j.$

$\Sigma a^6\beta^3$ $= -b^5c^3 + 3bc^4 + 3b^4cd - 9b^2c^2d - 3c^3d - 3b^3d^2 + 18bcd^2 - 6d^3$
$- 3b^3e + 12b^3ce - 9bc^2e - 9b^2de + 9be^2 + 3b^2f - 9b^2cf$
$+ 9c^2f - 9ef - 3b^3g + 9dg + 9b^2h - 9ch - 9bi + 9j.$

$\Sigma a^5\beta^4$ $= -bc^4 + 4b^2c^2d + c^3d - 2b^3d^2 - 7bcd^2 + 3d^3 - 4b^3ce + 3bc^2e$
$+ 13b^2de - 2cde - 11be^2 + 4b^4f - 7b^2cf - c^2f - 2bdf$
$+ 11ef - 9b^3g + 18bcg - 9dg + 9b^2h - 9ch - 9bi + 9j.$

$\Sigma a^7\beta\gamma$ $= -b^5d + 6b^4cd - 9b^2c^2d + 2c^3d - 6b^3d^2 + 12bcd^2 - 3d^3 + b^5e$
$- 5b^3ce + 5bc^2e + 11b^2de - 11cde - 5be^2 - b^4f + 4b^2cf$
$- 2c^2f - 10bdf + 9ef + b^3g - 3bcg + 9dg - b^2h + 2ch$
$+ bi - 9j.$

$\Sigma a^6\beta^2\gamma$ $= -b^4cd + 4b^3c^2d - 2c^3d + b^3d^2 - 7bcd^2 + 3d^3 + 3b^3e - 14b^3ce$
$+ 12bc^2e + 13b^2de - 4cde - 14be^2 - 3b^4f + 11b^2cf - 4c^2f$
$- 10bdf + 18ef + 3b^3g - 8bcg - 3b^2h + 4ch + 10bi - 18j.$

$\Sigma a^5\beta^3\gamma$ $= -b^3c^2d + 2c^3d + 2b^3d^2 - 4bcd^2 + 3d^3 + b^3ce - 2bc^2e - 5b^3de$
$+ 2cde + 6be^2 - 4b^4f + 15b^2cf - 8c^2f - 5bdf - 2ef$
$+ 4b^3g - 10bcg - 10b^2h + 18ch + 10bi - 18j.$

$\Sigma a^4\beta^4\gamma$ $= -c^3d + 3bcd^2 - 3d^3 + bc^2e - 5b^2de + 2cde + 5be^2 - b^2cf$
$+ c^2f + 6bdf - 11ef + 5b^3g - 14bcg + 9dg - 5b^2h$
$+ 9ch + 5bi - 9j.$

$\Sigma a^5\beta^2\gamma^2$ $= -b^3d^2 + 3bcd^2 - 3d^3 + 2b^3ce - 6bc^2e + 6cde + be^2 - 2b^4f + 6b^2cf$
$- 8bdf - ef + 2b^3g - 4bcg + 9dg - 2b^2h - 5bh + 9bi - 9j.$

$\Sigma a^4\beta^3\gamma^2$ $= -bcd^2 + 3d^3 + 2bc^2e + b^2de - 8cde + 2be^2 - 5b^2cf + 6c^2f$
$+ 2bdf - 2ef + 5b^3g - 4bcg - 11b^2h + 4ch + 18bi - 18j.$

$\Sigma a^3\beta^3\gamma^3$ $= -d^3 + 3cde - 3be^2 - 3c^2f + 3bdf + 3ef + 3bcg - 6dg$
$- 3b^2h + 3ch + 3bi - 3j.$

$\Sigma a^6\beta\gamma\delta$ $= -b^3e + 5b^3ce - 5bc^2e - 5b^2de + 5cde + 5be^2 + b^3f - 4b^2cf + 2c^2f$
$+ 4bdf - 9ef - b^3g + 3bcg - 3dg + b^2h - 2ch - bi + 9j.$

TABLES. 289

$\Sigma a^5 \beta^2 \gamma \delta$ $= -b^5ce + 3bc^2e + b^4de - 5cde - be^2 + 4b^3f - 15b^2cf + 6c^2f + 15bdf$
$- 7ef - 4b^3g + 11bcg - 9dg + 4b^2h - 6ch - 11bi + 27j.$

$\Sigma a^4 \beta^3 \gamma \delta$ $= -bc^3e + 2b^3de + cde - 5be^2 + b^2cf - 5bdf + 13ef - 5b^3g$
$+ 13bcg - 9dg + 11b^2h - 20ch - 11bi + 27j.$

$\Sigma a^4 \beta^2 \gamma^2 \delta$ $= -b^3de + 2cde + be^2 + 3b^2cf - 6c^2f - 2bdf + 3ef - 5b^3g$
$+ 12bcg - 9dg + 5b^2h + ch - 19bi + 27j.$

$\Sigma a^3 \beta^3 \gamma^2 \delta$ $= -cde + 3be^2 + 3c^2f - 4bdf - 7ef - 7bcg + 18dg + 12b^2h$
$- 13ch - 19bi + 27j.$

$\Sigma a^3 \beta^2 \gamma^2 \delta^2$ $= -be^2 + 2bdf + ef - 2bcg - 3dg + 2b^2h + 5ch - 9bi + 9j.$

$\Sigma a^5 \beta \gamma \delta \varepsilon$ $= -b^4f + 4b^2cf - 2c^2f - 4bdf + 4ef + b^3g - 3bcg + 3dg - b^2h$
$+ 2ch + bi - 9j.$

$\Sigma a^4 \beta^2 \gamma \delta \varepsilon$ $= -b^2cf + 2c^2f + bdf - 4ef + 5b^3g - 14bcg + 12dg - 5b^2h$
$+ 8ch + 12bi - 36j.$

$\Sigma a^3 \beta^3 \gamma \delta \varepsilon$ $= -c^2f + 2bdf - 2ef + bcg - 3dg - 6b^2h + 11ch + 6bi - 18j.$

$\Sigma a^3 \beta^2 \gamma^2 \delta \varepsilon$ $= -bdf + 4ef + 4bcg - 9dg - 9b^2h + 5ch + 30bi - 54j.$

$\Sigma a^2 \beta^2 \gamma^2 \delta^2 \varepsilon$ $= -ef + 3dg - 5ch + 7bi - 9j.$

$\Sigma a^4 \beta \gamma \delta \varepsilon \zeta$ $= -b^3g + 3bcg - 3dg + b^2h - 2ch - bi + 9j.$

$\Sigma a^3 \beta^2 \gamma \delta \varepsilon \zeta$ $= -bcg + 3dg + 6b^2h - 10ch - 13bi + 45j.$

$\Sigma a^2 \beta^2 \gamma^2 \delta \varepsilon \zeta$ $= -dg + 5ch - 14bi + 30j.$

$\Sigma a^3 \beta \gamma \delta \varepsilon \zeta \eta$ $= -b^2h + 2ch + bi - 9j.$

$\Sigma a^2 \beta^2 \gamma \delta \varepsilon \zeta \eta$ $= -ch + 7bi - 27j.$

$\Sigma a^2 \beta \gamma \delta \varepsilon \zeta \eta \theta$ $= -bi + 9j.$

X. Σa^{10} $= b^{10} - 10b^8c + 35b^6c^2 - 50b^4c^3 + 25b^2c^4 - 2c^5 + 10b^7d - 60b^5cd$
$+ 100b^3c^2d - 40bc^3d + 25b^2d^2 - 60b^2cd^2 + 15c^2d^2 + 10bd^3$
$- 10b^6e + 50b^4ce - 60b^2c^2e + 10c^3e - 40b^3de + 60bcde$
$- 10d^2e + 15b^2e^2 - 10ce^2 + 10b^5f - 40b^3cf + 30bc^2f$
$+ 30b^2df - 20cdf - 20bef + 5f^2 - 10b^4g + 30b^2cg$
$- 10c^2g - 20bdg + 10eg + 10b^3h - 20bch + 10dh - 10b^2i$
$+ 10ci + 10bj - 10k.$

$\Sigma a^9 \beta$ $= b^8c - 8b^6c^2 + 20b^4c^3 - 16b^2c^4 + 2c^5 - b^7d + 15b^5cd - 46b^3c^2d$
$+ 31bc^3d - 7b^4d^2 + 33b^2cd^2 - 15c^2d^2 - 7bd^3 + b^6e$
$- 14b^4ce + 33b^2c^2e - 10c^3e + 13b^3de - 42bcde + 10d^2e$
$- 6b^2e^2 + 10ce^2 - b^5f + 13b^3cf - 21bc^2f - 12b^2df + 20cdf$
$+ 11bef - 5f^2 + b^4g - 12b^2cg + 10c^2g + 11bdg - 10eg$
$- b^3h + 11bch - 10dh + b^2i - 10ci - bj + 10k.$

P P

290   TABLES.

$\Sigma_{\alpha^6 \beta^2}$ $= b^6c^2 - 6b^5c^3 + 9b^4c^4 - 2c^5 - 2b^7d + 12b^5cd - 12b^3c^3d - 8bc^4d$
$\quad - 13b^6d^2 + 28b^4cd^2 + c^3d^2 - 10bd^3 + 2b^5e - 10b^4ce$
$\quad + 4b^2c^3e + 6c^5e + 24b^3de - 28bcde + 10d^2e - 11b^4e^2$
$\quad + 2ce^2 - 2b^5f + 8b^3cf + 2bc^2f - 22b^2df + 4cdf + 20bef$
$\quad - 5f^2 + 2b^4g - 6b^2cg - 6c^2g + 20bdg - 10eg - 2b^3h$
$\quad + 4bch - 10dh + 2b^2i + 6ci - 10bj + 10k.$

$\Sigma_{\alpha^7 \beta^3}$ $= b^4c^3 - 4b^5c^4 + 2c^5 - 3b^5cd + 12b^3c^3d - 2bc^4d + 3b^4d^2 - 24b^2cd^2$
$\quad + 6c^2d^2 + 11bd^3 + 3b^5e - 15b^4ce + 18b^2c^3e - 10c^5e + 12b^3de$
$\quad + 3bcde - 11d^2e - 15b^3e^2 + 10ce^2 - 3b^3f + 12b^2cf - 9bc^2f$
$\quad - 9b^3df - cdf + 20bef - 5f^2 + 3b^4g - 9b^2cg + 10c^2g - bdg$
$\quad - 10eg - 3b^3h - bch + 11dh + 10b^2i - 10ci - 10bj + 10k.$

$\Sigma_{\alpha^6 \beta^4}$ $= b^4c^4 - 2c^5 - 4b^5c^3d + 8bc^4d + 2b^4d^2 - 9c^2d^2 + 2bd^3 + 4b^4ce$
$\quad - 12b^2c^3e + 10c^5e - 8b^3de + 12bcde - 2d^2e + 9b^4e^2 - 14ce^2$
$\quad - 4b^3f + 16b^3cf - 18bc^2f - 6b^2df + 20cdf - 4bef - 5f^2$
$\quad + 4b^4g - 6b^2cg - 2c^2g - 4bdg + 14eg - 10b^3h + 20bch$
$\quad - 10dh + 10b^2i - 10ci - 10bj + 10k.$

$\Sigma_{\alpha^5 \beta^5}$ $= c^5 - 5bc^3d + 5b^3cd^2 + 5c^2d^2 - 5bd^3 + 5b^3c^3e - 5c^5e - 5b^4de$
$\quad - 5bcde + 5d^2e + 5b^3e^2 + 5ce^2 - 5b^3cf + 10bc^2f + 10b^2df$
$\quad - 15cdf - 15bef + 10f^2 + 5b^4g - 15b^2cg + 5c^2g + 10bdg$
$\quad - 5eg - 5b^3h + 10bch - 5dh + 5b^2i - 5ci - 5bj + 5k.$

$\Sigma_{\alpha^6 \beta \gamma}$ $= b^7d - 7b^5cd + 14b^3c^3d - 7bc^3d + 7b^4d^2 - 21b^2cd^2 + 7c^2d^2$
$\quad + 7bd^3 - b^5e + 6b^4ce - 9b^2c^3e + 2c^5e - 13b^3de + 26bcde$
$\quad - 10d^2e + 6b^3e^2 - 6ce^2 + b^5f - 5b^3cf + 5bc^2f + 12b^2df$
$\quad - 12cdf - 11bef + 5f^2 - b^4g + 4b^2cg - 2c^2g - 11bdg + 10eg$
$\quad + b^3h - 3bch + 10dh - b^2i + 2ci + bj - 10k.$

$\Sigma_{\alpha^7 \beta^2 \gamma}$ $= b^5cd - 5b^3c^3d + 5bc^3d - b^4d^2 + 9b^2cd^2 - 7c^2d^2 - 4bd^3 - 3b^4e$
$\quad + 17b^4ce - 23b^2c^3e + 4c^5e - 16b^3de + 21bcde + d^2e + 17b^3e^2$
$\quad - 12ce^2 + 3b^3f - 14b^3cf + 12bc^2f + 13b^3df - 3cdf - 31bef$
$\quad + 10f^2 - 3b^4g + 11b^2cg - 4c^2g - 10bdg + 20eg + 3b^3h$
$\quad - 8bch - dh - 3b^2i + 4ci + 11bj - 20k.$

$\Sigma_{\alpha^6 \beta^3 \gamma}$ $= b^3c^3d - 3bc^3d - 2b^4d^2 + 6b^2cd^2 + 3c^2d^2 - 7bd^3 - b^4ce + 3b^2c^3e$
$\quad + 5b^3de - 15bcde + 13d^2e - 3b^3e^2 + 4ce^2 + 4b^3f - 19b^3cf$
$\quad + 18bc^2f + 15b^2df - 19cdf - 7bef + 10f^2 - 4b^4g + 15b^2cg$
$\quad - 8c^2g - 4bdg - 4eg + 4b^3h - 10bch - dh - 11b^2i + 20ci$
$\quad + 11bj - 20k.$

$\Sigma_{\alpha^5 \beta^4 \gamma}$ $= bc^3d - 3b^3cd^2 - c^2d^2 + 5bd^3 - b^3c^3e + 5b^3de - 8d^2e - 8b^3e^2$
$\quad + 4ce^2 + b^3cf - bc^2f - 12b^2df + 10cdf + 23bef - 15f^2$
$\quad - 5b^4g + 18b^2cg - 8c^2g - 7bdg - 4eg + 11b^3h - 31bch$
$\quad + 20dh - 11b^2i + 20ci + 11bj - 20k.$

TABLES. 291

$\Sigma a^6 \beta^2 \gamma^2$ = $b^4 d^2 - 4b^5 cd^2 + 2c^5 d^3 + 4bd^5 - 2b^4 ce + 8b^3 c^2 e - 4c^3 e - 8bcde$
$\qquad - 4d^2 e - b^3 e^2 + 10ce^2 + 2b^2 f - 8b^3 cf + 4bc^3 f + 10b^2 df$
$\qquad - 4cdf - 8bef + 5f^2 - 2b^4 g + 6b^2 cg - 8bdg - 2eg + 2b^3 h$
$\qquad - 4bch + 10dh - 2b^4 i - 6ci + 10bj - 10k.$

$\Sigma a^5 \beta^3 \gamma^2$ = $b^3 cd^2 - 2c^2 d^2 - bd^3 - 2b^3 c^2 e + 4c^3 e - b^3 de + 5bcde + d^2 e + b^3 e^2$
$\qquad - 12ce^2 + 5b^3 cf - 13bc^2 f - 4b^3 df + 17cdf + 10bef - 15f^2$
$\qquad - 5b^3 g + 15b^2 cg - 4c^3 g - 19bdg + 20eg + 5b^3 h - 3bch$
$\qquad - dh - 12b^3 i + 4ci + 20bj - 20k.$

$\Sigma a^4 \beta^4 \gamma^2$ = $c^3 d^2 - 2bd^3 - 2c^3 e + 4bcde + 2d^2 e - 3b^2 e^2 + 2ce^2 + 2bc^2 f$
$\qquad + 2b^3 df + 12edf + 4bef + 5f^2 - 6b^2 cg + 10c^2 g + 4bdg$
$\qquad - 14eg + 6b^3 h - 12bch + 10dh - 6b^2 i + 2ci + 10bj - 10k.$

$\Sigma a^4 \beta^3 \gamma^3$ = $bd^3 - 3bcde - d^2 e + 3b^2 e^2 + 2ce^2 + 3bc^2 f - 3b^2 df + cdf - 8bef$
$\qquad + 5f^2 - 3b^2 cg - 4c^2 g + 13bdg - 2eg + 3b^3 h + bch - 11dh$
$\qquad - 10b^2 i + 10ci + 10bj - 10k.$

$\Sigma a^7 \beta \gamma \delta$ = $b^5 e - 6b^4 ce + 9b^3 c^2 e - 2c^5 e + 6b^3 de - 12bcde + 3d^2 e - 6b^2 e^2$
$\qquad + 6ce^2 - b^5 f + 5b^3 cf - 5bc^2 f - 5b^3 df + 5cdf + 11bef - 5f^2$
$\qquad + b^4 g - 4b^2 cg + 2c^2 g + 4bdg - 10eg - b^3 h + 3bch - 3dh$
$\qquad + b^2 i - 2ci - bj + 10k.$

$\Sigma a^6 \beta^2 \gamma \delta$ = $b^4 ce - 4b^3 c^2 e + 2c^3 e - b^3 de + 7bcde - 3d^2 e + b^2 e^2 - 6ce^2 - 4b^3 f$
$\qquad + 19b^3 cf - 17bc^2 f - 19b^2 df + 15cdf + 18bef - 15f^2 + 4b^4 g$
$\qquad - 15b^2 cg + 6c^2 g + 15bdg - 6eg - 4b^3 h + 11bch - 9dh$
$\qquad + 4b^2 i - 6ci - 12bj + 30k.$

$\Sigma a^5 \beta^3 \gamma \delta$ = $b^3 c^2 e - 2c^3 e - 2b^3 de + 4bcde - 3d^2 e + 2b^2 e^2 + 2ce^2 - b^3 cf$
$\qquad + 2bc^2 f + 3b^2 df - 4cdf - 12bef + 10f^2 + 5b^4 g - 19b^2 cg$
$\qquad + 10c^2 g + 15bdg - 6eg - 5b^3 h + 13bch - 9dh + 12b^2 i$
$\qquad - 22ci - 12bj + 30k.$

$\Sigma a^4 \beta^4 \gamma \delta$ = $c^3 e - 3bcde + 3d^2 e + 3b^2 e^2 - 3ce^2 - bc^2 f + 2b^2 df + cdf - 8bef$
$\qquad + 5f^2 + b^2 cg - c^2 g - 3bdg + 9eg - 6b^3 h + 17bch - 15dh$
$\qquad + 6b^2 i - 11ci - 6bj + 15k.$

$\Sigma a^5 \beta^2 \gamma^2 \delta$ = $b^3 de - 3bcde + 3d^2 e - b^2 e^2 + 2ce^2 - 3b^3 cf + 9bc^2 f + 2b^3 df$
$\qquad - 13cdf - bef + 10f^2 + 5b^4 g - 17b^2 cg + 4c^2 g + 18bdg$
$\qquad - 18eg - 5b^3 h + 12bch - 9dh + 5b^2 i + 2ci - 21bj + 30k.$

$\Sigma a^4 \beta^3 \gamma^2 \delta$ = $bcde - 3d^2 e - 3b^2 e^2 + 4ce^2 - 3bc^2 f + 4b^2 df + 5cdf - 5f^2$
$\qquad + 7b^2 cg - 8c^2 g - 15bdg + 12eg - 12b^3 h + 21bch + 3dh$
$\qquad + 26b^2 i - 28ci - 42bj + 60k.$

$\Sigma a^3 \beta^3 \gamma^3 \delta$ = $d^2 e - 2ce^2 - cdf + 5bef - 5f^2 + 4c^2 g - 7bdg + 2eg - 4bch$
$\qquad + 11dh + 7b^2 i - 10ci - 7bj + 10k.$

$\Sigma a^4 \beta^2 \gamma^2 \delta^2$ = $b^2 e^2 - 2ce^2 - 2b^2 df + 4cdf - 5f^2 + 2b^2 cg - 4c^2 g + 10eg - 2b^3 h$
$\qquad + 4bch - 10dh + 2b^2 i + 6ci - 10bj + 10k.$

292   TABLES.

$\Sigma a^3\beta^2\gamma^2\delta^2 = ce^2 - 2cdf - bef + 5f^2 + 2c^2g + 3bdg - 9eg - 7bch + 6dh$
$\qquad + 7b^2i + ci - 15bj + 15k.$

$\Sigma a^6\beta\gamma\delta\epsilon = b^3f - 5b^2cf + 5bc^2f + 5b^2df - 5cdf - 5bef + 5f^2 - b^3g + 4b^2cg$
$\qquad - 2c^2g - 4bdg + 4eg + b^3h - 3bch + 3dh - b^2i + 2ci + bj - 10k.$

$\Sigma a^5\beta^2\gamma\delta\epsilon = b^2cf - 3bc^2f - b^2df + 5cdf + bef - 5f^2 - 5b^3g + 19b^2cg - 8c^2g$
$\qquad - 19bdg + 16eg + 5b^3h - 14bch + 12dh - 5b^2i + 8ci$
$\qquad + 13bj - 40k.$

$\Sigma a^4\beta^3\gamma\delta\epsilon = bc^2f - 2b^2df - cdf + 5bef - 5f^2 - b^2cg + 5bdg - 8eg + 6b^3h$
$\qquad - 16bch + 12dh - 13b^2i + 24ci + 13bj - 40k.$

$\Sigma a^4\beta^2\gamma^2\delta\epsilon = b^2df - 2cdf - bef + 5f^2 - 4b^2cg + 8c^2g + 3bdg - 12eg + 9b^3h$
$\qquad - 23bch + 18dh - 9b^2i + 4ci + 33bj - 60k.$

$\Sigma a^3\beta^3\gamma^2\delta\epsilon = cdf - 3bef + 5f^2 - 4c^2g + 6bdg + 9bch - 24dh - 21b^2i$
$\qquad + 28ci + 33bj - 60k.$

$\Sigma a^3\beta^2\gamma^2\delta^2\epsilon = bef - 5f^2 - 3bdg + 8eg + 5bch - 2dh - 7b^2i - 8ci + 31bj - 40k.$

$\Sigma a^2\beta^2\gamma^2\delta^2\epsilon^2 = f^2 - 2eg + 2dh - 2ci + 2bj - 2k.$

$\Sigma a^5\beta\gamma\delta\epsilon\zeta = b^3g - 4b^2cg + 2c^2g + 4bdg - 4eg - b^3h + 3bch - 3dh + b^2i$
$\qquad - 2ci - bj + 10k.$

$\Sigma a^4\beta^2\gamma\delta\epsilon\zeta = b^2cg - 2c^2g - bdg + 4eg - 6b^3h + 17bch - 15dh + 6b^2i$
$\qquad - 10ci - 14bj + 50k.$

$\Sigma a^3\beta^3\gamma\delta\epsilon\zeta = c^2g - 2bdg + 2eg - bch + 3dh + 7b^2i - 13ci - 7bj + 25k.$

$\Sigma a^3\beta^2\gamma^2\delta\epsilon\zeta = bdg - 4eg - 5bch + 12dh + 14b^2i - 12ci - 46bj + 100k.$

$\Sigma a^2\beta^2\gamma^2\delta^2\epsilon\zeta = eg - 4dh + 9ci - 16bj + 25k.$

$\Sigma a^4\beta\gamma\delta\epsilon\zeta\eta = b^3h - 3bch + 3dh - b^2i + 2ci + bj - 10k.$

$\Sigma a^3\beta^2\gamma\delta\epsilon\zeta\eta = bch - 3dh - 7b^2i + 12ci + 15bj - 60k.$

$\Sigma a^2\beta^2\gamma^2\delta\epsilon\zeta\eta = dh - 6ci + 20bj - 50k.$

$\Sigma a^2\beta\gamma\delta\epsilon\zeta\eta\theta = b^2i - 2ci - bj + 10k.$

$\Sigma a^2\beta^2\gamma\delta$ &c. $= ci - 8bj + 35k.$

$\Sigma a^2\beta\gamma$ &c. $= bj - 10k.$

Mr. Cayley has noticed a certain symmetry in the coefficients of the preceding formulæ which may be more easily exhibited by using Hirsch's notation. Let such a sum as $\Sigma a^3\beta^2\gamma^2\delta\epsilon\zeta$ be denoted $[32^21^2]$, and let the coefficients be a_1, a_2, &c. so that (32^21^2) will denote $a_3 a_2^2 a_1^2$; then the formulæ for the sums of the fourth order may be written

	$\widehat{(4)}$	$\widehat{(31)}$	$\widehat{(2^2)}$	$\widehat{(21^2)}$	$\widehat{(1^4)}$
$[4]$ = -4	$+4$	$+2$	-4	$+1$
$[31]$ = $+4$	-1	-2	$+1$	
$[2^2]$ = $+2$	-2	$+1$		
$[21^2]$ = -4	$+1$			
$[1^4]$ = $+1$				

The first line of which is to be read

$$\Sigma a^4 = -4a_4 + 4a_3 a_1 + 2a_2^2 - 4a_2 a_1^2 + a_1^4,$$

and so on for the rest. Now what Mr. Cayley has proved is that when the formulæ already given are written in this form, the figures are the same whether we read according to the rows or to the columns. The same thing holds for a set of formulæ given by Mr. Cayley (*Phil. Trans.*, 1856, p. 489) expressing the coefficients (4), (31), &c. in terms of the sums [4], [31], &c.

I add in conclusion the values of a few symmetric functions of the differences of the roots of the general equation written with binomial coefficients, as given by Mr. M. Roberts (*Quarterly Journal*, vol. IV.), in whose papers are to be found several interesting relations connecting the different covariants of binary quantics. Let

$$b^2 - ac = H, \quad ae - 4bd + 3c^2 = S, \quad ace + 2bcd - ad^2 - eb^2 - c^3 = T,$$
$$ag - 6bf + 15ce - 10d^2 = A, \quad ai - 8bh + 28cg - 5df + 35e^2 = P,$$
$$b^2 g - 2cd^2 + bde - 3bcf - acg + 3adf - 2ae^2 + 3c^2 f = M.$$

Then

$$a^2 \Sigma (a - \beta)^2 = n^2 (n - 1) H,$$
$$a^4 \Sigma (a - \beta)^4 = n^2 (n - 1) \{n^2 H^2 - \tfrac{1}{6}(n - 2)(n - 3) a^2 S\},$$
$$a^6 \Sigma (a - \beta)^6 = n^2 (n - 1) \{n^4 H^3 - \tfrac{1}{4} n^2 (n - 2)(n - 5) a^2 HS$$
$$\qquad - \tfrac{1}{4} n (n - 2)(7n - 15) a^3 T - \tfrac{1}{120}(n - 2)(n - 3)(n - 4)(n - 5) a^4 A\},$$
$$a^8 \Sigma (a - \beta)^8 = n^2 (n - 1) \Big\{ n^6 H^4 - \tfrac{1}{2} n^4 (n - 2)(n - 7) a^2 H^2 S$$
$$\qquad + 2n^3 (n - 2)(3n - 7) a^3 HT + \tfrac{1}{72} n^2 (n - 2)(n - 3)(n^2 + 8n - 21) a^4 S^2$$
$$\qquad - \tfrac{1}{90} n^2 (n - 2)(n - 3)(n - 4)(n - 21) a^4 HA$$
$$\qquad - \tfrac{1}{6} n (n - 2)(n - 3)(n - 4)(3n - 7) a^5 M$$
$$\qquad - \frac{(n - 2)(n - 3)(n - 4)(n - 5)(n - 6)(n - 7)}{2.3.4.5.6.7} a^6 P \Big\}.$$

By the help of these can be calculated the first few terms in the equation for the squares of the differences of the roots.

(294)

INDEX.

Aronhold, on symbolical methods, 130.
  On the invariants of a ternary cubic, 270.
  On the differential equations of invariants, 271.

Baltzer, on determinants, 266.
Bezout, on elimination, 66, 266, 282.
Binet, on determinants, 266.
Bezoutiants, 282.
Boole, on linear transformations, 94, 270.
  His form for the resultant of two quadratics, 148, 160.
Borchardt, proof that the equation of the secular inequalities has all its roots real, 36.
Bordered Hessians reduced, 15.
  Symmetrical determinants, value of, 29.
Brioschi, expression for differential equation of invariants in terms of roots, 51.
  On solution of the quintic, 201.
  On determinants, 266.
Burnside, investigation of radius of sphere circumscribing tetrahedron, 21.

Canonical forms, 133, 274.
Canonizants, 138.
Catalecticants, 169, 203.
Cauchy, on determinants, 266.
Cayley, (see also p. 270).
  His expression for relation connecting mutual distances of five points on a sphere, 21.
  of five points in space, 22.
  Application of skew determinants to the theory of orthogonal substitutions, 33, 266.
  Statement of Bezout's method of elimination, 68.
  General expression for resultants as quotients of determinants, 71.
  Notation for quantics, 83.
  Discovery of invariants, 94.
  On the number of invariants of a binary quartic, 116, 210.
  Definition of covariants, 119.
  Symbolical method of expressing invariants and covariants, 120.
  Identifies two forms of canonizant of equations of odd degree, 139.
  On discriminants of discriminants, 149.

Cayley, on tact-invariants, 153.
  Relation connecting covariants of cubic, 162.
  Solution of a cubic, 162.
  Solution of a quartic, 174.
  On covariants of system formed by quartic and its Hessian, 178.
  On covariants of quintic, 190.
  Tables of Sturmian functions, 192.
  Tables of symmetric functions, 285.
  On Tschirnhausen transformation, 278.
Clebsch, on symbolical methods, 130.
  On canonical form of ternary quartics, 134.
  His proof that every invariant may be symbolically expressed, 142.
  His form for resultant of two cubics, 274.
  Investigation of resultant of quadratic and general equation, 248.
  General expression for discriminant, 252.
Cockle, on the solution of the quintic, 201.
Combinants, 144, 273.
  Invariant of invariant of $U + \lambda V$ is a combinant, 167.
  Of a system of two quartics, 180.
Common roots determined, 76.
Commutants, 267.
Concomitants, 104.
Conditions that equations should have two common factors, 63, 82.
  For systems of equalities between roots, 109.
  That $U + \lambda V$ should have a cubic factor, 147.
  That two quartics should be differentials of same quintic, 184.
  That quartic should have two square factors, 173.
  That quintic should have two square, or a cubic factor, 189.
  That sextic should have two square, or a cubic factor, 205.
  That roots of sextic should be in involution, 210.
  See also Lesson XVIII.
Contragredience, 102.
Contravariants, 101.
Covariants, 99.
  Number of, for a binary quantic, 159.

INDEX. 295

Cramer, on determinants, 266.
Critical functions, 40.
Cubic discussed, 160.
Cubic quaternary, its canonical form, 144.
Derivatives of derivatives expressed symbolically, 243.
Differential coefficients of determinants, 28.
Of resultants with respect to quantities entering into all the quantics, 81.
Differential equation of functions of differences of roots, 47.
Of invariants, 113.
Differentiation mutual, of covariants and contravariants, 110.
Discriminant of binary quantics expressed as determinant, 20.
Of products of two quantics, 86.
Of discriminants, 148.
Enables us to distinguish whether equation has even or odd numbers of pairs of imaginary roots, 191.
General symbolical expression for, 246, 252.

Eisenstein, expression for general solution of quartic, 270.
Emanants, 99.
Equalities between roots of an equation, conditions for, 109.
Euler, on the theory of orthogonal substitutions, 36.
On elimination, 64, 268.
Evectants, 105.
When discriminant vanishes, 107.
Of discriminant of cubic, 161.

Faà de Bruno, calculates invariant of quintic, 186.
On elimination, 269, 282.
Forme-type of quintic, 278.

Gauss, on linear transformations, 266, 268.

Harley, on solution of a quintic, 201.
Hermite, law of reciprocity, 125.
Form for covariants of system formed by quartic and its Hessian, 178.
Discovery of skew invariant of quintic, 189.
Canonical form for quintic, 189.
Forme-type of quintic, 91, 278.
Expression by invariants of conditions of reality of roots, 193, 279.
Expression of invariants in terms of roots, 202.
Solution of quintics by elliptic functions, 200.
On Tschirnhausen transformation, 276.
Hessians, 15, 100, 268.
Contain all square factors of original quartic, 136.
Of Hessians, 178.
Hirsch's tables of symmetric functions, 285.

Invariants, 92.
Absolute, 96.
Skew, 115.
How many independent, 95, 272.
Relation connecting weight and order of, 113.
Involution, 145, 160.
Condition roots of sextic shall be in, 210.

Jacobi, on determinants and linear transformations, 266, 268.
Jacobian, of system of equations, 69, 101, 272.
Its discriminant discussed, 147.
Of quartic and its Hessian, 177.
Jerrard, transformation of a quintic, 188, 275.
Joachimsthal, expression for area of a triangle inscribed in an ellipse, 21.
Theorem on form of discriminant, 87.

Kronecker, solution of quintic by elliptic functions, 201.
Kümmer's resolution into sum of squares of discriminant of cubic which determines axes of a quadric, 43.

Lagrange, on the general solution of equations, 200.
On conditions that equation should have two pairs of equal roots, 269.
On linear transformations, 273.
Laplace, on determinants, 266.
On equation of secular inequalities, 36.
Leibnitz, his claim to invention of determinants, 266.
Linear covariants of cubic and quadratic, 163.
Of quintic, 190.
Of equations of odd degrees, 191, 274.

Minor determinants, 9, 23.
Of reciprocal system how related to those of original, 25.
Multiplication of determinants, 17.
Number of quadrics which can be described through five points to touch four planes, 226.
Of invariants of a binary quantic, 271.

Order of determinants, 6.
Of symmetric functions, 45.
Of invariants, 114.
Of systems of equations, 213.
Orthogonal transformations, 33.
Osculants, 156.

Poisson's method of forming symmetric functions of common roots of systems of equations, 268.

INDEX.

Quadratic forms, reducible to sum of squares, 136.
  Number of positive and negative squares fixed, 136.
  And equation of nth degree, general expression for resultant, 244.
Quadrinvariants of binary quantics, 112.
Quartic, theory of, 169.

Reciprocal determinants, 24.
Reciprocity, Hermite's law of, 125.
Reducing sextic for quintic, forms of, 201.
Resultant of two quadratics, 55.
  Two cubics, 63, 165, 283.
  Two quartics, 180, 284.
  Tables of, 283.
Roberts, Michael, on covariants, 117.
  On application of Sturm's theorem to quantics, 192, 268.
  On equation of squares of differences, 298.
Roberts, Samuel, on orders of systems of equations, 280.
Rodrigues, on orthogonal transformations, 36.

Skew symmetric determinants of even degree are perfect squares, 31.
Skew invariant of quintic, 189.
  Vanishes if quintic can be linearly transformed to the recurring form, 194.
  Or to one wanting alternate terms, 274.
  Of sextic, 210, 253.
  Of all quantics vanish when quantic wants alternate terms, 274.
Source of covariants, 117.
Sphere circumscribing tetrahedron, 21.
  Relations connecting mutual distances of points on, 21, 22.

Sturm's functions, Sylvester's expressions for, 37.
  In case of quartic, 175.
  of quintic, 192.
  Extensions of, 269.
Sylvester (see also p. 270).
  Umbral notation for determinants, 7.
  Proof that equation of secular inequalities has all real roots, 23.
  Expression for Sturm's functions in terms of roots, 37.
  Dialytic method of elimination, 65.
  Expression of resultant as determinants, 70.
  On nomenclature, 104.
  Canonical forms of odd and even degrees, 137, 141.
  Of quaternary cubic, 144.
  Expressions for discriminant with regard to variables which do not enter explicitly, 181.
  On osculants, 151.
  Investigation of expression by invariants of conditions for reality of roots of quintic, 193.
  On Bezoutiants, 282.
Symbolical expression for invariants, 121, 239.
Symmetric functions, 44.
  Their use in finding invariants, 108.
  Tables of, 285.

Tact-invariants, 153.
  Of complex curves, 155.
Tetrahedron, radius of circumscribing sphere, 21.
Tschirnhausen, transformation of equations, 276.

Vandermonde, on determinants, 266.

Warren, on resultant of two cubics, 169.

THE END.

W. Metcalfe, Printer, Green Street, Cambridge.

www.ingramcontent.com/pod-product-compliance
Lightning Source LLC
Chambersburg PA
CBHW022043230426
43672CB00008B/1050